# ENCYCLOPEDIA OF DINOSAURS

# ENCYCLOPEDIA OF DINOSAURS

Consultant editor: Carl Mehling

This edition first published in 2025

Copyright © 2017 Amber Books Ltd

All rights reserved. No part of this publication may be reproduced, stored in a retrieval system, or transmitted in any form or by any means, electronic, mechanical, photocopying, recording, or otherwise, without prior written permission of the copyright holder.

Published by
Amber Books Ltd
United House
North Road
London N7 9DP
United Kingdom

www.amberbooks.co.uk
Facebook: amberbooks
YouTube: amberbooksltd
Instagram: amberbooksltd
X(Twitter): @amberbooks

ISBN:978-1-83886-560-3

Project Editor: Sarah Uttridge
Editorial Assistant: Kieron Connolly
Picture Research: Terry Forshaw
Design: Graham Beehag

Printed in China

CONTRIBUTORS:

SEAN CALLERY has researched and written numerous reference books for adults and children. He specialises in history and science topics.

C.A. CURTIS is a freelance writer and editor with many years of experience working on reference books for both adults and children.

GERRIE McCALL obtained a Batchelor of Arts in English Literature and a minor in History from the University of Houston. She worked as a lecturer at the University of Texas and is now a successful freelance writer.

BRENDA R. LEWIS has written more than 85 books, including *Monarchy: The History of an Idea* and *Great Civilizations from Prehistoric Times to the Present Day*.

# CONTENTS

| | |
|---|---|
| Introduction | 8 |

**BEFORE THE DINOSAURS**

| | |
|---|---|
| Anomalocaris | 12 |
| Hallucigenia | 13 |
| Trilobite | 14 |
| Pterygotus | 15 |
| Ammonite | 16 |
| Cladoselache | 17 |
| Eusthenopteron | 18 |
| Ichthyostega | 19 |
| Dunkleosteus | 20 |
| Coelacanth | 22 |
| Eogyrinus | 23 |
| Hylonomus | 24 |
| Arthropleura | 25 |
| Eurypterid | 26 |
| Cacops | 27 |
| Diadectes | 28 |
| Eryops | 29 |
| Mesosaurus | 30 |
| Moschops | 31 |
| Platyhystrix | 32 |
| Scutosaurus | 33 |
| Seymouria | 34 |
| Youngina | 35 |
| Dimetrodon | 36 |
| Diplocaulus | 40 |
| Coelurosauravus | 42 |

**TRIASSIC**

| | |
|---|---|
| Eoraptor | 43 |
| Erythrosuchus | 44 |
| Euparkeria | 45 |
| Euskelosaurus | 46 |
| Henodus | 47 |
| Hyperodapedon | 48 |
| Kannemeyeria | 49 |
| Lagosuchus | 50 |
| Liliensternus | 51 |
| Lotosaurus | 52 |
| Melanorosaurus | 53 |
| Mussaurus | 54 |
| Nanchangosaurus | 55 |
| Nothosaurus | 56 |
| Pisanosaurus | 57 |
| Protoavis | 58 |
| Riojasaurus | 59 |
| Saltopus | 60 |
| Sellosaurus | 61 |
| Shansisuchus | 62 |
| Shonisaurus | 63 |
| Tanystropheus | 64 |
| Thecodontosaurus | 65 |
| Gracilisuchus | 66 |
| Postosuchus | 67 |
| Coelophysis | 68 |
| Cynognathus | 72 |
| Herrerasaurus | 76 |
| Lystrosaurus | 80 |
| Coloradisaurus | 82 |
| Desmatosuchus | 83 |
| Staurikosaurus | 84 |

**JURASSIC**

| | |
|---|---|
| Abrictosaurus | 85 |
| Ammosaurus | 86 |
| Anchisaurus | 87 |
| Barapasaurus | 88 |
| Emausaurus | 89 |
| Kotasaurus | 90 |
| Lufengosaurus | 91 |
| Lycorhinus | 92 |

*OMEISAURUS*

| | |
|---|---|
| Megapnosaurus | 93 |
| Dilophosaurus | 94 |
| Scelidosaurus | 95 |
| Cryolophosaurus | 96 |
| Yunnanosaurus | 98 |
| Rhoetosaurus | 99 |
| Datousaurus | 100 |
| Gasosaurus | 101 |
| Lapparentosaurus | 102 |
| Metriacanthosaurus | 103 |
| Omeisaurus | 104 |
| Piatnitzkysaurus | 105 |
| Proceratosaurus | 106 |
| Xiaosaurus | 107 |
| Huayangosaurus | 108 |
| Dimorphodon | 110 |
| Megalosaurus | 111 |
| Eustreptospondylus | 112 |
| Shunosaurus | 114 |
| Cetiosaurus | 118 |
| Cetiosauriscus | 119 |
| Lexovisaurus | 120 |
| Liopleurodon | 122 |
| Tuojiangosaurus | 123 |
| Bothriospondylus | 124 |
| Camptosaurus | 125 |
| Chialingosaurus | 126 |
| Coelurus | 127 |
| Dicraeosaurus | 128 |
| Diplodocus | 129 |
| Dracopelta | 130 |
| Elaphrosaurus | 131 |
| Euhelopus | 132 |
| Haplocanthosaurus | 133 |
| Othnielia | 134 |
| Szechuanosaurus | 135 |
| Ultrasauros | 136 |
| Yangchuanosaurus | 137 |
| Dacentrurus | 138 |

| | |
|---|---|
| Ornitholestes | 140 |
| Apatosaurus | 141 |
| Archaeopteryx | 142 |
| Compsognathus | 143 |
| Allosaurus | 144 |
| Brachiosaurus | 148 |
| Ceratosaurus | 152 |
| Kentrosaurus | 156 |
| Seimosaurus | 158 |
| Stegosaurus | 160 |
| Ophthalmosaurus | 164 |
| Dryosaurus | 165 |
| Dsungaripterus | 166 |

**EARLY CRETACEOUS**

| | |
|---|---|
| Gastonia | 167 |
| Afrovenator | 168 |
| Atlascopcosaurus | 169 |
| Becklespinax | 170 |
| Chilantaisaurus | 171 |
| Fulgurotherium | 172 |
| Gilmoreosaurus | 173 |
| Harpymimus | 174 |
| Hylaeosaurus | 175 |
| Leaellynasaura | 176 |
| Muttaburrasaurus | 177 |
| Pelicanimimus | 178 |
| Pelorosaurus | 179 |
| Polacanthus | 180 |
| Silvisaurus | 181 |
| Stenopelix | 182 |
| Tapejara | 183 |
| Tenontosaurus | 184 |
| Tropeognathus | 185 |

| | |
|---|---|
| Wuerhosaurus | 186 |
| Yaverlandia | 187 |
| Minmi | 188 |
| Sauropelta | 190 |
| Zephyrosaurus | 192 |
| Giganotosaurus | 193 |
| Hypsilophodon | 194 |
| Kronosaurus | 195 |
| Ouranosaurus | 196 |
| Psittacosaurus | 197 |
| Acrocanthosaurus | 198 |
| Amargasaurus | 202 |
| Baryonyx | 206 |
| Deinonychus | 208 |
| Iguanodon | 212 |
| Probactrosaurus | 216 |
| Pterodaustro | 217 |
| Utahraptor | 218 |
| Suchomimus | 222 |

**MID CRETACEOUS**

| | |
|---|---|
| Argentinosaurus | 223 |
| Carcharodontosaurus | 224 |
| Spinosaurus | 228 |

**LATE CRETACEOUS**

| | |
|---|---|
| Abelisaurus | 230 |
| Adasaurus | 231 |
| Aeolosaurus | 232 |
| Alamosaurus | 233 |
| Albertosaurus | 234 |
| Alectrosaurus | 235 |
| Alioramus | 236 |
| Alvarezsaurus | 237 |
| Anatotitan | 238 |
| Anchiceratops | 239 |
| Anserimimus | 240 |
| Antarctosaurus | 241 |
| Aralosaurus | 242 |
| Archaeornithomimus | 243 |
| Arrhinoceratops | 244 |
| Aublysodon | 245 |
| Avaceratops | 246 |
| Bagaceratops | 247 |
| Borogovia | 248 |
| Brachyceratops | 249 |
| Brachylophosaurus | 250 |
| Centrosaurus | 251 |
| Chasmosaurus | 252 |
| Chirostenotes | 253 |
| Conchoraptor | 254 |
| Corythosaurus | 255 |
| Diceratus | 256 |
| Dravidosaurus | 257 |

GRACILISUCHUS

| | |
|---|---|
| Dromaeosaurus | 258 |
| Dryptosaurus | 259 |
| Edmontosaurus | 260 |
| Einiosaurus | 261 |
| Elasmosaurus | 262 |
| Elmisaurus | 263 |
| Erlikosaurus | 264 |
| Euoplocephalus | 265 |
| Garudimimus | 266 |
| Goyocephale | 267 |
| Hadrosaurus | 268 |
| Homalocephale | 269 |
| Hypacrosaurus | 270 |
| Hypselosaurus | 271 |
| Indosuchus | 272 |
| Ingenia | 273 |
| Jaxartosaurus | 274 |
| Leptoceratops | 275 |
| Magyarosaurus | 276 |
| Majungasaurus | 277 |
| Mandschurosaurus | 278 |
| Microceratus | 279 |
| Montanoceratops | 280 |
| Nanotyrannus | 281 |
| Nanshiungosaurus | 282 |
| Nemegtosaurus | 283 |
| Neuquensaurus | 284 |
| Nipponosaurus | 285 |
| Noasaurus | 286 |
| Opisthocoelicaudia | 287 |
| Pachyrhinosaurus | 288 |
| Panoplosaurus | 289 |
| Parksosaurus | 290 |
| Pentaceratops | 291 |
| Pinacosaurus | 292 |
| Prenocephale | 293 |
| Prosaurolophus | 294 |
| Quaesitosaurus | 295 |
| Rhabdodon | 296 |
| Saurolophus | 297 |
| Saurornithoides | 298 |
| Secernosaurus | 299 |
| Shantungosaurus | 300 |
| Stegoceras | 301 |
| Struthiomimus | 302 |
| Stygimoloch | 303 |
| Talarurus | 304 |
| Therizinosaurus | 305 |
| Thescelosaurus | 306 |
| Titanosaurus | 307 |
| Tsintaosaurus | 308 |
| Tylocephale | 309 |
| Tylosaurus | 310 |
| Wannanosaurus | 311 |
| Ankylosaurus | 312 |
| Nodosaurus | 316 |
| Saichania | 318 |
| Struthiosaurus | 322 |
| Xiphactinus | 324 |
| Deinosuchus | 325 |
| Euoplocephalus | 326 |
| Gallimimus | 327 |
| Hesperornis | 328 |
| Lambeosaurus | 329 |
| Libonectes | 330 |
| Mononykus | 331 |
| Mosasaur | 332 |
| Parasaurolophus | 333 |
| Quetzalcoatlus | 334 |
| Saltasaurus | 335 |
| Tyrannosaurus rex | 336 |
| Carnotaurus | 340 |
| Deinocheirus | 344 |
| Edmontonia | 346 |
| Maiasaura | 350 |
| Oviraptor | 354 |
| Pachycephalosaurus | 358 |
| Protoceratops | 362 |
| Pteranodon | 366 |
| Tarbosaurus | 370 |
| Triceratops | 374 |
| Troodon | 378 |
| Velociraptor | 380 |
| Styracosaurus | 384 |

**NEOGENE**

| | |
|---|---|
| Borhyaena | 385 |
| Gastornis | 386 |
| Carcharocles Megalodon | 390 |
| Platybelodon | 391 |
| Thylacosmilus | 392 |

**TERTIARY (EOCENE)**

| | |
|---|---|
| Coryphodon | 393 |
| Hyracotherium | 394 |
| Mesonyx | 395 |
| Pristichampsus | 396 |
| Uintatherium | 397 |
| Andrewsarchus | 398 |
| Basilosaurus | 400 |
| Brontotherium | 404 |

**TERTIARY (OLIGOCENE)**

| | |
|---|---|
| Arsinoitherium | 408 |
| Mammalodon | 409 |
| Pyrotherium | 410 |
| Palaeocastor | 411 |

**TERTIARY (MIOCENE)**

| | |
|---|---|
| Amebelodon | 412 |
| Daeodon | 413 |
| Deinogalerix | 414 |
| Homalodotherium | 415 |
| Argentavis | 416 |
| Moropus | 418 |
| Borophagus | 419 |
| Syndyoceras | 420 |

**LATE TERTIARY (PLIOCENE)**

| | |
|---|---|
| Megatherium | 421 |
| Sivatherium | 422 |

**PLEISTOCENE**

| | |
|---|---|
| Colossochelys | 423 |
| Doedicurus | 424 |
| Smilodon | 426 |
| Woolly Mammoth | 430 |
| Homotherium | 434 |

**QUATERNARY (PLEISTOCENE)**

| | |
|---|---|
| Coelodonta | 435 |
| Diprotodon | 436 |
| Glyptodon | 437 |
| Megaloceros | 438 |

| | |
|---|---|
| Glossary | 439 |
| Fossil Sites Map | 440 |
| Index | 442 |

*Lexovisaurus*

INTRODUCTION

# Introduction

The history of life of Earth is an endlessly fascinating topic. It informs us about our origins and place in the colossal scope of Life. Our biological perspective is honed with every new discovery or reinterpretation. We are humbled by the immense time spans and infinite variety that evolution has spawned. Contemplating lost worlds and their bizarre inhabitants fires our imagination like no other subject. Chances are that you were drawn to this book for at least some of these reasons.

To reconstruct our planet's biological history is to be a detective. Aside from the genetic data stowed within every extant organism, fossils are what remain of the story of species long gone. Thus, they are the starting point for our elucidations and restorations of deep time. Remember that even the best-preserved fossils never tell the whole story. For example, an insect hermetically sealed in amber lacks a record of the rich behaviours that the living animal displayed. And the vast majority of fossils don't come close to the physical 'perfection' of a bug encased in fossil tree resin. Moreover, we have an infinitesimally small sample of the countless trillions of organisms that have populated the globe over nearly four billion years.

**Even the best fossils leave plenty of room for interpreting the appearance and behaviour of long-extinct organisms like these Brachiosaurs.**

# INTRODUCTION

**Were giant horned dinosaurs able to rear up on their hind legs? We are unlikely to know for sure but can make educated guesses.**

As more and better fossils are discovered, our grasp of the story improves bit by bit. This is certainly true of dinosaurs. As terrestrial and aerial vertebrates, they were less apt to be preserved in the fossil record than, say, a huge invertebrate reef community in a shallow sea. Most dinosaurs, on their way to becoming fossilized for our later scrutiny, were significantly scavenged and weathered prior to burial, leaving behind tantalizing but frustrating fragments. This holds true for many of the terrestrial animals outlined in this book.

While reading, you will notice repeated use of such qualifiers as 'probably', 'possibly' and 'maybe'. This is both intentional and critical. When paleontologists are trying to reconstruct organisms from fragmentary remains, great caution must be used. There is often little that we can state with certainty, particularly with regard to behaviour. Just consider how much remains to be learned about organisms living in our modern world, including ourselves, and it's easy to see how much more dense is the fog that surrounds those for which we have only fossils. Would we ever conclude by studying a fossilized skeleton that, in life, the animal passively drew up water from moist sand via its forelimbs and conducted it to the corners of its mouth by means of the capillary action of the gaps separating its scales? Such an animal exists today – the modern thorny devil, *Moloch horridus*. If snakes were an extinct group, known only from their fossil remains, who would ever imagine them capable of snatching bats from the air in pitch-black caves, as one Cuban species does? How could we ever guess that a prehistoric animal might intentionally break bones in its own toes so that they erupt through the skin to fashion 'claws', as does the modern frog *Trichobatrachus robustus*?

## The Meaning of Names

You may also observe that the etymologies for some genera featured herein lack certainty. It has long been standard practice to employ Latin or Greek words when naming organisms. Scholars of old were well-versed in these languages, and could safely assume that their readers would accurately perceive their significance: the dinosaur *Dilophosaurus* ('Two-crested lizard') had two crests on its skull, and *Gracilisuchus* ('Slender crocodile') was indeed a slender crocodile. Other names, however, were devised for reasons not immediately apparent from a straight translation of their components. A literal translation of *Hypsilophodon* means 'High-ridged tooth', but a deeper look reveals that the dinosaur's name was actually intended to mean '*Hypsilophus* tooth'. While the modern lizard *Hypsilophus* does sport a high ridge of scale spines along its back, *Hypsilophodon* was actually (and somewhat less obviously) named for the similarity of its teeth to those of its namesake. And then there are examples like *Emausaurus*, whose name translates to 'EMAU lizard', after Ernst-Moritz-Arndt-Universität in Greifswald, Germany, near where the fossil was found – a fluent knowledge of Latin and Greek wouldn't get you far in parsing that name! For some genera named long ago, and for which we have no etymologies, we have done our best to interpret the probable intent of the names. Fortunately, modern naming rules now require an explanation of newly named species.

People hunger for information about prehistoric animals and, for myriad and complex reasons, about dinosaurs in particular. And for equally complicated reasons, the word 'dinosaur' often gets misconstrued by laypeople to mean 'any huge, long-extinct animal known from fossils'. But scientists try to be more precise with their definitions. To them, the group shares more concrete, unique, evolutionarily significant features than size, extinction, or preservation, especially since we now recognize dinosaurs as including tiny, extant animals known from live specimens (modern birds). That said, the term 'dinosaur' is surprisingly hard to define.

**Everyone knows *Tyrannosaurus rex* is a dinosaur, but devising a clear definition of the term 'dinosaur' is a surprisingly difficult task.**

## The Definition of a Dinosaur

Currently, the two widely accepted definitions for the term dinosaur are (a) 'all descendants of the most recent common ancestor of *Triceratops* and modern birds' and (b) 'all of the descendants of the most recent common ancestor of *Megalosaurus* and *Iguanodon*'. (The latter refers to the first two non-avian dinosaurs that were scientifically described.) Either of these definitions brackets the same discrete group of animals, but what does that actually mean? The layperson certainly can't just look at an organism and determine that it's a descendant of the most recent common ancestor of *Triceratops* and birds or *Megalosaurus* and *Iguanodon*.

If we consider the species used above to define the group Dinosauria, they share the features exemplified by all members of the group. These include aspects of the humerus, ilium, tibia and astragalus, as well as an erect stance for the hind limbs, which, at first, would seem a bizarre and mundane set of criteria on which to define such a charismatic group. Bizarre and mundane, perhaps, but consistency across the group is essential in forming a solid definition. But, since fossils can never be relied upon to give us the whole picture, we are destined always to find fossils of organisms that have many but not all of the features we consider crucial in defining a *group* of organisms. This is especially true for fossils of organisms that fall close to the origin of a large evolutionary radiation of organisms, like the Dinosauria. This is why definitions using two bracketing taxa are favoured – new organisms either fit within them or do not. If a suite of characteristics are chosen to define a group, and we find an organism that is missing one or two, we are forced to decide to exclude the new organism or refine our definition ad infinitum.

## Moving Forward into the Past

The same is true for any group of organisms at any level in the hierarchy of organismal categories. Dinosaurs are a good group with which to begin teaching these issues because they possess such an immediate and vast mythology – they are well liked, if not well understood by the public. The concerns outlined in this introduction are the same across all of paleontology and should also be kept in mind when reading the non-dinosaurian entries in this book, included here as a springboard to other areas of paleontology outside of the Dinosauria. And as a temptation, it's intriguing to note that fully 83 per cent of the Earth's fossil record is non-dinosaurian.

We will close with one last illuminating issue that vexes people attempting to define Dinosauria: its history as a topic of scientific study. Until fairly recently, birds were not considered part of this group, and thus, we could speak of several givens: the total extinction of dinosaurs at the end of the Cretaceous, the fact that humans have never lived alongside dinosaurs, and the idea that no dinosaurs ever flew. But now that it is remarkably clear that birds are a lineage of dinosaurs, our definition must account for this, and should be crafted to encompass all that we now accept about the group. This necessity to reexamine and redefine applies to *all* areas of paleontology, and indeed, to all science. In the end, we must ask, 'Can one simple meaning suffice for such a diverse group?' Essentially, that question is the daily work of a paleontologist: define what is seen in the fossil record and then, hopefully, refine our definitions as new discoveries and methods become available.

Thus readers of this, or any other piece of writing on paleontology, should stay aware of this indispensable caveat: the ideas encountered here are as apt to change over time as are the organisms being described. Our ideas evolve to fit the environment of evidence we have accumulated, as the organisms themselves evolve to fit their ever-changing environment.

**Since the naming of *Dilophosaurus* in 1970, our concept of the Dinosauria has been refined more than in the term's entire history.**

CAMBRIAN

# Anomalocaris
• ORDER • Radiodonta  • FAMILY • Anomalocarididae  • GENUS & SPECIES • *Anomalocaris nathorsti* and *A. canadensis*

This huge arthropod is the largest known and possibly the most deadly animal swimming in the seas of the Cambrian Era. It took a century for it to be identified confidently.

## VITAL STATISTICS

| | |
|---|---|
| FOSSIL LOCATION | Canada, China and Australia |
| DIET | Carnivorous |
| PRONUNCIATION | an-OM-ah-low-KAR-is |
| WEIGHT | Unknown |
| LENGTH | Mostly 60cm (23½in), but up to 2m (6ft) |
| HEIGHT | Unknown |
| MEANING OF NAME | 'Strange shrimp' from the misinterpretation of an early fossil as an odd kind of prawn |

### WHERE IN THE WORLD?

Found in the Burgess Shale of the Canadian Rockies, the Chengjiang shales of China and Australia's Emu Bay Shale.

**MOUTH**
The disc-shaped mouth had a ring of sharp teeth to break the shells of its possible favourite food, trilobites, either by biting straight through or gripping and shaking until the hard shell cracked.

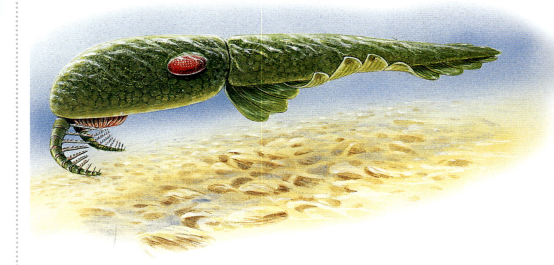

## FOSSIL EVIDENCE

J.F. Whiteaves' study of segments found from this creature in 1892 suggested it was a kind of shrimp (he thought he had a lobster or prawn tail: in fact it was an arm). In 1911 mouth fossils were found but mistakenly identified as primitive jellyfish. Only in the 1980s did palaeontologists realize that that these were parts of the same animal, ten times bigger than anything else living at the time. The following decade saw identification of several new species with a variety of claw and eating appendages from fossils in China.

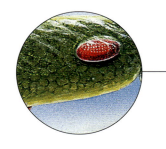

**EYES**
With its large eyes and swimming ability, *Anomalocaris* was a formidable hunter, flexing its segmented body to zoom through the oceans after prey.

### HOW BIG IS IT?

PREHISTORIC ANIMAL

CAMBRIAN

**TIMELINE (millions of years ago)**

| 540 | 505 | 438 | 408 | 360 | 280 | 248 | 208 | 146 | 65 | 1.8 to today |

CAMBRIAN

# Hallucigenia

• ORDER • Scleronychophora • FAMILY • Hallucigeniidae • GENUS & SPECIES • *Hallucigenia sparsa, H. fortis*

*Hallucigenia* was a tiny worm-like creature that scurried about on the seabed on pairs of stilt-like legs, but which end was up and which was down?

| VITAL STATISTICS | |
|---|---|
| FOSSIL LOCATION | Canada, China |
| DIET | Dead sea creatures |
| PRONUNCIATION | hal-OO-suh-JEEN-ee-uh |
| WEIGHT | Unknown |
| LENGTH | 0.5–3cm (⅕–1⅕in) |
| HEIGHT | Unknown |
| MEANING OF NAME | 'Strange and dream-like' because it is so bizarre |

**TENTACLES**
Recent specimens from China suggest there is a second set of tentacles, paired with the first, with claws at the ends. These are now thought to be what it walked on.

### WHERE IN THE WORLD?

The Burgess Shale of Southwestern Canada and the Cambrian Maotianshan shales of China.

### FOSSIL EVIDENCE

How can you make head or tail of *Hallucigenia*? There are dark stains at both ends of the fossils, either of which might be a head. So we still don't know for sure how it ate – some palaeontologists argue it took food down its tentacles. These have also caused confusion as early reconstructions of *Hallucigenia* were upside down. What are now seen as defensive spikes were initially identified as legs. Meanwhile the seven dorsal tentacles originally on its back are now depicted as pairs of clawed legs – although no second set has yet been found.

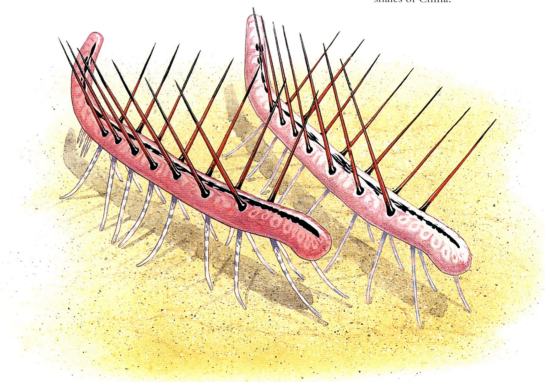

**HOW BIG IS IT?**

**SPINES**
It is not known what the spines were made of. They have not been preserved well, suggesting they were soft and therefore not capable of offering much protection.

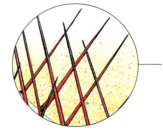

PREHISTORIC ANIMAL

CAMBRIAN

**TIMELINE (millions of years ago)**

CAMBRIAN TO PERMIAN

# Trilobite

• ORDER • Numerous • FAMILY • Numerous • GENUS & SPECIES • Numerous

| VITAL STATISTICS | |
|---|---|
| FOSSIL LOCATION | Worldwide |
| DIET | Varied between species; mainly scavengers |
| PRONUNCIATION | TRI-low-bite |
| WEIGHT | Unknown |
| LENGTH | 5mm–80cm (⅙in–32in) |
| HEIGHT | Unknown |
| MEANING OF NAME | 'Three lobes' because the body is in three parts: its three longitudinal lobes |

**FOSSIL EVIDENCE**
So many trilobites lived that some fossils have been preserved with complete exoskeletons. Many show signs of attack by their predators, including fish, orthocones and other arthropods. As predatory jawed fish evolved, trilobites developed more protective features, such as spines. They were one of the first creatures to develop eyes, made from dozens of small crystal lenses. As each species is easy to date, trilobites are 'index fossils' used to determine the age of the rocks around them.

PREHISTORIC ANIMAL

CAMBRIAN TO PERMIAN

More than 15,000 species have been identified of this long-surviving arthropod group. In fact, the Cambrian Period is sometimes known as 'The Age of Trilobites'.

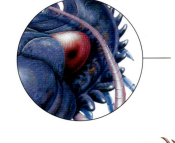

**EYES**
Trilobite eyes were often complex and delicate, like those of modern insects, being made up of many crystals. They were sensitive to movement and some offered stereoscopic vision.

**WHERE IN THE WORLD?**

Trilobites are found everywhere around the world, always in rocks containing other saltwater animals.

**LEGS**
Legs were jointed and covered in spines, and used for walking, grabbing prey and pushing it into the mouth. The legs also had filaments for extracting oxygen from the water.

**HOW BIG IS IT?**

**TIMELINE (millions of years ago)**

| 540 | 505 | 438 | 408 | 360 | 280 | 248 | 208 | 146 | 65 | 1.8 to today |

SILURIAN

# Pterygotus

• ORDER • Eurypterida • FAMILY • Pterygotidae • GENUS & SPECIES • Various species within the genus *Pterygotus*

### VITAL STATISTICS

| | |
|---|---|
| FOSSIL LOCATION | Worldwide |
| DIET | Carnivore |
| PRONUNCIATION | TARE-ih-GO-tus |
| WEIGHT | Unknown |
| LENGTH | 2.1m (7ft) |
| HEIGHT | Unknown |
| MEANING OF NAME | 'Wing fish' because 'wing' is often used for fins or paddles, and *Pterygotus* was first thought to be a fish; or 'Wing ear' because of the shape of one of the parts on its underside |

*Pterygotus* was an enormous, predacious eurypterid that evolved formidable pincers, or 'chelae'. In addition to its last pair of legs, which were flattened swimming paddles, it also had a round, flattened terminal segment (the 'telson'), which helped propel it through water.

**EYES**
A pair of huge compound eyes mounted on the forward edge of its exoskeleton strongly suggests that *Pterygotus* was a visual predator.

### WHERE IN THE WORLD?

Fossils of *Pterygotus* have been found in all continents except for Antarctica. This is not surprising because that continent was apparently land-locked at the time.

### FOSSIL EVIDENCE

Fossils of *Pterygotus*, known from a variety of species, are relatively common, yet complete skeletons are rare. This is because arthropods periodically moult, and the shedded exoskeletons disarticulate rapidly. It is also very hard for animals the size of *Pterygotus* to be preserved, exposed, and then recovered whole. It flourished in the Silurian Period, persisted into the Devonian, and is among the largest known eurypterids and one of the largest arthropods of all time.

**TEXTURE**
The distinctive scale-like texture of the exoskeleton allows identification of the often disarticulated (disjointed) segments of pterygotids.

### HOW BIG IS IT?

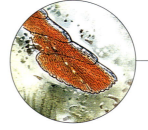

**TIMELINE** (millions of years ago)

| 540 | 505 | 438 | 408 | 360 | 280 | 248 | 208 | 146 | 65 | 1.8 to today |

DEVONIAN

# Ammonite

• **ORDERS** • Ammonitida, Goniatitida and Ceratitida • **FAMILY** • Numerous • **GENUS & SPECIES** • Numerous

These early molluscs had no vertebrae and were protected by a hard shell as they hunted down other marine animals. Their closest modern relatives are octopus, squid and cuttlefish.

| VITAL STATISTICS | |
|---|---|
| FOSSIL LOCATION | Worldwide |
| DIET | Carnivorous |
| PRONUNCIATION | AM-uh-nite |
| WEIGHT | Unknown |
| LENGTH | 2.5cm–2m (1in–6ft) diameter |
| HEIGHT | Unknown |
| MEANING OF NAME | Their resemblance to ram's horns inspired ammonites to be named after the Egyptian god Amun, who was often depicted wearing horns |

**FOSSIL EVIDENCE**
Ammonite fossils are usually found in rocks and are often used as an index fossil to date the surrounding rock because each species lived only in certain periods. The earliest are Devonian and survived until finally dying out with the non-avian dinosaurs. Their often spiral forms inspired the early belief that they were snakes turned to stone. One group, called the heteromorphids, had uncoiled shells like twisted wire.

PREHISTORIC ANIMAL
DEVONIAN–CRETACEOUS

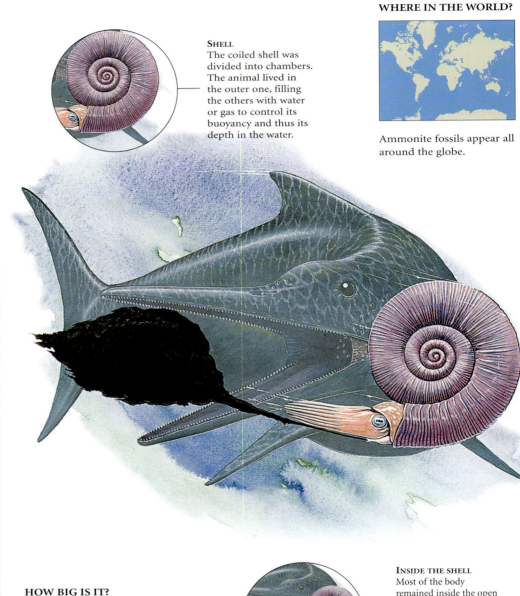

**SHELL**
The coiled shell was divided into chambers. The animal lived in the outer one, filling the others with water or gas to control its buoyancy and thus its depth in the water.

**WHERE IN THE WORLD?**

Ammonite fossils appear all around the globe.

**HOW BIG IS IT?**

**INSIDE THE SHELL**
Most of the body remained inside the open chamber at the end part of the shell. Only the tentacles, mouth, eyes and a thin tube, called a hyponome, would have been visible. This tube squirted water to propel the mollusc through the seas.

**TIMELINE (millions of years ago)**

| 540 | 505 | 438 | 408 | 360 | 280 | 248 | 208 | 146 | 65 | 1.8 to today |

DEVONIAN

# Cladoselache

• **ORDER** • Cladoselachiformes • **FAMILY** • Unattributed • **GENUS & SPECIES** • *Cladoselache fyleri*

**This primitive shark was a fast and agile swimmer. It was highly successful, surviving for 100 million years, but how it mated remains a mystery to palaeontologists.**

### VITAL STATISTICS

| | |
|---|---|
| FOSSIL LOCATION | North America |
| DIET | Carnivorous: fish and marine animals |
| PRONUNCIATION | KLAD-oh-sel-LAK-ee |
| WEIGHT | Unknown |
| LENGTH | Up to 2m (6ft) |
| HEIGHT | Unknown |
| MEANING OF NAME | 'Branch shark' because of its triple-pointed teeth |

**BODY SHAPE**
This early shark had a simple and dynamic body shape with a flexible, light skeleton that helped it move fast.

### WHERE IN THE WORLD?

The warm seas of North America.

### FOSSIL EVIDENCE
Cartilaginous skeletons lack bones and decompose quickly, so few became fossils. Intact fossils of bony fish have been found in the stomachs of some specimens, usually positioned to indicate that they had been eaten tail-first. This implies *Cladoselache* outpaced its prey and pulled them back into its mouth, which was mounted at the front of the head. This shark's fin spines were shorter and sharper than the spikier spines of its relatives. One enigma was that males had no claspers, used by sharks for mating, so no one is sure how these creatures reproduced.

PREHISTORIC ANIMAL

DEVONIAN

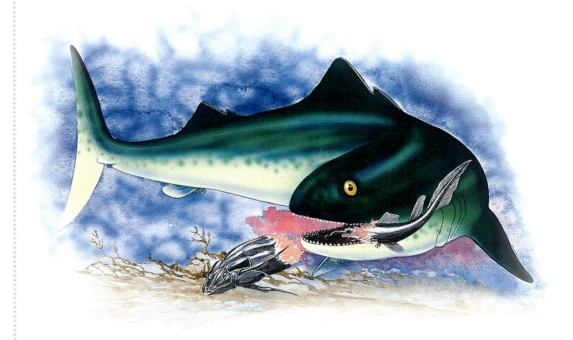

**HOW BIG IS IT?**

**TEETH**
*Cladoselache* teeth were smooth and multi-cusped, good for grabbing prey.

**TIMELINE (millions of years ago)**

| 540 | 505 | 438 | 408 | 360 | 280 | 248 | 208 | 146 | 65 | 1.8 to today |

DEVONIAN

# Eusthenopteron

• **SUPERORDER** • Osteolepidida • **FAMILY** • Tristichopteridae • **GENUS & SPECIES** • Numerous

*Eusthenopteron* is famous for being the 'fish with legs', but it lived in the open seas and there is no evidence that it ever hauled itself onto land, even if it could.

| VITAL STATISTICS | |
|---|---|
| FOSSIL LOCATION | Canada, Europe |
| DIET | Carnivorous |
| PRONUNCIATION | YOOSS-thin-OP-ter-on |
| WEIGHT | Unknown |
| LENGTH | 2.7cm–1m (1in–3ft) |
| HEIGHT | Unknown |
| MEANING OF NAME | 'Good strong fin' because of the sturdy internal bones of its ventral fins. |

### WHERE IN THE WORLD?

Thousands of fossils have been found in the Frasnian Escuminac Formation of the province of Québec, in Canada, making this one of the most studied specimens in the world. There are also remains in Scotland and Russia.

**TEETH**
There are small teeth on both the upper and lower jaws, with fangs a little farther back in the mouth. This was a nasty predator.

### FOSSIL EVIDENCE

The numerous fossils found make this one of the most studied animals in history. Its fin endoskeleton has a distinct set of muscular lobes that look like limbs, and drawings once showed it, quite wrongly, on land, because it was a marine animal. It shares other features with tetrapods (four-legged creatures), such as similar teeth and nostrils. Palaeontologists speculate it also had strong lungs. Specimens come in many sizes, but studies of the skeleton bone structures show this fish usually went through two growth spurts in its lifetime.

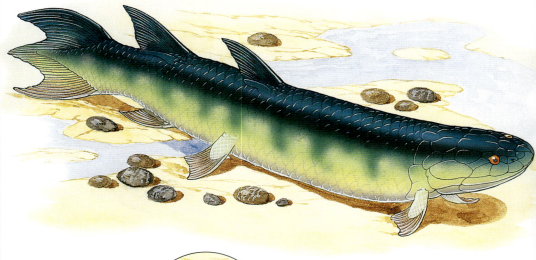

**FINS**
The median fins are far back on the body and shaped like sails, allowing the fish to sprint and surprise its prey.

| PREHISTORIC ANIMAL |
|---|
| DEVONIAN |

**HOW BIG IS IT?**

**TIMELINE (millions of years ago)**

| 540 | 505 | 438 | 408 | 360 | 280 | 248 | 208 | 146 | 65 | 1.8 to today |

# DEVONIAN

# Ichthyostega

• **ORDER** • Ichthyostegalia • **FAMILY** • Ichthyostegidae • **GENUS & SPECIES** • Various species within *Ichthyostega*

*Ichthyostega* was a large four-limbed, aquatic vertebrate. It had a skull with many features in common with lobe-finned fishes of the time. It also had a deep, flattened, finned tail that would have been used to propel it through water. Its sharp teeth mark it as a predator.

## VITAL STATISTICS

| | |
|---|---|
| FOSSIL LOCATION | Greenland |
| DIET | Carnivorous |
| PRONUNCIATION | IK-thee-o-STEG-ah |
| WEIGHT | Unknown |
| LENGTH | 1.5m (5ft) |
| HEIGHT | Unknown |
| MEANING OF NAME | 'Fish roof' because the bones of its skull roof are very fish-like |

**TOES**
The hind feet of *Ichthyostega* each had seven digits. It is currently unknown what the hands looked like, but they probably had more than today's vertebrate maximum of five digits.

## WHERE IN THE WORLD?

All *Ichthyostega* specimens have been unearthed in east Greenland. The first fossils were found in the late 1920s.

## FOSSIL EVIDENCE

*Ichthyostega*, an early tetrapod that lived in the Upper Devonian Period, is one of the earliest and best-known and tetrapods (along with *Acanthostega*, also from Greenland). Abundant specimens collected in the late 1920s and 30s were the first Devonian tetrapods to be found and described. Most of the skeleton is well represented, except, frustratingly, the hands. Knowing what the hands were like is essential to understanding the tetrapods' first transition to a terrestrial life. Devonian tetrapods are traditionally referred to as 'amphibians', but the first true amphibians appeared in the Carboniferous Period.

**RIBS**
Some of *Ichthyostega*'s ribs broadened dramatically in the middle and extensively overlapped adjacent ribs. This evidently restricted flexibility and may have been related to breathing.

PREHISTORIC ANIMAL

DEVONIAN

**HOW BIG IS IT?**

**TIMELINE (millions of years ago)**

| 540 | 505 | 438 | 408 | 360 | 280 | 248 | 208 | 146 | 65 | 1.8 to today |

DEVONIAN

# Dunkleosteus

| VITAL STATISTICS | |
|---|---|
| Fossil Location | Worldwide |
| Diet | Carnivorous |
| Pronunciation | Dunk-lee-OH-stee-us |
| Weight | 1 ton (1.1 tonnes) |
| Length | 6m (20ft) |
| Height | Unknown |
| Meaning of name | 'Dunkle's bone', after the former curator of the Cleveland Museum of Natural History |

*Dunkleosteus* came right at the top of its food chain and was one of the fiercest marine predators that ever lived. Its size and weight meant that it lacked speed, but it had a bite that was both fast and powerful, and almost four times as strong as *Tyrannosaurus rex's*. It was so powerful, in fact, that it has never been equalled. Capable of exerting a pressure of 5600kg per sq cm (80,000lb per sq in), it could tear its prey in two with just a single bite. And its prey was whatever it chose.

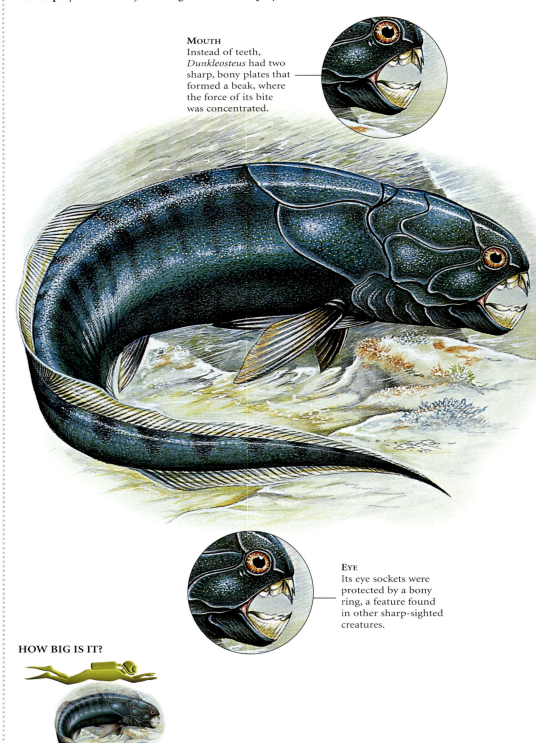

**Mouth**
Instead of teeth, *Dunkleosteus* had two sharp, bony plates that formed a beak, where the force of its bite was concentrated.

**Eye**
Its eye sockets were protected by a bony ring, a feature found in other sharp-sighted creatures.

**HOW BIG IS IT?**

### FOSSIL EVIDENCE

The Cleveland Museum of Natural History holds the most famous *Dunkleosteus* specimen. The armoured frontal sections of this predator primarily have survived in the fossil record, so reconstructions are based on smaller members of the order Arthrodira. Fossils are often found with bone mass and the partially eaten remains of other fish, suggesting that it tended to regurgitate bones – or that it suffered from indigestion. Armour with unhealed bite marks suggests that *Dunkleosteus* may also have turned to cannibalism.

| PREHISTORIC ANIMAL |
| DEVONIAN |

# DEVONIAN

• **ORDER** • Arthrodira • **FAMILY** • Dinichthyidae • **GENUS & SPECIES** • *Dunkleosteus terrelli*

### WHERE IN THE WORLD?

One of the most successful aquatic predators of the Devonian period, *Dunkleosteus* was found worldwide.

**HUNTING**
*Dunkleosteus* could open its mouth very quickly, creating a powerful suction that pulled prey into its mouth.

### REPRODUCTION

*Dunkleosteus* may have been among the first creatures in which females carried the young inside their bodies rather than laying eggs. Evidence for this comes from the fossilized remains of another placoderm, *Materpiscis attenboroughi*. Found in the Gogo Formation in Australia, this female had died while giving birth. Before the fossil's discovery, it was assumed that the first creature to nourish its young inside its body appeared some 200 million years after the placoderms.

### SUCCESSFUL, BUT SHORT-LIVED

*Dunkleosteus* belonged to a class of armour-plated fish known as placoderms. Only their heads and thoraxes were protected by armour; the rest of the body was either naked or covered with scales. One of the earliest fish to have jaws, it dominated its environment but survived for only about 50 million years, leaving no relatives. Sharks, by contrast, have existed for 400 million years.

**TIMELINE (millions of years ago)**

| 540 | 505 | 438 | 408 | 360 | 280 | 248 | 208 | 146 | 65 | 1.8 to today |

DEVONIAN TO TODAY

# Coelacanth

• ORDER • Coelacanthiformes • FAMILY • Numerous • GENUS & SPECIES • Numerous

The coelacanth is a primitive bony fish once known only through fossils. More recently, live specimens have been caught in fisherman's nets in the Indian Ocean. The coelacanth has hardly changed in 360 million years.

## VITAL STATISTICS

| | |
|---|---|
| FOSSIL LOCATION | Worldwide |
| DIET | Carnivorous. Cephalopods, such as cuttlefish, squid and octopus |
| PRONUNCIATION | seel-a-canth |
| WEIGHT | Up to 82kg (180lb) |
| LENGTH | Up to 2m (6½ft) |
| HEIGHT | Unknown |
| MEANING OF NAME | The hollow spines supporting the tail fin: 'coela' means hollow and 'acanth' is spine |

**EYES**
The large eyes have a reflective layer behind the retina. The coelacanth is highly sensitive to light.

### WHERE IN THE WORLD?

The first fossils were discovered in the oceans off Africa but Coelacanth fossils have been found Worldwide.

## FOSSIL EVIDENCE

One mystery about the coelacanth is that no fossils have been found covering the last 65 million years. This may be because it moved to habitats near steep volcanic islands where fossils are unlikely to form. These ancient creatures have many characteristics that distinguish them from other fish, including a 'rostral organ' in the snout (with which it seems to position itself upright to use on the seabed), and a unique hinge that allows the mouth to open very wide. It also has a three-lobed tail.

**FINS**
It can't walk, but this ancient fish does have something in common with humans: unlike other fish, it moves its opposite fins alternately, as we move our legs, rather than in unison.

PREHISTORIC ANIMAL
DEVONIAN TO TODAY

**HOW BIG IS IT?**

**TIMELINE (millions of years ago)**

| 540 | 505 | 438 | 408 | 360 | 280 | 248 | 208 | 146 | 65 | 1.8 to today |

CARBONIFEROUS

# Eogyrinus

• **ORDER** • Anthracosauria • **FAMILY** • Eogyrinidae • **GENUS & SPECIES** • *Eogyrinus attheyi*

### VITAL STATISTICS

| | |
|---|---|
| Fossil Location | England |
| Diet | Fish |
| Pronunciation | EE-oh-ji-RINE-us |
| Weight | 560kg (half a ton) |
| Length | 4.6m (15ft) |
| Height | Unknown |
| Meaning of name | 'Early frog' because it was a primitive amphibian |

*Eogyrinus* was an alligator-like swamp-dwelling tetrapod whose very long tail would have made it a powerful swimmer capable of pouncing on prey in or close to the water.

### WHERE IN THE WORLD?

Fossils have been found in coal seams in England.

**JAWS**
The deep, narrow head had long jaw muscles that equipped *Eogyrinus* with a strong bite similar to a crocodile's today.

### FOSSIL EVIDENCE

*Eogyrinus* was larger than any animal living on land at the time it was alive. With twice as many vertebrae as most amphibians or reptiles, *Eogyrinus* was very flexible and able to move its long tail and body easily through swampland. Its short limbs were fine for paddling and steering, but meant that it would have been unable to lift its belly up when moving around on dry land. It lived a mainly aquatic life, lurking in shallow waters and bursting up to snap its long jaws around its prey.

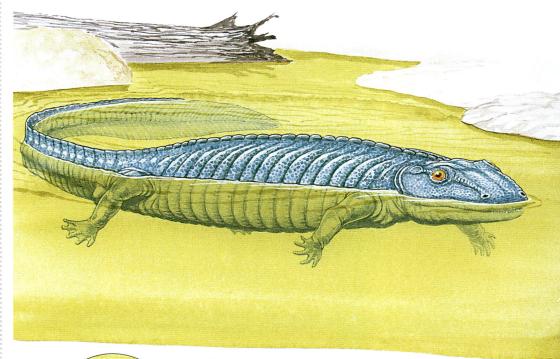

**TEETH**
For its size, *Eogyrinus* was comparatively light, allowing quick, agile movements. It is speculated it could have drowned its victims with a lethal 'death roll' under the water.

### HOW BIG IS IT?

PREHISTORIC ANIMAL

CARBONIFEROUS

**TIMELINE (millions of years ago)**

| 540 | 505 | 438 | 408 | 360 | 280 | 248 | 208 | 146 | 65 | 1.8 to today |

CARBONIFEROUS

# Hylonomus

• ORDER • Captorhinida • FAMILY • Protorothyrididae • GENUS & SPECIES • *Hylonomus lyelli*

| VITAL STATISTICS | |
|---|---|
| FOSSIL LOCATION | Nova Scotia, Canada |
| DIET | Carnivorous |
| PRONUNCIATION | High-lo-NO-muss |
| WEIGHT | Unknown |
| LENGTH | 25cm (10in) |
| HEIGHT | Unknown |
| MEANING OF NAME | 'Forest wanderer' because its fossils are found in ancient forest |

In life, *Hylonomus* probably looked much like a modern lizard. Its small, sharp teeth in a single row along the jawline indicate that it probably ate small invertebrates, such as millipedes or early insects.

**WHERE IN THE WORLD?**

Fossils of *Hylonomus* have been found in the remains of fossilized tree stumps in Joggins, Nova Scotia. Fossil tracks found in New Brunswick have also been attributed to *Hylonomus*.

**EARS**
Details of the ear region of *Hylonomus* suggest that it was not very good at hearing airborne sounds.

**FOSSIL EVIDENCE**
*Hylonomus* was an early reptile that lived during the Carboniferous Period. The first fossils were discovered by Canadian geologist Sir William Dawson in the 1800s. The specimens occur in fossilized lycopod tree stumps at the famous Joggins site in Nova Scotia. The stumps, hollowed out by rot, acted as traps for small vertebrates. One stump was found to contain seventeen small skeletons. Some scientists have recently suggested the trunks were dens rather than traps. Today, these trees can still be seen eroding out of the cliffs on to the beach below.

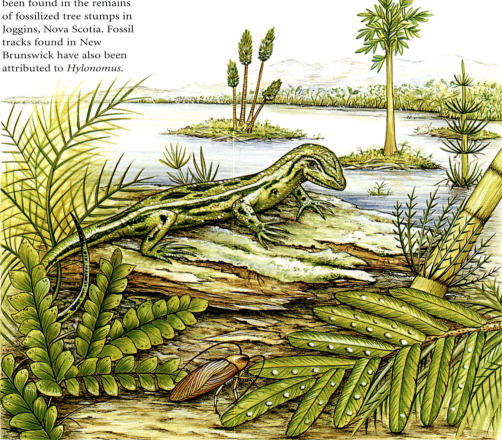

PREHISTORIC ANIMAL

CARBONIFEROUS

**HOW BIG IS IT?**

**TAIL**
The tail of *Hylonomus* and other animals built like it, such as modern lizards, was used as a balancing organ for these swift and agile runners.

**TIMELINE (millions of years ago)**

| 540 | 505 | 438 | 408 | 360 | 280 | 248 | 208 | 146 | 65 | 1.8 to today |

LATE CARBONIFEROUS

# Arthropleura

**ORDER** • Arthropleurida • **FAMILY** • Arthropleuridae • **GENUS & SPECIES** • Various

**Arthropleura** was a huge arthropod that had a flattened, many-segmented body with bumpy ornamentation. It is the largest known land invertebrate of all time.

## VITAL STATISTICS

| | |
|---|---|
| Fossil Location | North America and Europe (England, Scotland, Germany, the Netherlands) |
| Diet | Herbivorous |
| Pronunciation | AR-thro-PLUR-ah |
| Weight | Unknown |
| Length | 1.8m (6ft) |
| Height | Unknown |
| Meaning of name | 'Jointed sides' because of its jointed body armour |

## FOSSIL EVIDENCE

*Arthropleura* was an Upper Carboniferous relative of centipedes and millipedes. It is known mostly from isolated pieces of its segmented body, but rare, nearly complete specimens are also known. Some of its immense trackways are also known. *Arthropleura* may have evolved to be so large because of a local lack of large terrestrial predators. However, its giant size might also have been related to the comparatively oxygen-enriched atmosphere of its time.

PREHISTORIC ANIMAL
LATE CARBONIFEROUS

**TRACKWAYS**
Fossil tracks attributable to *Arthropleura* have been given the name *Diplichnites cuithensis* and sometimes indicate a sinuous walking habit.

## WHERE IN THE WORLD?

Body and trace fossils of *Arthropleura* have been found in the USA (Ohio, Pennsylvania, Illinois, Kansas and New Mexico), Canada, and Europe.

**LEGS**
One of the things that separates *Arthropleura* from centipedes and millipedes is the large number of segments in its legs.

## HOW BIG IS IT?

**TIMELINE (millions of years ago)**

| 540 | 505 | 438 | 408 | 360 | 280 | 248 | 208 | 146 | 65 | 1.8 to today |

ORDOVICIAN TO PERMIAN

# Eurypterid

• ORDER • Xiphosura • FAMILY • Various • GENUS & SPECIES • Various

Eurypterids, as arthropods, had a segmented external skeleton and jointed legs. Their tails were either pointed spines (as in their modern relatives, the scorpions) or flattened swimming paddles.

| VITAL STATISTICS | |
|---|---|
| Fossil Location | Worldwide |
| Diet | Varied |
| Pronunciation | ya-RIP-ta-rid |
| Weight | 5.5 tonnes (5 tons) |
| Length | 10cm–2.4m (4in–over 8ft) |
| Height | Unknown |
| Meaning of name | 'Broad wing' because of the flattened swimming paddles that many had |

### WHERE IN THE WORLD?

Fossils of eurypterids are found all over the world. The Silurian deposits in New York State are especially rich and represent a broad sample of eurypterid diversity.

**PADDLES**
Many eurypterids had two paddles as their rearmost set of legs. This shows they were swimming animals.

### FOSSIL EVIDENCE

Eurypterid fossils come from rocks of the Paleozoic Era between the Ordovician and the Permian Era. Trackways suggest they may have been around as early as the Cambrian. Complete specimens can be common in some beds, but disarticulated body elements are much more common because all arthropods moult their exoskeletons from time to time in order to grow. The anatomy and appearance of some of the largest eurypterids must be reconstructed from isolated parts: the larger an animal becomes, the harder it is to preserve the entire creature.

PREHISTORIC ANIMAL

ORDOVICIAN – PERMIAN

### HOW BIG IS IT?

**SEXES**
Many eurypterid species are well-preserved and abundant so scientists are able to tell the difference between males and females.

**TIMELINE (millions of years ago)**

| 540 | 505 | 438 | 408 | 360 | 280 | 248 | 208 | 146 | 65 | 1.8 to today |

PERMIAN

# Cacops

• **ORDER** • Temnospongdyli • **FAMILY** • Dissorophidae • **GENUS & SPECIES** • *Cacops aspidephorus, C. aspidophorus*

*Cacops* was a large amphibian that was well protected from its predators by its heavy body armour. It might have been able to locate prey in the dark using its superb hearing.

## VITAL STATISTICS

| | |
|---|---|
| Fossil Location | USA |
| Diet | Carnivorous |
| Pronunciation | KAY-kops |
| Weight | Unknown |
| Length | 40cm (16in) |
| Height | Unknown |
| Meaning of name | 'Bad Face' |

**Head**
A large head over a wide, stocky body held up by heavy limbs gave *Cacops* a sprawling stance. Its movements are likely to have been fairly clumsy.

### WHERE IN THE WORLD?

Poorly preserved remains of *Cacops* are found in the Cacops Bone Bed in Baylor County, Texas, and at Fort Sill, Oklahoma.

## FOSSIL EVIDENCE

Much of the interest in this armoured amphibian centres on its otic notch. This is an opening behind each eye enclosed by a bony bar. It was covered by thin membrane, which might have vibrated like an eardrum and picked up sounds. Such equipment might have allowed the *Cacops* to pick out small land animals in the dark and snap them up in its large jaws. It seems likely it was a nocturnal and land-based hunter, able to enter water for defence and to seek other prey.

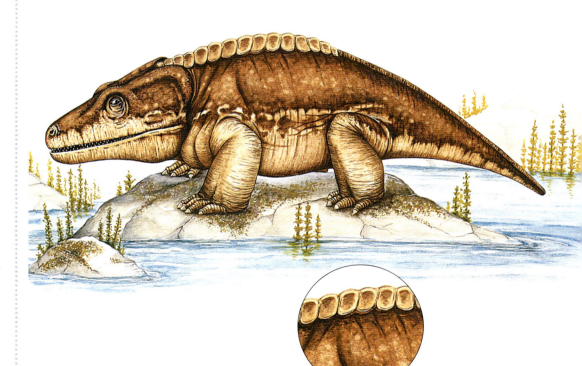

PREHISTORIC ANIMAL

PERMIAN

**HOW BIG IS IT?**

**Protection**
Bony plates covered its body, offering good protection from predators. *Cacops* also sported doubled bony plates along its backbone. Clearly it was not a fast mover and so needed other ways of keeping itself off the local menu

**TIMELINE (millions of years ago)**

| 540 | 505 | 438 | 408 | 360 | 280 | 248 | 208 | 146 | 65 | 1.8 to today |

PERMIAN

# Diadectes

• **ORDER** • Diadectomorpha • **FAMILY** • Diadectidae • **SPECIES** • *Diadectes absitus, D. maximus, D. tenuitectus*

| VITAL STATISTICS | |
|---|---|
| FOSSIL LOCATION | North America and Europe |
| DIET | Herbivorous |
| PRONUNCIATION | Die-ah-Deck-teez |
| WEIGHT | 100kg (220lb) |
| LENGTH | Up to 3m (10ft) |
| HEIGHT | Unknown |
| MEANING OF NAME | 'Through biter' because of its cutting teeth |

*Diadectes* was a giant, hauling a stocky 3-metre (10-ft) body through the Permian forest and rooting out plants. *Diadectes* was among the earliest vertebrates with a herbivorous diet.

**TEETH**
*Diadectes* had eight peg-like front teeth that nipped off mouthfuls of vegetation to be ground up by the molar-like cheek teeth.

### WHERE IN THE WORLD?

Fossils have been found across North America, especially in Texas, and also in Europe.

### FOSSIL EVIDENCE

The combination of amphibian-like skull and reptilian body led early palaeontologists to believe that *Diadectes* was the link between amphibians and reptiles. But it isn't. There was also speculation that it burrowed underground like a mole with its strong fingers. However, it is now thought that it had claws to dig up plants. Unlike many far more advanced reptiles, it was able to breathe and chew at the same time because it had a partial secondary palate. This was useful because such a large animal needs to eat and digest a lot of vegetation to stay alive.

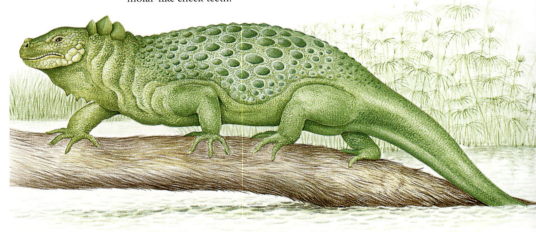

**LAND-BASED**
The thick-boned skull, heavy vertebrae and ribs and strong, stocky limbs clearly indicate that this was a land-based animal.

### HOW BIG IS IT?

| PREHISTORIC ANIMAL |
|---|
| PERMIAN |

**TIMELINE (millions of years ago)**

| 540 | 505 | 438 | 408 | 360 | 280 | 248 | 208 | 146 | 65 | 1.8 to today |

PERMIAN

# Eryops

• ORDER • Temnospondyli • FAMILY • Unknown • GENUS & SPECIES • *Eryops megacephalus*

One of the largest land animals of its time, *Eryops* was a fierce predatory amphibian equipped with a mouthful of sharp teeth to grab prey, which was then flung straight down its throat.

### VITAL STATISTICS

| | |
|---|---|
| FOSSIL LOCATION | USA |
| DIET | Carnivorous |
| PRONUNCIATION | Air-ee-ops |
| WEIGHT | 90kg (200lb) |
| LENGTH | 1.5–2m (5–6ft) |
| HEIGHT | About 35cm (14in) |
| MEANING OF NAME | 'Drawn-out face' because most of the skull was in front of the eyes |

### WHERE IN THE WORLD?

Mostly in Texas, southern USA, but also in eastern USA and New Mexico.

**HEAD**
The skull was broad and flat and could be 60cm (24in) long. The jaw opened to reveal fang-like teeth for gripping prey, but it couldn't chew so it would throw the victim whole down its throat.

### FOSSIL EVIDENCE

*Eryops* is a good example of a creature that adapted from an aquatic to a land-based environment. It developed strong limbs to carry its weight out of water and a sturdy backbone to prevent its body sagging. Its fish jawbones evolved into a simple ear mechanism. *Eryops* had a short tail, suggesting it was not a fast swimmer through the swamp; it probably hunted by stealth, staying low or partially submerging itself like a modern crocodile or alligator.

**LEGS**
With its body weight pushed forwards in front of stubby legs, *Eryops* could only take short strides, so it was no runner. Fossilized prints of close relatives show a slow and clumsy gait.

**HOW BIG IS IT?**

PREHISTORIC ANIMAL

PERMIAN

**TIMELINE (millions of years ago)**

| 540 | 505 | 438 | 408 | 360 | 280 | 248 | 208 | 145 | 65 | 1.8 to today |

PERMIAN

# Mesosaurus

• ORDER • Mesosauria • FAMILY • Mesosauridae • GENUS & SPECIES • *Mesosaurus tenuidens* and *M. brasiliensis*

| VITAL STATISTICS | |
|---|---|
| FOSSIL LOCATION | South America and South Africa |
| DIET | Piscivorous or planktivorous |
| PRONUNCIATION | MEE-zo-SAW-rus |
| WEIGHT | Unknown |
| LENGTH | 0.95m (3ft) |
| HEIGHT | Unknown |
| MEANING OF NAME | 'Middle lizard' because of its multiple evolutionary affinities |

*Mesosaurus* was a freshwater swimmer, as evidenced by its webbed feet and its long flattened tail. Additionally, its nostrils were located high on the skull, allowing it to breathe with only the top of its head breaking the water's surface.

### WHERE IN THE WORLD?

Fossils of *Mesosaurus* are found in southern Africa and in southern South America. They only occur in rocks deposited in fresh water and are often complete.

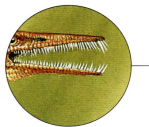

**TEETH**
Some scientists think that its numerous, thin teeth are a classic indicator that *Mesosaurus* was a fish-eater.

## FOSSIL EVIDENCE

*Mesosaurus* was one of the first tetrapods to return to a life in the water after its ancestors had adapted to life on land. It is common in many freshwater exposures of continents flanking the southern Atlantic. This fact, along with the observation that such an animal could not have crossed an entire ocean, helped add weight to the theory of Continental Drift when it was first mooted in the early twentieth century. As it turns out, when *Mesosaurus* was alive, the Atlantic Ocean hadn't yet formed, and its habitats were continuous and part of a single super-continent: Pangaea.

**TAIL**
The flattened, flexible tail of *Mesosaurus* was one of its main propulsive adaptations.

### HOW BIG IS IT?

PREHISTORIC ANIMAL

PERMIAN

**TIMELINE (millions of years ago)**

| 540 | 505 | 438 | 408 | 360 | 280 | 248 | 208 | 146 | 65 | 1.8 to today |

PERMIAN

# Moschops

• ORDER • Therapsida • FAMILY • Tapinocephalidae • GENUS & SPECIES • Various species within the genus *Moschops*

### VITAL STATISTICS

| | |
|---|---|
| FOSSIL LOCATION | South Africa |
| DIET | Herbivorous |
| PRONUNCIATION | MAH-skops |
| WEIGHT | Over a ton |
| LENGTH | 5m (16ft) |
| HEIGHT | Unknown |
| MEANING OF NAME | 'Calf face' |

*Moschops* was a bulky, herbivorous synapsid with splayed forelimbs and shorter, erect hind limbs. The different lengths of the front and hind limbs gave the back a slope that went down towards the tail, which was very short.

### WHERE IN THE WORLD?

*Moschops* is found in South Africa but others in its family, the Tapinocephalidae, are also known from Russia.

**TEETH**
*Moschops*' short, peg-like, chisel-edged teeth are all of one type throughout the mouth. They show that *Moschops* was a plant-eater.

### FOSSIL EVIDENCE

*Moschops* is known from a large number of skulls and associated skeletons from Permian rocks from the Karoo region of South Africa. The massive, heavy-boned skulls are sturdy and thus more readily preserved. Over the years, the fossils have been given many names and with further study, many have been shown to be synonymous: *Moschoides*, *Agnosaurus*, *Moschognathus* and *Pnigalion*.

**TRACKWAYS**
Fossil trackways of close relatives of *Moschops* were described recently from the Karoo rocks of South Africa. They come from an area that was a floodplain back in the Permian Era.

### HOW BIG IS IT?

PREHISTORIC ANIMAL

PERMIAN

**TIMELINE** (millions of years ago)

| 540 | 505 | 438 | 408 | 360 | 280 | 248 | 208 | 146 | 65 | 1.8 to today |

PERMIAN

# Platyhystrix

• ORDER • Temnospondyli • FAMILY • Dissorophidae • GENUS & SPECIES • *Platyhystrix rugosus*

| VITAL STATISTICS | |
|---|---|
| Fossil Location | USA (Texas, Utah, New Mexico and Colorado) |
| Diet | Carnivorous |
| Pronunciation | PLAT-ee-HISS-trix |
| Weight | Unknown |
| Length | 0.9m (3ft) |
| Height | Unknown |
| Meaning of name | 'Flat porcupine' because of its flat, vertical, vertebral prongs |

*Platyhystrix* was a terrestrial temnospondyl with a distinctive sail along its back. The sail seems to have evolved from rows of overlapping bony plates that sat flat over the spines of its ancestors. They may have served to strengthen the back or offer armour protection.

**ARMOUR**
The late Pennsylvanian to early Permian dissorophids had well-developed limbs and solid vertebrae.

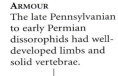

### WHERE IN THE WORLD?

*Platyhystrix* has only been found in the USA. The family it belongs to is known from Europe as well and was likely more widespread.

### FOSSIL EVIDENCE

Animals with a series of elongated, vertical bony projections above their backbones have a long history, spanning over 300 million years. They include dinosaurs, their close relatives, some 'amphibians', and synapsids (mammals and their ancestors). Tall, narrow series were most likely used to support skin sails, which are thought to have regulated body temperature. Heavier series might have been hump-support structures as found in modern bison. *Platyhystrix's* dramatically curved and textured vertebral extensions may have served a different, as yet unknown, function.

### HOW BIG IS IT?

**OSTEODERMS**
Roughened osteoderms (bony plates), which possibly had a protective function, were stuck to its ribs.

| PREHISTORIC ANIMAL |
| PERMIAN |

**TIMELINE (millions of years ago)**

540 | 505 | 438 | 408 | 360 | 280 | 248 | 208 | 146 | 65 | 1.8 to today

PERMIAN

# Scutosaurus

• **ORDER** • Procolophonia • **FAMILY** • Pareiasauridae • **GENUS & SPECIES** • *Scutosaurus karpinskii*

| VITAL STATISTICS | |
|---|---|
| FOSSIL LOCATION | Russia |
| DIET | Herbivorous |
| PRONUNCIATION | SKOO-toe-SAW-rus |
| WEIGHT | Unknown |
| LENGTH | perhaps more than 3.3m (11ft) |
| HEIGHT | Unknown |
| MEANING OF NAME | 'Shield reptile' because it was covered with bony armour |

*Scutosaurus* was a large, armoured, plant-eating pareiasaur. Pareiasaurs lived during the Permian Era and were synapsids, as are mammals. They had stocky bodies with short tails and often highly sculpted and adorned skulls.

**EARS**
A slender inner ear bone (the stapes) found in some pareiasaurs suggests they were able to hear high-frequency airborne sounds.

**WHERE IN THE WORLD?**

*Scutosaurus* is well known from Russia's Permian rocks. Examples of the group it comes from, the pareiasaurs, are also well represented in Permian rocks of South Africa.

**FOSSIL EVIDENCE**
The Permian beds of Russia are renowned for their abundant vertebrate fossils. In one site, several *Scutosaurus* skeletons were found together in a flood channel sand deposit. Some *Scutosaurus* skeletons have been found in a standing position, suggesting that the animals became mired in the mud and swamps, possibly while feeding on vegetation. Some scientists have used this observation to assert that scutosaurus was probably an aquatic animal, but animals adapted to living in water are less likely to be mired in mud than those that spend most of their time on land.

PREHISTORIC ANIMAL

PERMIAN

**HOW BIG IS IT?**

**LEGS**
Unlike most reptiles, *Scutosaurus* held its legs directly underneath its body to support its great weight. Today, only endothermic vertebrates do this.

**TIMELINE (millions of years ago)**

| 540 | 505 | 438 | 408 | 360 | 280 | 248 | 208 | 146 | 65 | 1.8 to today |

PERMIAN

# Seymouria

• ORDER • Seymouriamorpha • FAMILY • Seymouriidae •
GENUS & SPECIES • *Seymouria baylorensis, S. sanjuanensis, S. grandis, S. agilis*

| VITAL STATISTICS | |
|---|---|
| Fossil Location | North America |
| Diet | Omnivorous |
| Pronunciation | See-MOOR-ee-ah |
| Weight | 15kg (23lb) |
| Length | 60cm (24in) |
| Height | Unknown |
| Meaning of name | 'Seymour' is the name of the town in Texas where it was first discovered |

With its amphibian skull on a reptilian body, *Seymouria* is an enigma: which is it? Many palaeontologists believe it is a reptilomorph, part of a group of amphibians from which reptiles evolved.

**NASAL GLAND**
*Seymouria* could spend more time on land because it may have had a gland in its nose through which it could excrete excess salt from its blood. Some modern reptiles are also able to do this.

### WHERE IN THE WORLD?

First found in Texas. Other specimens have been located in Utah, Oklahoma and New Mexico, and there is a European branch in Germany.

**FOSSIL EVIDENCE**
Several three-dimensional fossil skeletons have survived, allowing for accurate reconstruction of the skull. Skulls believed to be male were particularly thick and it is speculated it may have used this to batter rivals in mating contests. It may have had dry skin giving it the ability to conserve moisture allowed it to spend long periods away from water, but its short forelimbs and shank suggest it was not a fast mover. Therefore, this tetrapod was well adapted to roam the land in search of food in the dry Permian climate, hardly returning to the water apart from possibly laying its eggs.

| PREHISTORIC ANIMAL |
|---|
| PERMIAN |

**HOW BIG IS IT?**

**LEGS**
Long, muscular legs allowed *Seymouria* to travel far from water seeking its diet of possibly insects, small vertebrates and carrion.

**TIMELINE** (millions of years ago)

540 | 505 | 438 | 408 | 360 | 280 | 248 | 208 | 146 | 65 | 1.8 to today

PERMIAN

# Youngina

• ORDER • Eosuchia • FAMILY • Younginidae • GENUS & SPECIES • *Youngina capensis*

*Youngina* was a lizard-like reptile of the Late Permian Period. It had a powerful bite and snapped its jaws, which were armed with tiny blade-like teeth.

### VITAL STATISTICS

| | |
|---|---|
| FOSSIL LOCATION | South Africa |
| DIET | Snails and insects |
| PRONUNCIATION | Yung-in-nah |
| WEIGHT | Unknown |
| LENGTH | 30–45cm (12–18in) |
| HEIGHT | Unknown |
| MEANING OF NAME | Named after Scottish fossil collector John Young |

**SNOUT**
Its long, slender snout and slim body suggest that *Youngina* was a burrower, possibly living underground or retreating there as refuge from predators.

### WHERE IN THE WORLD?

*Youngina* remains have been found in the Karoo Beds of South Africa.

### FOSSIL EVIDENCE

*Youngina* had many features seen in modern lizards, such as a broad breastbone, short neck, long tail and thin digits. These long toes and fingers may have been useful for burrowing, or possibly climbing trees. *Youngina* is part of the diapsid group, which features skull openings allowing for the attachment of strong muscles that could have given it a formidable jaw. *Youngina* died out in the mass extinction of the late Permian period, but its close relatives could include the lizards and snakes of today.

PREHISTORIC ANIMAL

PERMIAN

### HOW BIG IS IT?

**TEETH**
*Youngina* had a mighty bite with sharp, blade-like teeth that could cut through tough skin.

**TIMELINE (millions of years ago)**

| 540 | 505 | 438 | 408 | 360 | 280 | 248 | 208 | 146 | 65 | 1.8 to today |

35

PERMIAN

# Dimetrodon

| VITAL STATISTICS | |
|---|---|
| Fossil Location | USA and Europe |
| Diet | Carnivorous |
| Pronunciation | dy-MEE-tro-don |
| Weight | Unknown |
| Length | 3.3m (11ft) |
| Height | 1.2m (4ft) |
| Meaning of name | 'Two-measures of teeth', because it had both shearing and sharp canine teeth |

*Dimetrodon* was a large, predatory, early synapsid with a huge sail on its back. Synapsids are the group of vertebrates that includes mammals and their ancestors. Other related synapsids, both herbivorous and carnivorous, sported similar sails: *Edaphosaurus*, *Ianthasaurus* and, to lesser degree, *Sphenacodon*.

**Sail**
The most striking feature of this animal was its back sail, thought by many to have had a function in regulating the creature's body temperature.

**FOSSIL EVIDENCE**
Fossils of *Dimetrodon* are common in the Permian beds of Texas, USA, where the first fossils of this genus were found in the 1870s and named by the famous palaeontologist Edward Drinker Cope. It is now known from as many as 15 species in the USA and Europe. The bony prongs that supported the sail, which are extensions of the vertebrae, are extremely fragile once mineralized; it is very difficult to find them intact and equally hard to extract them intact. Because of this, extra care must be taken to exhibit this iconic animal.

PREHISTORIC CREATURE

PERMIAN

**HOW BIG IS IT?**

**A DOMINANT PREDATOR**
*Dimetrodon* was the dominant predator of its time and its enemies attacked it at their peril. This may have been why *Dimetrodon* did not have to be fast on its feet, unlike more vulnerable creatures which needed speed if they were going to escape their enemies. *Dimetrodon*'s diet included prey much smaller than itself, such as other vertebrates.

36

# PERMIAN

• **ORDER** • Pelycosauria • **FAMILY** • Sphenacodontidae • **GENUS & SPECIES** • Various species within the genus *Dimetrodon*

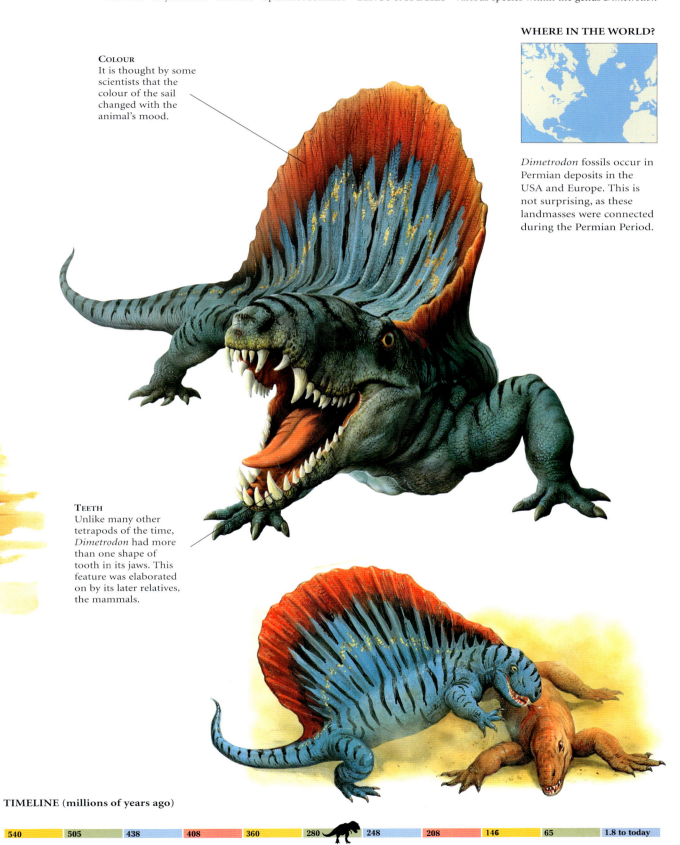

**COLOUR**
It is thought by some scientists that the colour of the sail changed with the animal's mood.

**WHERE IN THE WORLD?**

*Dimetrodon* fossils occur in Permian deposits in the USA and Europe. This is not surprising, as these landmasses were connected during the Permian Period.

**TEETH**
Unlike many other tetrapods of the time, *Dimetrodon* had more than one shape of tooth in its jaws. This feature was elaborated on by its later relatives, the mammals.

**TIMELINE (millions of years ago)**

| 540 | 505 | 438 | 408 | 360 | 280 | 248 | 208 | 146 | 65 | 1.8 to today |

PERMIAN
# Dimetrodon

PERMIAN

• **ORDER** • Pelycosauria • **FAMILY** • Sphenacodontidae • **GENUS & SPECIES** • Various species within the genus *Dimetrodon*

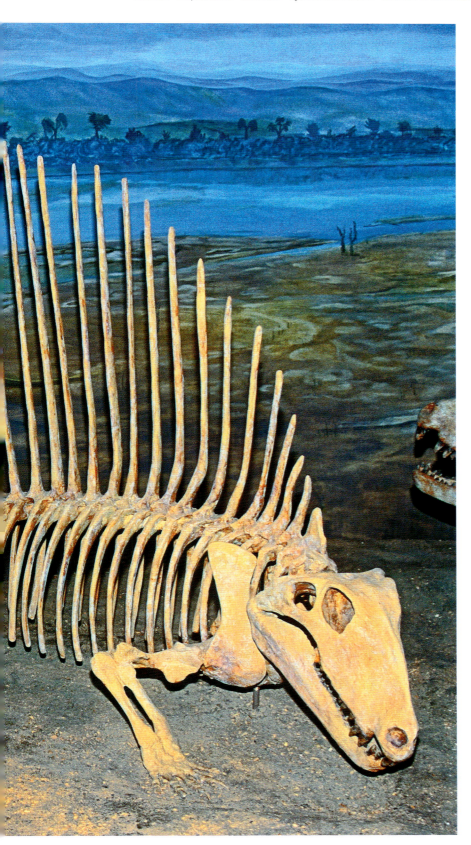

### DIMETRODON SAILS

*Dimetrodon* has often been popularly classed as a dinosaur. In fact, it lived 40 million years before the first dinosaurs appeared on Earth and is classed among the pelycosaurs, a group of vertebrates – animals with spines – that carried sails on their backs. As well as possibly regulating *Dimetrodon's* body temperature, the sails could have been used for display to impress a female *Dimetrodon* during the mating ritual. The sail served, too, as a warning signal to other predators by making *Dimetrodon* look a great deal bigger than it actually was.

Scientists believe that the sail worked like this: the sail, which was covered by skin, absorbed heat from the sun, which warmed the body. If the animal was in danger of overheating, it might have angled the sail away from direct sun to cool off. This was an efficient system that enabled *Dimetrodon* to remain active for longer during the day, so increasing its ability to find food and fend off enemies.

PERMIAN

# Diplocaulus

| VITAL STATISTICS | |
|---|---|
| Fossil Location | USA |
| Diet | Carnivorous |
| Pronunciation | DIF-low-caw-us |
| Weight | Unknown |
| Length | 1m (3ft) |
| Height | Unknown |
| Meaning of name | 'Two-fold stem' after the wing-shaped bones in its head |

*Diplocaulus* preceded dinosaurs by some 20 million years. It is remarkable for its boomerang-shaped head, which it may have used for defence, by punching sideways. Just 1m (3ft) long, *Diplocaulus* crawled on short, weak legs, but moved quickly through water, propelled by its tail. An amphibian, it may have retained its gills into adulthood, enabling it to spend long periods submerged, watching for prey – probably fish and insects, which it sliced with its sharp teeth. It laid its eggs in water – or, at least, damp locations – and these probably hatched into larvae.

**HEAD**
Up to six times wider than it was long, the head of *Diplocaulus* would have been difficult to swallow by even the largest predator.

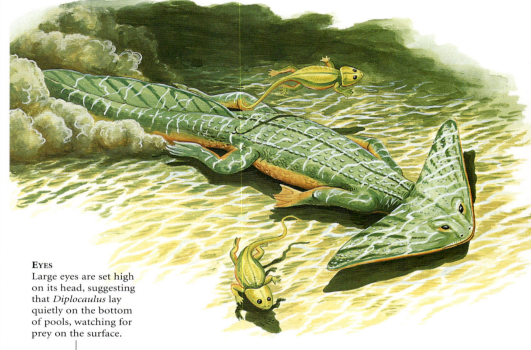

**FOSSIL EVIDENCE**
Several specimens of *Diplocaulus* have been found in the Texas Red Beds. Until the discovery of fossils here, almost no record existed of the Palaeozoic Era, a period that began about 540 million years ago and ended some 250 million years ago. Discovered in 1877, the Texas Red Beds have yielded thousands of specimens, revealing many species. *Diplocaulus* fossils are a common find.

**EYES**
Large eyes are set high on its head, suggesting that *Diplocaulus* lay quietly on the bottom of pools, watching for prey on the surface.

PREHISTORIC ANIMAL

PERMIAN

**HOW BIG IS IT?**

**AHEAD OF THE GAME?**
Young *Diplocaulus* did not have the curious head shape of adults. The bones that formed the boomerang grew slowly, its ever-increasing size being a sign of maturity. In the same period, *Diploceraspis* was another amphibian with a vastly expanded skull, suggesting that an oddly shaped head offered advantages. It may have acted like a hydrofoil, enabling it to swim against the current.

PERMIAN

• ORDER • Nectridea • FAMILY • Keraterpetonidae • GENUS & SPECIES • *Diplocaulus salamandroides*

**WHERE IN THE WORLD?**

*Diplocaulus* was found in what are now the US states of Texas and Oklahoma.

**TAIL**
*Diplocaulus* probably moved through the water with a side-to-side movement, propelled by a short but relatively powerful tail.

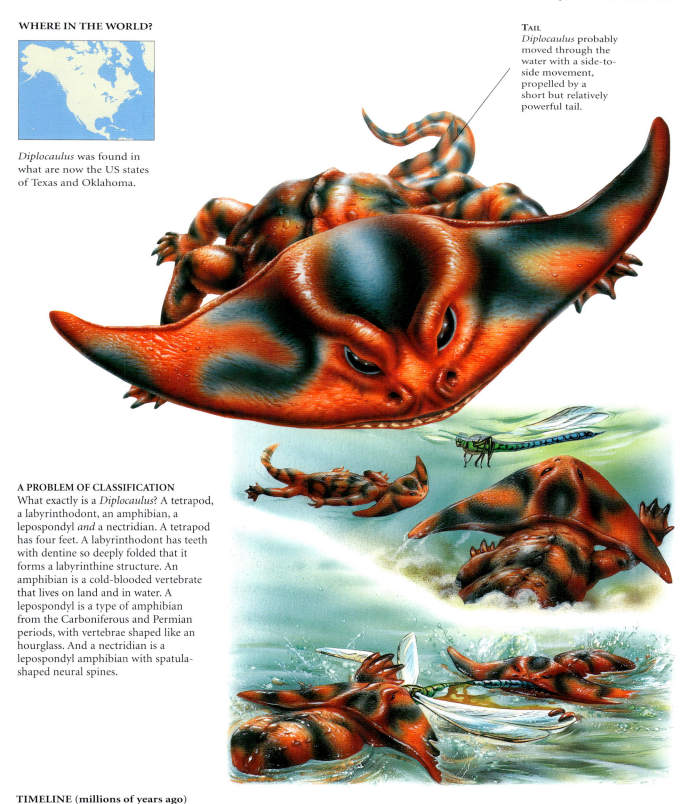

**A PROBLEM OF CLASSIFICATION**
What exactly is a *Diplocaulus*? A tetrapod, a labyrinthodont, an amphibian, a lepospondyl *and* a nectridian. A tetrapod has four feet. A labyrinthodont has teeth with dentine so deeply folded that it forms a labyrinthine structure. An amphibian is a cold-blooded vertebrate that lives on land and in water. A lepospondyl is a type of amphibian from the Carboniferous and Permian periods, with vertebrae shaped like an hourglass. And a nectridian is a lepospondyl amphibian with spatula-shaped neural spines.

**TIMELINE (millions of years ago)**

| 540 | 505 | 438 | 408 | 360 | 280 | 248 | 208 | 146 | 65 | 1.8 to today |

PERMIAN

# Coelurosauravus

• ORDER • [unranked]
• FAMILY • Coelurosauravidae • GENUS & SPECIES • *Coelurosauravus jaekeli, C. elivensis*

## VITAL STATISTICS

| | |
|---|---|
| FOSSIL LOCATION | Europe and Madagascar |
| DIET | Probably insects and small animals |
| PRONUNCIATION | Seel-LUR-ah-sawr-AY-vus |
| WEIGHT | Unknown |
| LENGTH | 30cm (12in) |
| HEIGHT | Unknown |
| MEANING OF NAME | 'Grandfather of Coelurosaur' |

### WHERE IN THE WORLD?

Fossils have been found in Germany, England (which were linked in the Permian Period) and Madagascar.

This was one of the earliest gliding reptiles known. Its wing design was unique and died with it about 250 million years ago.

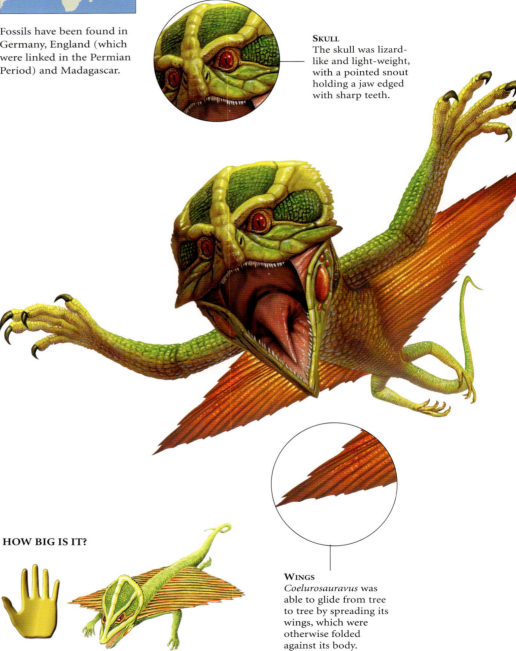

**SKULL**
The skull was lizard-like and light-weight, with a pointed snout holding a jaw edged with sharp teeth.

### FOSSIL EVIDENCE

*Coelurosauravus* had retractable wings formed from skin that grew on each side of its body. They were supported by rod-like bones that were, surprisingly, not attached to the ribcage These would have opened up like curved paper fans when it launched itself from a tree. It was kept stable – and possibly steered – by using its long, slender tail, and it probably also stretched its arms to aid direction when gliding. These arms ended with flexible claws enabling it to grab and hold onto its new perch.

PREHISTORIC ANIMAL
PERMIAN

### HOW BIG IS IT?

**WINGS**
*Coelurosauravus* was able to glide from tree to tree by spreading its wings, which were otherwise folded against its body.

**TIMELINE (millions of years ago)**

| 540 | 505 | 438 | 408 | 360 | 280 | 248 | 208 | 146 | 65 | 1.8 to today |

TRIASSIC

# Eoraptor

• ORDER • Saurischia • FAMILY [unranked] • GENUS & SPECIES • *Eoraptor lunensis*

## VITAL STATISTICS

| | |
|---|---|
| FOSSIL LOCATION | Argentina |
| DIET | Carnivorous, possibly also a scavenger |
| PRONUNCIATION | EE-oh-RAP-tor |
| WEIGHT | 10kg (22lb) |
| LENGTH | 1m (3ft) |
| HEIGHT | 50cm (20in) |
| MEANING OF NAME | 'Dawn thief' because it was an early carnivorous dinosaur |

Only identified in 1993, *Eoraptor* was one of the earliest dinosaurs, a meat-eating beast that stood up on two long hind legs and tore its prey apart with its large claws.

**MOUTH**
*Eoraptor's* mouth had dozens of teeth that were clearly designed for a carnivorous diet.

**WHERE IN THE WORLD?**

Discovered in 1991 in the Ischigualasto Badlands in northwestern Argentina. This area was a river valley at the time *Eoraptor* existed but is now desert.

## FOSSIL EVIDENCE

This is one of the few early dinosaurs for which an entire skeleton has been found. With its light, hollow bones and long, thin legs ending in three-toed feet, *Eoraptor* was built for speed. Staying stable by keeping its tail out behind it, it simply outran its victims and then ripped them apart with its claws or teeth. *Eoraptor* was able to maintain an upright posture on two legs because of the structural strength provided by fused vertebrae in its hip region.

**HOW BIG IS IT?**

**Hands**
*Eoraptor* had five digits at the end of each 'hand', three of them forming long claws. The other two digits were shorter.

DINOSAUR

TRIASSIC

**TIMELINE (millions of years ago)**

| 540 | 505 | 438 | 408 | 360 | 280 | 248 | 208 | 143 | 65 | 1.8 to today |

43

TRIASSIC

# Erythrosuchus

• ORDER • [unranked] • FAMILY • Erythrosuchidae • GENUS & SPECIES • *Erythrosuchus africanus*

| VITAL STATISTICS | |
|---|---|
| Fossil Location | South Africa |
| Diet | Carnivorous |
| Pronunciation | Ee-RITH-row-SOOK-us |
| Weight | Unknown |
| Length | 5m (16ft) |
| Height | Unknown |
| Meaning of name | 'Red crocodile' because the rocks in which it is found stains fossils red. |

This brutally strong beast is the largest known predator of its time, about the size of a crocodile today, and equally able to overpower and devour prey.

**HEAD**
The head was up to 1m (40in) long and featured a massive jaw furnished with a set of sharp conical teeth.

### WHERE IN THE WORLD?

Found in South Africa's *Cynognathus* Assemblage Zone.

## FOSSIL EVIDENCE

*Erythrosuchus* is an early archosaur related to dinosaurs. It probably ate plant-eaters, such as dicynodonts, which were abundant in the mid-Triassic period. It walked on all fours on limbs positioned semi-vertically under its body, and may also have been able to swim well. Such is the size and power of its jaw that it would have taken only one savage bite to kill or disable a victim.

PREHISTORIC ANIMAL

TRIASSIC

**HOW BIG IS IT?**

**STRENGTH**
It would have been hard to power this stocky body with any speed, so *Erythrosuchus* would have relied on surprise and sheer strength to overcome prey.

**TIMELINE** (millions of years ago)

| 540 | 505 | 438 | 408 | 360 | 280 | 248 | 208 | 146 | 65 | 1.8 to today |

TRIASSIC

# Euparkeria

• ORDER • [unranked] • FAMILY • Euparkeriidae • GENUS & SPECIES • *Euparkeria capensis*

## VITAL STATISTICS

| | |
|---|---|
| FOSSIL LOCATION | South Africa |
| DIET | Carnivorous |
| PRONUNCIATION | YOO-park-air-ree-ah |
| WEIGHT | 9kg (20lb) |
| LENGTH | 55cm (22in) |
| HEIGHT | Unknown |
| MEANING OF NAME | Named after English morphologist and naturalist W. Kitchen Parker, and meaning 'Parker's good animal' |

This was a light, lean reptile that probably fed on insects and small animals but may have been able to flee larger predators by switching from four legs to two for a self-preserving sprint.

### WHERE IN THE WORLD?

First found in South Africa in 1913 and 1924.

**TEETH**
The teeth were tiny and needle-like, and were shed periodically to allow room for sharper replacements.

## FOSSIL EVIDENCE

*Euparkeria* skulls have the hole in front of the eye socket that is characteristic of all archosaurs (ruling reptiles). This group includes the dinosaurs, pterosaurs and modern crocodiles. It was one of the earliest reptiles with the ability to move on two legs (bipedal). The speed this generated would have been rare in the Early Triassic period so was a good defence strategy.

**LEGS**
*Euparkeria* walked on four legs most of the time, but may have risen up on its long hind legs to run at greater speed when necessary.

### HOW BIG IS IT?

PREHISTORIC ANIMAL

TRIASSIC

## TIMELINE (millions of years ago)

| 540 | 505 | 438 | 408 | 360 | 280 | 248 | 208 | 146 | 65 | 1.8 to today |

TRIASSIC

# Euskelosaurus

• ORDER • Saurischia • FAMILY • Plateosauridae • GENUS & SPECIES • *Euskelosaurus browni*

One of the largest dinosaurs of the Triassic period, *Euskelosaurus* munched its way through enormous amounts of vegetation to keep its huge frame healthy.

| VITAL STATISTICS | |
|---|---|
| FOSSIL LOCATION | South Africa |
| DIET | Herbivorous |
| PRONUNCIATION | yoo-SKEL-oh-SORE-us |
| WEIGHT | 1,600kg (1.8 tons) |
| LENGTH | 10m (30ft) |
| HEIGHT | 3m (10ft) |
| MEANING OF NAME | 'Well-limbed lizard', after its long thigh bone |

### WHERE IN THE WORLD?

Lesotho in South Africa, and Zimbabwe, although it is similar to specimens found in South America. In the Triassic period these continents were joined.

**LEGS**
Like other plateosaurs, *Euskelosaurus* was probably able to rise up onto its hind legs to reach higher branches.

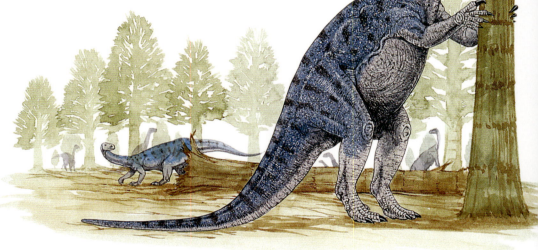

**FOSSIL EVIDENCE**
The hundreds of bones and 16 partial skeletons of *Euskelosaurus* that have been found suggest it was a very common dinosaur. However, the absence of any finds of head, hands or feet make it difficult to be clear about its appearance and eating habits. It seems safe to say it walked mostly on all fours and ate all the vegetation it could reach with its long neck. It was first described in 1866, making it one of the first dinosaurs to be identified in Africa.

**THIGH**
Fossils reveal that the shaft of its long thigh bone (from which it gets its name) is twisted. Some palaeontologists suggest this would have made it bow-legged – highly unusual for a dinosaur.

### HOW BIG IS IT?

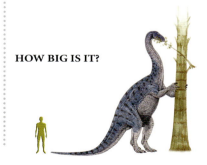

| DINOSAUR |
| TRIASSIC |

**TIMELINE (millions of years ago)**

| 540 | 505 | 438 | 408 | 360 | 280 | 248 | 208 | 146 | 65 | 1.8 to today |

46

TRIASSIC

# Henodus

• **ORDER** • Placodontia • **FAMILY** • Henodontidae • **GENUS & SPECIES** • *Henodus chelyops*

Seeing *Henodus* crawling along the floor of a lagoon would surely convince you it was a turtle, but it was only a distant cousin living a similar life.

| VITAL STATISTICS | |
|---|---|
| FOSSIL LOCATION | Germany |
| DIET | Shellfish |
| PRONUNCIATION | hen-O-dus |
| WEIGHT | Unknown |
| LENGTH | 1m (3ft) |
| HEIGHT | Unknown |
| MEANING OF NAME | 'Single tooth', as it was first thought to have had only one tooth on each side of its skull |

**TEETH**
*Henodus* had two teeth, one on each side of its mouth towards the rear. But a new specimen shows tiny teeth at the front of the jaws, too. The lower jaws were entirely toothless.

## FOSSIL EVIDENCE

*Henodus* was a placodont of the Late Triassic Period. It is known from several specimens, including a well-preserved, complete, articulated skeleton. Placodonts were a diverse group of aquatic reptiles of the Triassic. They had crushing dentition probably used for a molluscivorous diet. Some placodonts adopted a body form similar to modern marine iguana, while others evolved armour and became turtle-like. Some have suggested that the long-tailed, armoured varieties had a ray-like lifestyle. Details of the placodont vertebral column and limbs appear ill-suited for moving around on land, where they likely spent little time.

PREHISTORIC ANIMAL

TRIASSIC

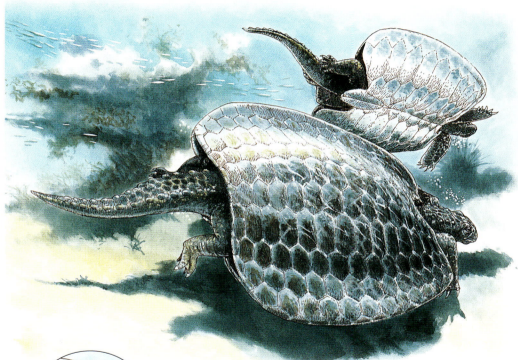

**SHELL**
The protective shell of *Henodus* was made up of tightly packed bony plates fused to the ribs as in turtles. However, unlike turtles, the limb girdles stayed outside of the ribcage.

**HOW BIG IS IT?**

**WHERE IN THE WORLD?**

Currently, the only *Henodus* fossils have been unearthed in Tübingen, Germany. But placodont fossils have turned up elsewhere in Europe as well as in the Middle East and North Africa.

**TIMELINE (millions of years ago)**

| 540 | 505 | 438 | 408 | 360 | 280 | 248 | 208 | 146 | 65 | 1.8 to today |

TRIASSIC

# Hyperodapedon

• ORDER • Rhynchosauria • FAMILY • Rhynchosauridae
• GENUS & SPECIES • Various species within the genus *Hyperodapedon*

| VITAL STATISTICS | |
|---|---|
| FOSSIL LOCATION | Worldwide |
| DIET | Herbivorous |
| PRONUNCIATION | HY-per-o-DAP-eh-don |
| WEIGHT | Unknown |
| LENGTH | 1.8m (6ft) |
| HEIGHT | Unknown |
| MEANING OF NAME | Many teeth in a pavement, because of its numerous subcylindrical palatal teeth |

**The terrifying beak of *Hyperodapedon*, looking much like a bony, oversized staple remover, was actually the tool of a herbivore.**

The firm clamp and scissor-like action of *Hyperodapedon*'s beak would have enabled it to cut up tough plant material. Beak wear patterns and the nature of the jaw joint, suggest a precision shear bite.

**HIND FEET**
The hind feet were equipped with massive claws that may have been used for digging up roots by backward scratching.

**TEETH**
The numerous teeth in the maxilla and palate were modified into broad tooth plates that served to crush the vegetation brought into the mouth by the shearing beak.

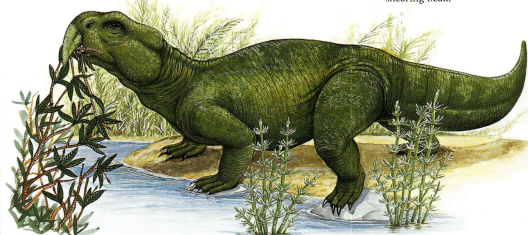

## FOSSIL EVIDENCE

*Hyperodapedon* is a genus of rhynchosaur from the Triassic Period. In some fossil localities, rhynchosaurs account for up to 60 per cent of specimens found. As part of the herbivorous fauna, this is not surprising; herbivores normally outnumber carnivores in terrestrial ecosystems.
*Hyperodapedon* likely ate tough vegetation, as evidenced by its huge, scissoring beak and giant jaw muscle attachments. Some of its diet might have been the subterranean parts of plants, which they dug up by scratch-digging with their robust hind limbs.

PREHISTORIC ANIMAL

TRIASSIC

**HOW BIG IS IT?**

**WHERE IN THE WORLD?**

As was common for many of the terrestrial life forms of the time, rhynchosaurs were distributed all around the world. This is because all major landmasses were conjoined as the supercontinent Pangaea.

**TIMELINE (millions of years ago)**

| 540 | 505 | 438 | 408 | 360 | 280 | 248 | 208 | 146 | 65 | 1.8 to today |

# Kannemeyeria

TRIASSIC

• ORDER • Therapsida • FAMILY • Kannemeyeriidae •
GENUS & SPECIES • Various species within genus *Kannemeyeria*

## VITAL STATISTICS

| | |
|---|---|
| FOSSIL LOCATION | South Africa, Argentina, India, China, Australia |
| DIET | Herbivorous |
| PRONUNCIATION | KAN-ah-MAY-er-ee-ah |
| WEIGHT | Unknown |
| LENGTH | 3m (10ft) |
| HEIGHT | Unknown |
| MEANING OF NAME | Named after a South African paleontologist D. Kannemeyer. |

This huge reptile was one of the first herbivorous vertebrates of the Triassic Period. Distributed worldwide, it probably roamed the plains mainly in herds for protection.

**HEAD**
The head was large, but actually quite light because of the large eye sockets and nasal cavity, so it could lift and stretch up easily to reach for food.

### WHERE IN THE WORLD?

Findings in South Africa, Argentina, India, Australia and possibly China suggest *Kannemeyeria* lived all around the world.

## FOSSIL EVIDENCE

*Kannemeyeria* had a broad, blunt snout and powerful jaws equipped with a toothless beak, so it could rip up the plants that formed its diet. It may have used its claws and strong front legs to dig up plants. Limb girdles formed massive plates of bone to carry its heavy body, which contained a large gut required to digest plant matter. It looks very similar to *Protoceratops*, but they are not closely related. *Kannemeyeria* would have been safer from its main predators, such as *Cynognathus*, in the company of the herd.

**JAW**
*Kannemeyeria* had a powerful beak and strong jaw muscles to help it chomp its way through plants.

PREHISTORIC ANIMAL

TRIASSIC

**HOW BIG IS IT?**

## TIMELINE (millions of years ago)

| 540 | 505 | 438 | 408 | 360 | 280 | 248 | 208 | 146 | 65 | 1.8 to today |

TRIASSIC

# Lagosuchus

• FAMILY • Lagosuchidae • GENUS & SPECIES • *Lagosuchus talanpayenisis*

**This small, fast and ferocious bipedal archosaur is thought to be a very close relative of the dinosaurs that were to follow, partly because of its well-developed ankle structure.**

### VITAL STATISTICS

| | |
|---|---|
| FOSSIL LOCATION | Argentina |
| DIET | Carnivorous |
| PRONUNCIATION | LAG-o-SOOK-us |
| WEIGHT | 200g (25oz) |
| LENGTH | 30–40cm (12-16in) |
| HEIGHT | Unknown |
| MEANING OF NAME | 'Rabbit crocodile' because it was small, lightly built and originally thought to be crocodylian. |

### WHERE IN THE WORLD?

Four skeletons have been found on the Ischigualasto Formation in Argentina.

**JAW**
An opening behind each eye allowed for the attachment of jaw muscles, strengthening the biting and holding power of this miniature 'dinosaur'.

### FOSSIL EVIDENCE

As a close relative, *Lagosuchus* looks like a dinosaur, with its upright stance (on two hind legs), short arms and long tail. Indeed, its feet were highly developed, with long bones and only the toes touching the ground, allowing a longer stride. This feature is similar to pterosaurs, birds and other dinosaurs. *Lagosuchus* was an agile predator, darting about on long legs that gave it the advantage of speed over its prey (probably insects and small reptiles) and its attackers.

DINOSAUR

TRIASSIC

### HOW BIG IS IT?

**SHINS**
With shins longer than its thighs, *Lagosuchus* could sprint across clearings in the South American forests.

**TIMELINE (millions of years ago)**

| 540 | 505 | 438 | 408 | 360 | 280 | 248 | 208 | 146 | 65 | 1.8 to today |

TRIASSIC

# Liliensternus

• ORDER • Saurischia • SUPERFAMILY • Coelphysoidea • GENUS & SPECIES • *Lilensternus liliensterni*

This small, fast and ferocious bipedal archosaur is thought to be a very close relative of the dinosaurs that were to follow, partly because of its well-developed ankle structure.

## VITAL STATISTICS

| | |
|---|---|
| FOSSIL LOCATION | Europe |
| DIET | Carnivorous |
| PRONUNCIATION | LIL-ee-in-STER-nus |
| WEIGHT | 130–400kg (290–880lb) |
| LENGTH | 3–7m (10–22ft) |
| HEIGHT | Unknown |
| MEANING OF NAME | Named after German palaeontologist Hugo Rühle von Lilienstern |

### WHERE IN THE WORLD?

Fossils have been found in Germany and France.

### FOSSIL EVIDENCE

This early theropod dinosaur was a quick sprinter equipped with vicious claws on both hands and feet. This basic body form of bipedal, clawed, carnivorous dinosaur went from the Triassic all the way to the Cretaceous. Meat-eating birds today are bipedal descendants of these early hunters. The pack would surround the victim and kill it with a series of flesh-ripping strikes to wound the larger foe so that it bled to death. *Lilensternus* inhabited the forests on the banks of rivers in an otherwise barren continent.

**CREST**
It is possible that *Lilensternus*'s crest, which was similar to its close relative *Dilophosaurus*, was brightly coloured.

**HANDS**
*Lilensternus* had five-fingered hands, the first and fifth digits being smaller than the central ones.

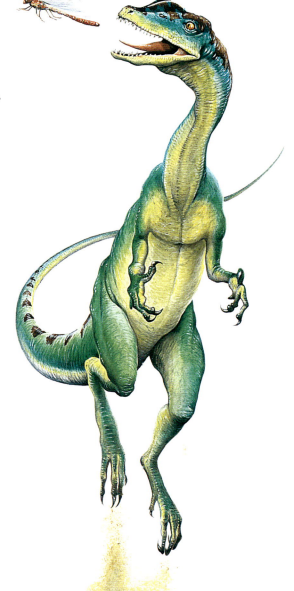

### HOW BIG IS IT?

DINOSAUR

TRIASSIC

**TIMELINE (millions of years ago)**

| 540 | 505 | 438 | 408 | 360 | 280 | 248 | 208 | 146 | 65 | 1.8 to today |

TRIASSIC

# Lotosaurus

• ORDER • Rauisuchia • FAMILY • Ctenosauriscidae • GENUS & SPECIES • *Lotosaurus adentus*

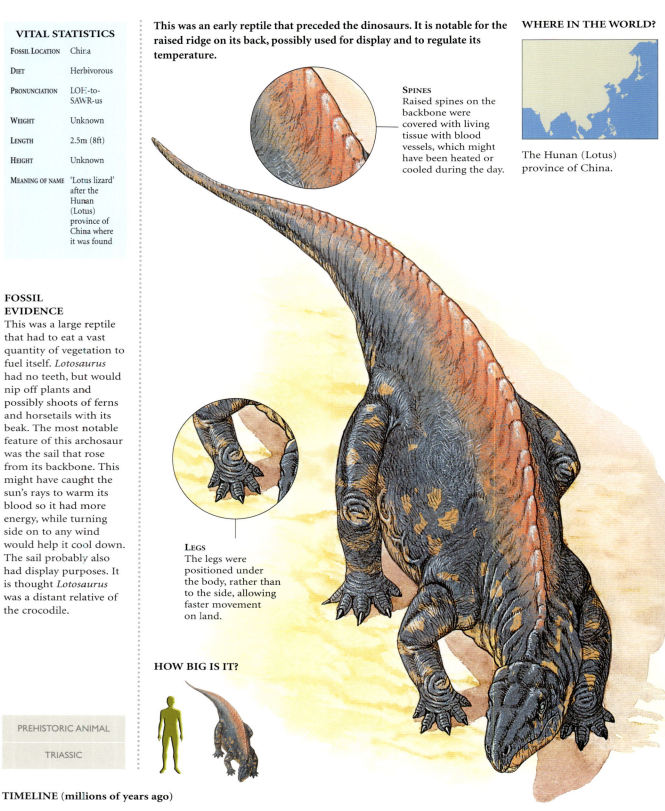

| VITAL STATISTICS | |
|---|---|
| Fossil Location | China |
| Diet | Herbivorous |
| Pronunciation | LOE-to-SAWR-us |
| Weight | Unknown |
| Length | 2.5m (8ft) |
| Height | Unknown |
| Meaning of name | 'Lotus lizard' after the Hunan (Lotus) province of China where it was found |

This was an early reptile that preceded the dinosaurs. It is notable for the raised ridge on its back, possibly used for display and to regulate its temperature.

**SPINES**
Raised spines on the backbone were covered with living tissue with blood vessels, which might have been heated or cooled during the day.

**WHERE IN THE WORLD?**

The Hunan (Lotus) province of China.

**LEGS**
The legs were positioned under the body, rather than to the side, allowing faster movement on land.

**FOSSIL EVIDENCE**
This was a large reptile that had to eat a vast quantity of vegetation to fuel itself. *Lotosaurus* had no teeth, but would nip off plants and possibly shoots of ferns and horsetails with its beak. The most notable feature of this archosaur was the sail that rose from its backbone. This might have caught the sun's rays to warm its blood so it had more energy, while turning side on to any wind would help it cool down. The sail probably also had display purposes. It is thought *Lotosaurus* was a distant relative of the crocodile.

**HOW BIG IS IT?**

PREHISTORIC ANIMAL

TRIASSIC

**TIMELINE (millions of years ago)**

| 540 | 505 | 438 | 408 | 360 | 280 | 248 | 208 | 146 | 65 | 1.8 to today |

TRIASSIC

# Melanorosaurus

• ORDER • Saurischia • FAMILY • Melanorosauridae • GENUS & SPECIES • *Melanorosaurus readi, M. thabanensis*

### VITAL STATISTICS

| | |
|---|---|
| FOSSIL LOCATION | South Africa |
| DIET | Herbivorous |
| PRONUNCIATION | MEH-luh-nor-oh-SAW-rus |
| WEIGHT | Unknown |
| LENGTH | 12m (39ft) |
| HEIGHT | 4.3m (14ft) |
| MEANING OF NAME | 'Black Mountain lizard' after the location of the site where it was found |

This giant herbivore was probably forced to move on four legs rather than two because of its size and weight. It was one of the largest early dinosaurs.

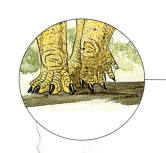

**CLAWED THUMB**
Like all prosauropods, *Melanorosaurus* had small fingers and a large clawed thumb. This may be been using for digging out food and for defence.

### WHERE IN THE WORLD?

Found on the Thaba 'Nyama (Black Mountain) in Transkei, South Africa.

### FOSSIL EVIDENCE

The hips have four sacral vertebrae, and the thigh bone is straight. These features were needed to allow the animal to carry its enormous weight on all four pillar-like legs, probably removing the option of walking on its hind legs. However, its rear legs were still longer than its front legs. The spinal bones had hollows to reduce their weight. So far no head has been found, but sauropod teeth were not designed for chewing so they may have swallowed stones to aid digestion. It had a long neck and tail.

**HOW BIG IS IT?**

**DIGESTING FOOD**
One way large herbivorous dinosaurs may have coped with digesting large amounts of plant matter was to swallow stones that would grind up the contents of the stomach. These stones are known as gastroliths.

DINOSAUR

TRIASSIC

**TIMELINE (millions of years ago)**

| 540 | 505 | 438 | 408 | 360 | 280 | 248 | 208 | 146 | 65 | 1.8 to today |

TRIASSIC

# Mussaurus

• ORDER • Saurischia • FAMILY • [unranked] • GENUS & SPECIES • *Mussaurus patagonicus*

| VITAL STATISTICS | |
|---|---|
| Fossil Location | Argentina |
| Diet | Herbivorous |
| Pronunciation | Moc-SAWR-us |
| Weight | 120kg (260lb) |
| Length | 3–5m (10–16ft) |
| Height | Unknown |
| Meaning of name | 'Mouse lizard' because of the miniature size of the only fossils |

**Describing an adult *Mussaurus* requires even more guesswork than is usually involved with dinosaurs, because the only evidence is a nest of hatchlings only a few days old.**

**Teeth**
The animal had weak, leaf-shaped teeth for tearing at tough leaves, and short claws. Neither offered much protection against predators.

**Brain**
Measurement of the tiny brain relative to huge estimated adult body weight suggests the *Mussaurus* was of very low intelligence.

## FOSSIL EVIDENCE
Ten incomplete juvenile fossils and some eggs are all we have to go on to describe this dinosaur. The specimens are among the smallest dinosaur fossils so far found and would fit in your hand. The skull is only 32mm (1¼in) long – slightly bigger than the eggs. From the shape of the skull we know it was a prosauropod and had a long neck and tail, a small head with a long snout, and large, five-fingered hands with a large thumb claw. It walked on four legs, possibly moving in migratory herds. It has been suggested that *Coloradisaurus* may be the adult version of *Mussaurus*.

**WHERE IN THE WORLD?**

Southern Argentina in South America, which in Late Triassic times was almost a desert.

**HOW BIG IS IT?**

DINOSAUR

TRIASSIC

**TIMELINE (millions of years ago)**

| 540 | 505 | 438 | 408 | 360 | 280 | 248 | 208 | 146 | 65 | 1.8 to today |

TRIASSIC

# Nanchangosaurus

• ORDER • Ichthyosauria • FAMILY • Nanchangosauridae • GENUS & SPECIES • *Nanchangosaurus suni*

## VITAL STATISTICS

| | |
|---|---|
| Fossil Location | China |
| Diet | Carnivorous |
| Pronunciation | Nan-CHANG-oh-SAWR-us |
| Weight | Unknown |
| Length | 1m (3ft) |
| Height | Unknown |
| Meaning of name | 'Nanchang lizard' after the part of China where it was discovered |

This marine reptile seems to have been an ancient evolutionary throwback – a seven-toed tetrapod that lived 100 million years after its ancestors developed limbs with five digits.

### WHERE IN THE WORLD?

The only remains have been found in the Jiangxi province of southern China.

**Eyes**
The head features large eyes, suggesting good eyesight, and a long toothless snout for snapping up its marine diet.

## FOSSIL EVIDENCE

Mystery surrounds why this reptile had seven digits on its forelimbs and six on its hind limbs. Early four-legged animals *Ichthyostega* and *Acanthostega* had seven or eight digits per limb, but their descendants lost them as they evolved. The *Nanchangosaurus*' extra digits may have allowed for stronger and more flexible limbs that made it a better swimmer. It remains possible that such reptiles form the 'missing link' between aquatic ichthyosaurs and their terrestrial archosaur ancestors.

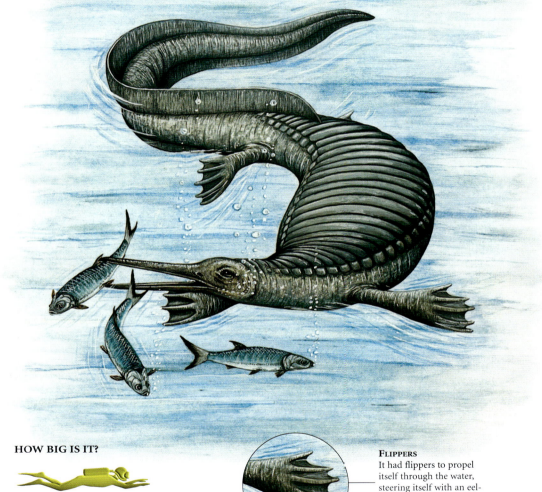

**HOW BIG IS IT?**

**Flippers**
It had flippers to propel itself through the water, steering itself with an eel-like tail that probably allowed for fast turning.

PREHISTORIC ANIMAL

TRIASSIC

**TIMELINE (millions of years ago)**

| 540 | 505 | 438 | 408 | 360 | 280 | 248 | 208 | 146 | 65 | 1.8 to today |

TRIASSIC

# Nothosaurus

• ORDER • Nothosauria • FAMILY • Nothosauridae
• GENUS & SPECIES • Various species within the genus *Nothosaurus*

| VITAL STATISTICS | |
|---|---|
| FOSSIL LOCATION | Europe, Middle East, Asia and North Africa |
| DIET | Fish and sea creatures |
| PRONUNCIATION | no-thoh-SAWR-us |
| WEIGHT | Unknown |
| LENGTH | 3m (10ft) |
| HEIGHT | Unknown |
| MEANING OF NAME | 'False lizard', since it only resembled a lizard |

This seal-like creature and its group were one of the most successful vertebrae groups of its time, surviving for more than 30 million years before being usurped by newer and faster marine reptiles.

**PADDLES**
Paddle-like limbs with five webbed fingers and toes made it a good swimmer, but its leg anatomy indicates that *Nothosaurus* was probably also comfortable on land.

**WHERE IN THE WORLD?**

Many good specimens in Germany, Italy, Netherlands, Switzerland, North Africa, China, Israel and Russia.

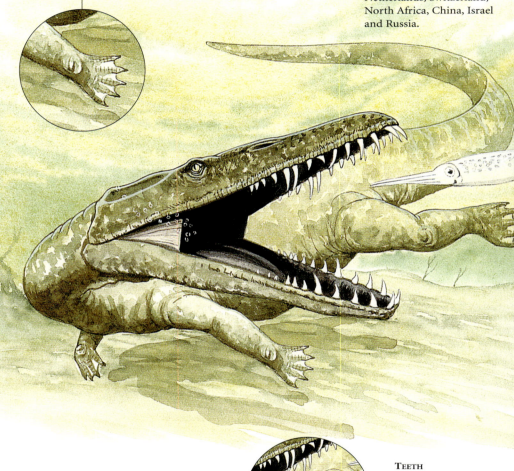

**FOSSIL EVIDENCE**
*Nothosaurus* was a reptile with a lifestyle somewhat like a modern seal. It would probably hunt in shallow tropical seas, propelling its streamlined body towards shoals of fish before pouncing with its teeth-lined jaw. It likely returned to the shore to rest and lay its eggs. Like all *Northosaurs*, the front teeth were bigger than those at the back, while the nostrils were at the top end of the snout to inhale air quickly when it surfaced.

**TEETH**
The broad, flat skull held dozens of sharp, interlocking teeth that allowed it to hold on to its prey.

**HOW BIG IS IT?**

PREHISTORIC ANIMAL

TRIASSIC

**TIMELINE (millions of years ago)**

| 540 | 505 | 438 | 408 | 360 | 280 | 248 | 208 | 146 | 65 | 1.8 to today |

TRIASSIC

# Pisanosaurus

• ORDER • Ornithischia • GENUS & SPECIES • *Pisanosaurus mertii*

| VITAL STATISTICS | |
|---|---|
| FOSSIL LOCATION | South America |
| DIET | Herbivorous |
| PRONUNCIATION | pe-ZAN-oh-SAWR-us |
| WEIGHT | 2–9kg (5–20lb) |
| LENGTH | 1m (3ft) |
| HEIGHT | 30cm (12in) |
| MEANING OF NAME | 'Pisano's lizard', named in honour of Argentinian palaeontologist Juan A. Pisano |

*Pisanosaurus* is a conundrum: an ornithischian head on a saurischian-like body. It is thought by many to be the first bird-hipped dinosaur, but the fossil evidence is so scant that some doubt the claim.

**LEGS**
*Pisanosaurus* walked on two long legs, possibly using its short arms to pull plants into its mouth.

**WHERE IN THE WORLD?**

The fragmented remains of a skeleton were found in La Rioja, Argentina in 1967.

**FOSSIL EVIDENCE**
*Pisanosaurus* was unusual for its time, being a dinosaur in a world dominated by other reptiles. This bipedal, lightly built animal grazed on low-level plants. It was able to bite with canine-like teeth to chew with its leaf-shaped cheek teeth in a similar way to the ornithischians that did not appear for another 25 million years. The pelvis and ankle joints are more characteristic of saurischians. Its long, slender feet would have helped it flee carnivorous predators, such as the primitive *Herrerasaurus*, remains of which were found in the same rock formation.

**HOW BIG IS IT?**

**TAIL**
The shape and length of the tail is pure guesswork based on other early ornithischians, but no tail fossil has been recovered so far.

DINOSAUR

TRIASSIC

**TIMELINE (millions of years ago)**

| 540 | 505 | 438 | 408 | 360 | 280 | 248 | 208 | 146 | 65 | 1.8 to today |

# TRIASSIC

## Protoavis

• ORDER • Saurischia (disputed) • FAMILY • Protoavidae • GENUS & SPECIES • *Protoavis texensis*

### VITAL STATISTICS

| | |
|---|---|
| Fossil Location | USA |
| Diet | Carnivorous |
| Pronunciation | PRO-toe-A-vis |
| Weight | 350g (12oz) |
| Length | 35cm (14in) |
| Height | Unknown |
| Meaning of name | 'Early bird' because it was thought to be a bird |

Is it a bird? Is it a theropod dinosaur? This is one of the most hotly disputed creatures among palaeontologists – if it really was avian, it means birds first appeared several million decades earlier than previously thought.

### WHERE IN THE WORLD?

Found in a mixed-up cache of crushed dinosaur bones in a Texan quarry.

### FOSSIL EVIDENCE

The bones attributed to *Protoavis* were found in a pile of various dinosaur fossils. They comprise a section of the skull (a partly toothless jaw) and some limb bones, from which a creature has been proposed with an avian-type skeleton featuring a tail, dinosaur-like rear legs, and hollow bones. However, most palaeontologists dispute whether the bones are from the same creature at all. If it was a primitive bird, *Protoavis* predates the famous *Archaeopteryx* by 75 million years, meaning that the first birds lived among the first dinosaurs.

DINOSAUR

TRIASSIC

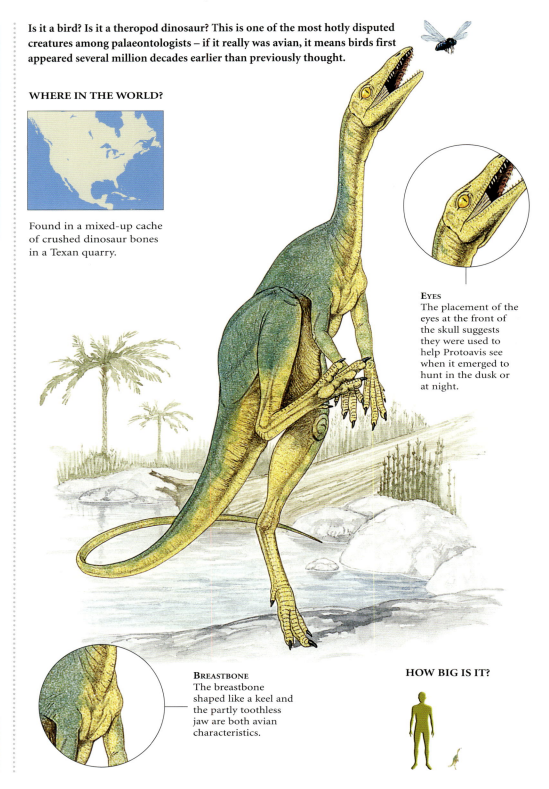

**Eyes**
The placement of the eyes at the front of the skull suggests they were used to help Protoavis see when it emerged to hunt in the dusk or at night.

**Breastbone**
The breastbone shaped like a keel and the partly toothless jaw are both avian characteristics.

### HOW BIG IS IT?

### TIMELINE (millions of years ago)

| 540 | 505 | 438 | 408 | 360 | 280 | 248 | 208 | 146 | 65 | 1.8 to today |

TRIASSIC

# Riojasaurus

• **ORDER** • Saurischia • **FAMILY** • Riojasauridae • **GENUS & SPECIES** • *Riojasaurus incertus*

*Riojasaurus* was a massive prosauropod with a long neck and tail sandwiching a bulky, heavy body. Its sheer size probably protected it from predators, especially if in a herd.

| VITAL STATISTICS | |
|---|---|
| FOSSIL LOCATION | Argentina |
| DIET | Herbivorous |
| PRONUNCIATION | Ree-OH-ha-SAWR-us |
| WEIGHT | 1 tonne (1 ton) |
| LENGTH | 10m (33ft) |
| HEIGHT | 4.9m (16ft) |
| MEANING OF NAME | 'Rioja lizard' after the La Rioja area of Argentina where it was found |

### WHERE IN THE WORLD?

Several almost complete skeletons and a skull have been discovered in the foothills of the Andes in Argentina.

**TEETH**
The teeth were spoon-shaped and serrated, with five at the front of the upper jaw and 24 in a row behind.

**FOSSIL EVIDENCE**

This was a giant of a dinosaur, probably only able to move on all fours without rising up on hind legs. The first finds were without skulls, but palaeontologists' guess that the head on the end of the long neck was small was later confirmed. Inside was a tiny brain, so this animal would have been slow-thinking as it was slow moving. It was among the heaviest animals before the sauropods, surviving by eating vegetation that grew higher than most other herbivores could reach. It was saurischian, or lizard-hipped.

**BACKBONE**
The backbone was hollow, reducing the body weight of this bulky creature with its thick, solid legs ending in claws.

**HOW BIG IS IT?**

DINOSAUR
TRIASSIC

**TIMELINE (millions of years ago)**

| 540 | 505 | 438 | 408 | 360 | 280 | 248 | 208 | 146 | 65 | 1.8 to today |

TRIASSIC

# Saltopus

• ORDER • [not classified – taxonomy disputed]
• FAMILY • [not classified – taxonomy disputed] • GENUS & SPECIES • *Saltopus elginensis*

## VITAL STATISTICS

| | |
|---|---|
| FOSSIL LOCATION | Scotland |
| DIET | Carnivorous |
| PRONUNCIATION | SALT-oh-pus |
| WEIGHT | 1kg (2lb) |
| LENGTH | 60cm (24in) |
| HEIGHT | 20cm (8in) at the hips |
| MEANING OF NAME | 'Leaping foot', because it was originally thought to have been able to jump |

### WHERE IN THE WORLD?

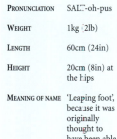

The scant remains were found at Lossiemouth in Scotland in 1910.

The fragmentary remains discovered of this animal suggest a cat-sized, agile meat-eater that may have been capable of leaping. However, it is still unclear whether it can be called a dinosaur at all.

**LONG HEAD**
*Saltopus* had a long head equipped with large eyes and dozens of small sharp teeth, indicating that it possessed keen eyesight.

**HIND LEGS**
Strong hind legs would have made it a fast mover, and it may have been able to leap to escape attackers or surprise prey.

### FOSSIL EVIDENCE

*Saltopus* was about the size of a cat and probably capable of fast movement on its two rear legs. Its hands had five fingers, the fourth and fifth digits being very small. It would have been able to spring on prey and pull it apart with its sharp teeth, and relied on speed and agility to dodge predators. It may also have scavenged and eaten carcasses of dead animals. *Saltopus* could be an early dinosaur, or a lagosuchid (a primitive reptile) or an ornithosuchian (closely related to dinosaurs).

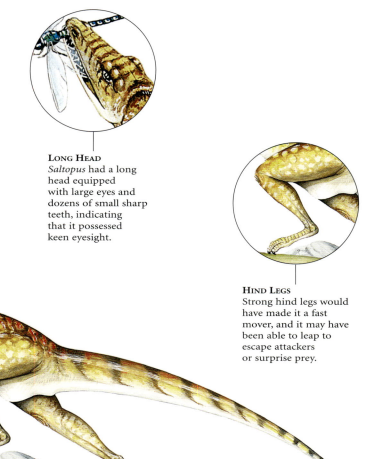

### HOW BIG IS IT?

PREHISTORIC ANIMAL

TRIASSIC

**TIMELINE (millions of years ago)**

| 540 | 505 | 438 | 408 | 360 | 280 | 248 | 208 | 146 | 65 | 1.8 to today |

# Sellosaurus

• ORDER • Saurischia • FAMILY • Plateosauridae • GENUS & SPECIES • Sellosaurus gracilis

*Sellosaurus* was a medium-sized herbivorous prosauropod that may have fed on the primitive conifers of the European Triassic deserts.

## VITAL STATISTICS

| | |
|---|---|
| FOSSIL LOCATION | Germany |
| DIET | Herbivorous |
| PRONUNCIATION | SELL-oh-SAWR-us |
| WEIGHT | 100–400kg (220–880lb) |
| LENGTH | 6.5m (21ft) |
| HEIGHT | 1.7m (6ft) |
| MEANING OF NAME | 'Saddle lizard' because of the saddle-shape of part of its spine |

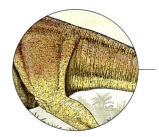

**WIDE TAIL**
Unlike other prosauropods, the tail vertebrae are wide and flat like a saddle, from which feature *Sellosaurus* gets its name.

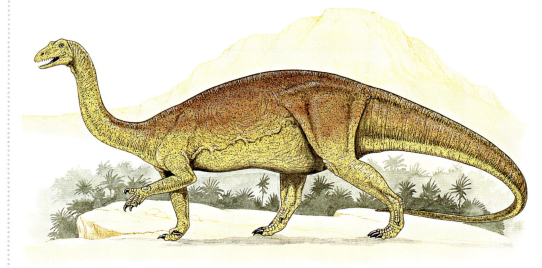

## FOSSIL EVIDENCE

*Sellosaurus* had a large gut, so would have been far more comfortable on four legs, but it was capable of rising onto its hind legs, possibly to run. It scraped leaves from branches with its serrated teeth and was able to eat plant food and possibly had cheeks to keep food from dropping out of its mouth. As with other prosauropods, it had a thumb claw that might have been for defence or to hook over branches to pull down leaves, but its tail vertebrae were unusually shaped.

**LONG TAIL**
*Sellosaurus* may have been able to prop itself up on its long tail so that it could stand on its hind legs to reach high foliage.

## HOW BIG IS IT?

## WHERE IN THE WORLD?

More than 20 skeletons have been found at Nordwürttemberg in Germany.

DINOSAUR

TRIASSIC

**TIMELINE (millions of years ago)**

| 540 | 505 | 438 | 408 | 360 | 280 | 248 | 208 | 146 | 65 | 1.8 to today |

TRIASSIC

# Shansisuchus

• ORDER • [unranked] • FAMILY • Erythrosuchidae
• GENUS & SPECIES • Various species within the genus *Shansisuchus*

**This was an archosaur, and a close relative of the ancestor of the dinosaurs. With its predatory skills it inhabited the world 200 million years ago in the middle Triassic Period.**

| VITAL STATISTICS | |
|---|---|
| FOSSIL LOCATION | China |
| DIET | Carnivorous |
| PRONUNCIATION | SHAN-see-SOO-kus |
| WEIGHT | Unknown |
| LENGTH | 2.2m (7ft) |
| HEIGHT | 50cm (20in) |
| MEANING OF NAME | 'Shansi crocodile' after the Chinese province where it was discovered |

### WHERE IN THE WORLD?

The remains were found in China's Shansi (Shanxi) province, west of Taihung Mountain.

**JAWS**
*Shansisuchus* had powerful jaws able to deliver a ferocious bite with which it tore off chunks of its victims.

## FOSSIL EVIDENCE

*Shansisuchus* was a formidable hunter. It was probably fast and had strong jaws. Like most archosaurs, it was thecodont, meaning its teeth were set in sockets and so were less likely to be torn loose when it was feeding. The legs were positioned under the body rather than to the side, which may have given it the advantage of speed, while the back legs were longer than the front ones. Few creatures, such as large therapsids of the time, would stand much chance against such a foe.

**HOW BIG IS IT?**

**LEGS**
The legs hung directly under the hip joints, possibly allowing it to move quickly despite its large size, attacking less agile reptiles.

PREHISTORIC ANIMAL

TRIASSIC

**TIMELINE (millions of years ago)**

| 540 | 505 | 438 | 408 | 360 | 280 | 248 | 208 | 146 | 65 | 1.8 to today |

TRIASSIC

# Shonisaurus

• **ORDER** • Ichthyosauria • **FAMILY** • Shastasauridae • **GENUS & SPECIES** • *Shonisaurus popularis & S. sikanniensis*

| VITAL STATISTICS | |
|---|---|
| Fossil Location | North America |
| Diet | Carnivorous |
| Pronunciation | SHO-ni-SAW-rus |
| Weight | Unknown |
| Length | as much as 21m (70 ft) |
| Height | Unknown |
| Meaning of name | 'Shoshone Mountain lizard' because it was first discovered there |

### WHERE IN THE WORLD?

*Shonisaurus* has been found in Nevada, USA and British Columbia, Canada. Another ichthyosaur found in the Himalayan Mountains, called *Himalayasaurus*, may actually be the same animal as Shonisaurus.

By far the largest marine reptile ever known, *Shonisaurus* likely surpassed the modern sperm whale in length and bulk.

**Vertebrae**
When *Shonisaurus* vertebrae were first unearthed in Nevada in the late 1800s, the miners working at the site used the huge discs as plates off which they ate their meals.

### FOSSIL EVIDENCE

*Shonisaurus* fossils were first discovered in the Triassic Luning Formation in Nevada. Dozens of skeletons were found all together, hence this species' name: *Shonisaurus popularis*. This was the largest species of ichthyosaur at the time at 15m (50ft) long. *Shonisaurus sikanniensis* fossils, at up to 21m (70ft) long, were first found in the Triassic Pardonet Formation of British Columbia in 1991. Mosquitoes, bears, floods and general inaccessibility led to a dig that took years to complete. A helicopter was needed to airlift the 2-ton blocks of stone out of the site before the fossil could be cleaned, described and published.

PREHISTORIC ANIMAL

TRIASSIC

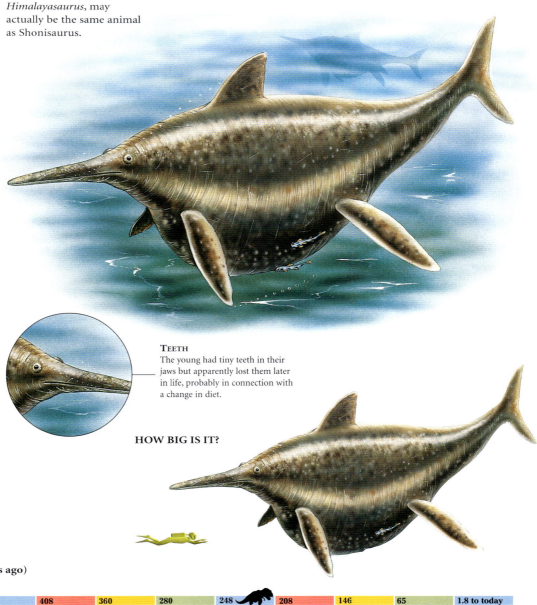

**Teeth**
The young had tiny teeth in their jaws but apparently lost them later in life, probably in connection with a change in diet.

### HOW BIG IS IT?

**TIMELINE (millions of years ago)**

| 540 | 505 | 438 | 408 | 360 | 280 | 248 | 208 | 146 | 65 | 1.8 to today |

TRIASSIC

# Tanystropheus

• ORDER • Prolacertiformes • FAMILY • Tanystrophidae
• GENUS & SPECIES • Various species within the genus *Tanystrophaeus*

| VITAL STATISTICS | |
|---|---|
| Fossil Location | Europe and Middle East |
| Diet | Carnivorous |
| Pronunciation | TAN-ee-STRO-fee-us |
| Weight | Unknown |
| Length | 6m (20ft) (most of which was neck) |
| Height | Unknown |
| Meaning of name | 'long-necked one' |

**FOSSIL EVIDENCE**
This improbable reptile is best known from the rich Middle Triassic deposits that span the Swiss/Italian border. Some were preserved so well that one specimen found in 2006 showed traces of skin with non-rectangular, overlapping scales. The aquatic sediments it comes from, plus the ungainly proportions of its neck, have led some to conclude that *Tanystropheus* was aquatic. However, details of the feet seem adapted to a terrestrial lifestyle. It is possible that juveniles, which had teeth of more varied nature, had a more land-based existence and that they moved into the water when adults. This move would have been connected with a change in diet.

PREHISTORIC ANIMAL
TRIASSIC

Since it was first discovered, the absurdly long neck of *Tanystropheus* has confused scientists about how this animal led its life.

**VERTEBRAE**
*Tanystropheus*' neck had eight to 12 extremely elongated vertebrae, while a possible Chinese tanystropheid had as many as 27.

**TAIL BONES**
Extra bones in the tail of certain specimens demonstrate differences in males and females. Some interpret this structure as a support for a brood pouch to raise young.

**WHERE IN THE WORLD?**

*Tanystropheus* fossils have been found in Italy, Israel, Germany and Switzerland, but others in the family have been found as far off as China and the USA.

HOW BIG IS IT?

TIMELINE (millions of years ago)

| 540 | 505 | 438 | 408 | 360 | 280 | 248 | 208 | 146 | 65 | 1.8 to today |

TRIASSIC

# Thecodontosaurus

· ORDER · Saurischia · FAMILY · Thecodontosauridae · GENUS & SPECIES · *Thecodontosaurus antiquus*

*Thecodontosaurus* was one of the earliest prosauropod dinosaurs that were closely related to the huge long-necked plant-eating sauropods.

| VITAL STATISTICS | |
|---|---|
| Fossil Location | England, Wales, Argentina |
| Diet | Herbivorous, possibly omnivorous |
| Pronunciation | THEE-co-DON'T-oh-SAWR-us |
| Weight | 70kg (154lb) |
| Length | 2.1m (7ft) |
| Height | 1.2m (4ft) |
| Meaning of name | 'Socket tooth lizard' because its teeth are set in sockets |

## FOSSIL EVIDENCE

Despite hundreds of *Thecodontosaurus* fossils being found, its diet and size – some palaeontologists believe it was smaller than the measurements given – are still in dispute. The serrated teeth were cone-shaped at the front, but those in the back of the jaw were flat. This suggests that the animal added meat to its herbivorous diet. As with other prosauropods, each five-fingered hand ended in a large, curved thumb claw. This was one of the earliest fossil dinosaurs to be named, in 1843. Some specimens in Bristol were destroyed by bombs in World War II.

**Teeth**
The different shapes of its serrated teeth have led palaeontologists to suggest that *Thecodontosaurus* ate meat as well as vegetation.

### WHERE IN THE WORLD?

Remains mostly in southern England, Wales and northern Argentina.

**Feet and Legs**
The feet and legs were slender and it could walk on (mostly) two or four legs, using its long tail to balance the elongated neck.

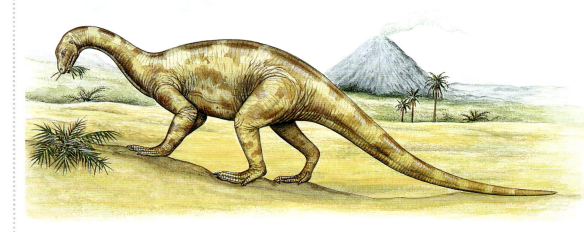

### HOW BIG IS IT?

DINOSAUR

TRIASSIC

**TIMELINE (millions of years ago)**

TRIASSIC

# Gracilisuchus

• ORDER • [unranked] • FAMILY • [unranked] • GENUS & SPECIES • *Gracilisuchus stipanicicorum*

| VITAL STATISTICS | |
|---|---|
| FOSSIL LOCATION | Argentina |
| DIET | Carnivorous |
| PRONUNCIATION | GRAS-i-li-SOOK-us |
| WEIGHT | Unknown |
| LENGTH | 30cm (12in) |
| HEIGHT | Unknown |
| MEANING OF NAME | 'Slender crocodile' because of its build |

**WHERE IN THE WORLD?**

Remains were found in Argentina in the 1970s.

It looks like a dinosaur, but actually it's a small early ancestor of the crocodile, even though it walked on two legs and lived on the land.

**HEAD**
The head is disproportionately large compared to its slender body, suggesting it captured prey with its jaws rather than its claws.

**FOSSIL EVIDENCE**

It is understandable that *Gracilisuchus* was originally thought to be a dinosaur when it was discovered in the 1970s. After all, it could run very fast on two legs and had a formidable set of teeth set in a short snout, which gave it the profile of a theropod dinosaur. However, a decade of study revealed that its anatomy was more like that of an early crocodile. Its large eyes suggest it hunted by sight, and it was equipped with razor-sharp claws and teeth to rip apart the victims it sprang on when it launched itself.

**BONY PLATES**
It was protected by a double row of bony interlocking plates down its backbone to the tip of the tail.

**HOW BIG IS IT?**

| PREHISTORIC ANIMAL |
|---|
| TRIASSIC |

**TIMELINE** (millions of years ago)

| 540 | 505 | 438 | 408 | 360 | 280 | 248 | 208 | 146 | 65 | 1.8 to today |

TRIASSIC

# Postosuchus

ORDER • Rauisuchia • FAMILY • Rauisuchidae • GENUS & SPECIES • *Postosuchus kirkpatricki, P. alisonae*

### VITAL STATISTICS

| | |
|---|---|
| FOSSIL LOCATION | USA |
| DIET | Carnivorous |
| PRONUNCIATION | POST-oh-SOOK-us |
| WEIGHT | 680kg (1500lb) |
| LENGTH | 5m (16½ft) |
| HEIGHT | 1.2m (3ft) |
| MEANING OF NAME | 'Crocodile from Post' after the Post Quarry (named for the town of Post) where it was first found |

### FOSSIL EVIDENCE

*Postosuchus* was a reptile that likely fed on dinosaurs and was probably in the enviable position of having no predators. Its skull was taller than its width, giving it a massively powerful bite using large, dagger-like teeth. Although it was flat-footed, it was fast because its legs were tucked under the body, giving it good agility. It could also possibly rise up on its hind legs, helping it attack larger animals. Proof of its terrible power is evident in one specimen where the gut contents were preserved: it contained the remains of at least four other creatures.

PREHISTORIC ANIMAL

TRIASSIC

To animals of the late Triassic Period, *Postosuchus* was their worst nightmare: a heavily armoured killing machine with a big appetite to satisfy its long body.

**BACKBONE**
The backbone was lined with rows of small bony plates to protect it from the teeth and claws of rivals – it probably had no predators.

### WHERE IN THE WORLD?

Specimens found at the Post Quarry in Texas, and at other locations in the USA.

**LEGS**
The legs were columnar, meaning they did not sprawl to the sides, with the hind legs slightly longer than those at the front. This was uncommon in reptiles.

### HOW BIG IS IT?

**TIMELINE (millions of years ago)**

| 540 | 505 | 438 | 408 | 360 | 280 | 248 | 208 | 146 | 65 | 1.8 to today |

TRIASSIC

# Coelophysis

| VITAL STATISTICS | |
|---|---|
| Fossil Location | USA |
| Diet | Carnivorous |
| Pronunciation | SEE-low-FIE-sis |
| Weight | Up to 45kg (100lb) |
| Length | 2.5–3m (8–10ft) |
| Height | Unknown |
| Meaning of name | 'Hollow form' in reference to its light, hollow bones |

One of the earliest dinosaurs, *Coelophysis* was a carnivore. It may have been a scavenger, but it was certainly an efficient predator. With powerful hind legs and a long, slender body, it was built to be fast. Virtually hollow leg bones, and a skull with large fenestrae (openings), reduced its body weight, further increasing its speed. A long, flexible neck, with a pronounced curve, helped it to grab its prey. *Coelophysis* may also have hunted in packs, which would have enabled it to take on large prey.

**Jaw**
*Coelophysis* had more than 100 dagger-like teeth, for ripping into flesh. A double hinge let it move its jaw back and forth, sawing its food.

**FOSSIL EVIDENCE**
Many specimens of *Coelophysis* have been discovered at the Ghost Ranch, New Mexico. This is a site rich in fossils, and includes a 'graveyard' containing hundreds of specimens – possibly victims of a flash flood. The remains of adult *Coelophysis* differ, one form being 'robust', the other 'gracile'; this may represent males and females. It was once thought that *Coelophysis* cannibalized its young: the abdomens of many adult specimens seemed to contain young *Coelophysis*, but the fossils were later re-analyzed and shown to be other reptiles.

DINOSAUR

TRIASSIC

**HOW BIG IS IT?**

**Hands**
Strong three-fingered hands were used to clasp its prey.

# TRIASSIC

• **ORDER** • Saurischia • **FAMILY** • Coelophysidae • **GENUS & SPECIES** • *Coelophysis bauri*

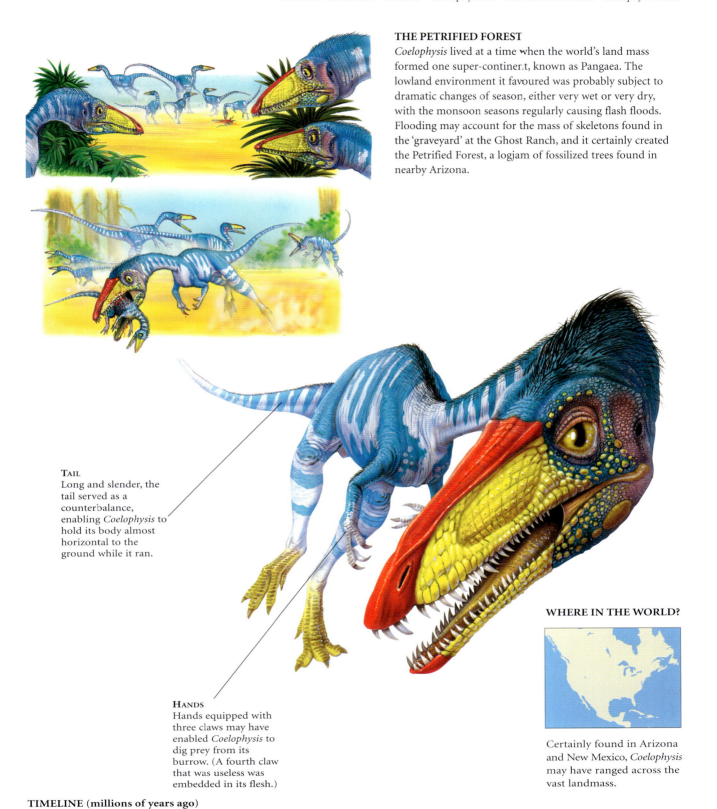

### THE PETRIFIED FOREST
*Coelophysis* lived at a time when the world's land mass formed one super-continent, known as Pangaea. The lowland environment it favoured was probably subject to dramatic changes of season, either very wet or very dry, with the monsoon seasons regularly causing flash floods. Flooding may account for the mass of skeletons found in the 'graveyard' at the Ghost Ranch, and it certainly created the Petrified Forest, a logjam of fossilized trees found in nearby Arizona.

**TAIL**
Long and slender, the tail served as a counterbalance, enabling *Coelophysis* to hold its body almost horizontal to the ground while it ran.

**HANDS**
Hands equipped with three claws may have enabled *Coelophysis* to dig prey from its burrow. (A fourth claw that was useless was embedded in its flesh.)

### WHERE IN THE WORLD?

Certainly found in Arizona and New Mexico, *Coelophysis* may have ranged across the vast landmass.

**TIMELINE (millions of years ago)**

| 540 | 505 | 438 | 408 | 360 | 280 | 248 | 208 | 146 | 65 | 1.8 to today |

TRIASSIC
# Coelophysis

# TRIASSIC

• **ORDER** • Saurischia • **FAMILY** • Coelophysidae • **GENUS & SPECIES** • *Coelophysis bauri*

**CHANGING EARTH**

Earth as *Coelophysis* and its contemporary dinosaurs knew it in the Late Triassic Period was very different from the planet we know today. North America, where the remains of *Coelophysis* were found in such abundance, was still part of Pangaea, where the Earth's future land masses were joined together in one giant super-continent. Sea levels in the Late Triassic were generally low and the climate in which *Coelophysis* lived could be extremely dry and hot. Much of *Coelophysis*' home territory was desert. The look of the land was changing, too. Conifers were starting to appear where forests had managed to gain a foothold. Ferns and two types of seed plants, the cycads and the bennettitales, grew in profusion on the forest floors. In America, animals who shared this environment with *Coelophysis* included *Massospondylus*, one of the first known prosauropods and *Protosuchus*, one of the earliest crocodiles (both from Arizona). Then, suddenly, a mass extinction 200 million years ago wiped out more than half of all animal species then living on Earth and dinosaurs like *Coelophysis* and their descendants became the principal land vertebrates on the planet.

TRIASSIC

# Cynognathus

| VITAL STATISTICS | |
|---|---|
| Fossil Location | Argentina, South Africa, China, Antarctica |
| Diet | Carnivorous |
| Pronunciation | Sy-nog-NAY-thus |
| Weight | Unknown |
| Length | 1.5m (5ft) |
| Height | Unknown |
| Meaning of name | 'Dog jaw' after its canine teeth |

### WHERE IN THE WORLD?

*Cynognathus* was found in Argentina, the Karoo Basin of South Africa, China and Antarctica.

A fierce predator that looked like a cross between a wolf and a lizard, *Cynognathus* was a key predator in its environment. Hunting in a pack, it attacked larger animals. One of its possible prey animals, *Kannemeyeria*, was a herbivore twice its length. Its head was up to 45cm (18in) long, almost one-third of its entire body length, and its large jaws had a powerful bite. A secondary palate in its mouth enabled it to breathe and swallow at the same time.

**Jaws**
With more than one form of tooth, including incisors and sharp canines, *Cynognathus* could process its food before swallowing it.

### FOSSIL EVIDENCE

The fossil record demonstrates that *Cynognathus* shared many features with modern mammals. The lack of ribs in the stomach region suggests that it had a diaphragm, a muscle strongly associated with mammals. Noting the fossil evidence of differentiated teeth and the presence of whiskers, some palaeontologists assume that *Cynognathus* was warm-blooded. There is no record of what its skin was like, but one fossil throws up the possibility that it was covered with fur: around a single possible *Cynognathus* paw print, the impression of fur can be seen.

### THE EVOLUTION OF MAMMALS

A member of the cynodont ('dog-tooth') family preceding the dinosaurs, *Cynognathus* has helped palaeontologists to understand how mammals evolved some tens of million years after it. This reptile-like mammal shares many features with modern mammals. Its legs were positioned under its body (rather than sticking out of the sides); it seems to have been warm-blooded; it may have given birth to live young (rather than laying eggs); and it had whiskers, which means that it probably also had fur.

PREHISTORIC ANIMAL

TRIASSIC

### HOW BIG IS IT?

TRIASSIC

• **ORDER** • Therapsida • **FAMILY** • Cynognathidae • **GENUS & SPECIES** • *Cynognathus crateronatus*

## ONE WORLD, ONE CONTINENT

Specimens of *Cynognathus* have been found on four continents, in Argentina, South Africa, China and Antarctica. Since these continents are now far apart, what can account for this range? In 1912, scientist Alfred Wegener suggested that the continents had once been joined together but had drifted apart, and pointed to the location of fossil finds, including *Cynognathus*, in support of his theory.

**SNOUT**
Pits and canals on its snout may have contained the sensory nerves of whiskers – a feature shared with some modern mammals.

**LEGS**
*Cynognathus* ran swiftly on short, muscular legs – but with a waddle, since its backbone moved more from side to side.

**TIMELINE** (millions of years ago)

| 540 | 505 | 438 | 408 | 360 | 280 | 248 | 208 | 146 | 65 | 1.8 to today |

TRIASSIC
# Cygnognathus

# TRIASSIC

**• ORDER •** Therapsida **• FAMILY •** Cynognathidae **• GENUS & SPECIES •** *Cynognathus crateronatus*

**CYNOGNATHUS IN THE TRIASSIC**

Earth in Triassic times, when *Cynognathus* was alive, saw many changes. Pangaea, the supercontinent comprising all Earth's land area, was in the process of breaking up. Corals made their appearance. The skies were full of new life, for the pterosaurs, or 'winged reptiles', had taken to the air. The seas also contained new inhabitants in the thalattosaurs ("ocean lizards") and the ichthyosaurs ("fish lizards"). Lycophytes and cycads grew on land, as did the ginkgophyta – represented today by the *Ginkgo biloba* – and most prolific of all, seed plants began to cover the ground. The climate in Triassic times was equatorial and mainly desert and therefore hot and dry. There was no ice and creatures were able to live in the moist, temperate zones around the poles. *Cynognathus* was one of the important predators during the early part of the Triassic Period, but after that, the archosaur reptiles began to arrive and ultimately became dominant. Mammals also appeared, but for a time remained small nocturnal creatures, probably living on insects. But the world of the Triassic was not stable and ended in a mass extinction possibly caused by tremendous volcanic eruptions, though global cooling was also a possible culprit.

TRIASSIC

# Herrerasaurus

| VITAL STATISTICS | |
|---|---|
| Fossil Location | Argentina |
| Diet | Carnivorous |
| Pronunciation | huh-RARE-ah-SAWR-us |
| Weight | 200–350kg (440–780lb) |
| Length | 3m (10ft) |
| Height | Unknown |
| Meaning of name | 'Herrera's lizard' after Victorino Herrera, a rancher who discovered the fossil |

One of the earliest carnivorous dinosaurs, *Herrerasaurus* lived in South America, at a time when dinosaurs were recently evolved and rare. Its large serrated teeth could penetrate bone, and it was probably a fast runner: powerful hind legs and a long stiff tail for balance are adaptations that enhance speed. This predator did not, however, have things all its own way: puncture wounds in a skull suggest that it fell prey to *Saurosuchus*, a giant reptile.

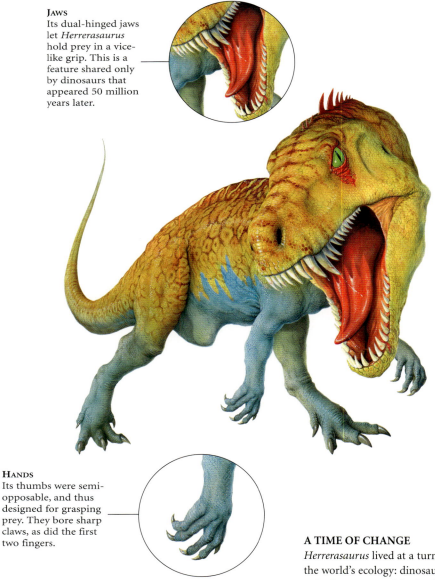

**Jaws**
Its dual-hinged jaws let *Herrerasaurus* hold prey in a vice-like grip. This is a feature shared only by dinosaurs that appeared 50 million years later.

**Hands**
Its thumbs were semi-opposable, and thus designed for grasping prey. They bore sharp claws, as did the first two fingers.

**FOSSIL EVIDENCE**
The best specimens of *Herrerasaurus* were discovered by chance: when spotted, close to where the first fragmentary fossil was found, erosion had only just begun to expose the two, almost complete, skeletons. The specimens were so well preserved that tiny earbones were visible, as well as plates in the iris of the eyes. With the discovery of these findings, *Herrerasaurus* could be reconstructed for the first time. Partly healed toothmarks found on the skulls suggests that it lived in packs, fighting to establish its place.

DINOSAUR

TRIASSIC

**HOW BIG IS IT?**

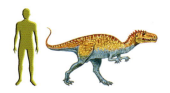

**A TIME OF CHANGE**
*Herrerasaurus* lived at a turning point in the world's ecology: dinosaurs account for only six per cent of the fossils from its time period. By the end of the Triassic Period, however, dinosaurs were beginning to dominate on land. Indeed, it tended to be the dinosaurs that survived a mass extinction event that closed the period, accompanied by volcanic eruptions breaking up the world's landmass.

# TRIASSIC

• **ORDER** • Saurischia • **FAMILY** • Herrerasauridae • **GENUS & SPECIES** • *Herrerasaurus ischigualastensis*

### WHERE IN THE WORLD?

To date, *Herrerasaurus* has been located only in northwest Patagonia, Argentina.

### CONFUSED EVOLUTION

It is difficult to draw up a family tree for *Herrerasaurus*. It shares features with dinosaurs from the much later Jurassic Period; its conical teeth were unique among its contemporaries; and its arms, much shorter than its hind legs, are designed for capturing prey, like later dinosaurs. Growing to lengths of up to 3m (10ft), it reached a size not seen again until the dinosaurs of the Jurassic.

**BODY**
The colour of any dinosaur is a matter of conjecture, but *Herrerasaurus* may have had camouflage markings for hiding in undergrowth.

**TIMELINE (millions of years ago)**

| 540 | 505 | 438 | 408 | 360 | 280 | 248 | 208 | 146 | 65 | 1.8 to today |

TRIASSIC
# Herrerasaurus

# TRIASSIC

• **ORDER** • Saurischia • **FAMILY** • Herrerasauridae • **GENUS & SPECIES** • *Herrerasaurus ischigualastensis*

**HERRERASAURUS DISCOVERY**
Herrerasaurus was discovered in the Ischigualasto Formation of northwest Argentina in 1958. This first find comprised a partial skeleton, but it was augmented in 1988 by the discovery of a complete skeleton, a skull, and some additional fragments. This second find enabled a reconstruction of Herrerasaurus that revealed some, though not all, of its dinosaur characteristics. One was the structure of Herrerasaurus' pelvis; another was its body shape, which was much like that of a carnivorous dinosaur. But the arrangement of the hip bones and the leg bones were markedly archosaur. At their peak, the archosaurs were more numerous, more evolved, and developed more species than any other creature on Earth. However, after some 50 million years at the top, a change in the climate caused a mass extinction in their ranks. Only a few archosaurs survived, enabling the dinosaurs, their descendants, to assume dominance.

TRIASSIC

# Lystrosaurus

| VITAL STATISTICS | |
|---|---|
| Fossil Location | Antarctica, South Africa, Russia, India, China |
| Diet | Herbivorous |
| Pronunciation | LY-stro-SAWR-us |
| Weight | 91kg (200lb) |
| Length | 1.5m (5ft) |
| Height | Unknown |
| Meaning of name | 'Shovel lizard' for its shovel-like beak |

About the size of a pig, *Lystrosaurus* lived near lakes and swamps, its wide feet enabling it to walk across boggy ground. Named 'shovel lizard', it may have been an amphibious feeder, shovelling aquatic plants out of lake and river beds. Its weak jaw moved backwards and forwards with a shearing action suitable for eating vegetation. With powerful forelimbs, it was possibly a powerful digger, and may have lived in burrows. It was one of the most common land vertebrates of the early triassic and represents up to 95 per cent of the specimens found in some fossil beds.

**Mouth**
Its sharp beak let it slice up tough vegetation, while two tusks may have enabled it to dig up plants.

**Nostrils**
*Lystrosaurus* had nostrils set high above its mouth, letting it root in pools and swamps without having to lift its head to breathe.

## FOSSIL EVIDENCE

Although specimens of *Lystrosaurus* have been found around the world, most have been found in the Karoo region of South Africa. The number of species recognized has been a matter for debate. Between the 1930s and the 1970s, palaeontologists recognized up to 23, a number revised down to six between the 1980s and 1990s, then further reduced in 2006 to four. One species, *Lystrosaurus maccaigi*, did not survive the Permian–Triassic extinction event. Appearing suddenly in the fossil record, it may have emigrated into the Karoo Basin.

PREHISTORIC ANIMAL

TRIASSIC

**HOW BIG IS IT?**

## A VERY SUCCESSFUL GENUS

At least one species of Lystrosaurus survived the Permian–Triassic extinction. Possibly because of the lack of predator and competitors lost to the extinction, Lystrosaurus diversity boomed at this time. Lystrosaurus fossils account for up to 95% of the total individuals in some Early Triassic fossil beds. Success to a degree of a single genus of land vertebrae has not been recorded elsewhere in the fossil record.

# TRIASSIC

**• ORDER** • Therapsida **• FAMILY** • Lystrosauridae **• GENUS & SPECIES** • Various species within the genus *Lystrosaurus*

### WHERE IN THE WORLD?

*Lystrosaurus* was found across the world, in Antarctica, South Africa, India, Russia, Mongolia, China and Australia.

**LEGS**
*Lystrosaurus* was once thought to have lived a wholly aquatic life, but its hind legs were strong and agile, better suited for land.

### A SURVIVOR

Why did *Lystrosaurus* survive the Permian–Triassic extinction event? There are several theories. A semi-aquatic creature, it could exploit both water and land. A specialized anatomy enabled it to cope with the increase of carbon dioxide in the atmosphere. It was protected by its relatively small size, since it is the larger creatures that are most vulnerable to an extinction event. Unfortunately, there are counter-arguments to all these theories, and luck may have played the largest part.

**TIMELINE (millions of years ago)**

| 540 | 505 | 438 | 408 | 360 | 280 | 248 | 208 | 146 | 65 | 1.8 to today |

LATE TRIASSIC

# Coloradisaurus

• ORDER • Saurischia • FAMILY • Massospondylidae • GENUS & SPECIES • Colaradisaurus brevis

| VITAL STATISTICS | |
|---|---|
| FOSSIL LOCATION | Argentina |
| DIET | Herbivorous |
| PRONUNCIATION | kohl-oh-RAHD-uh-SAR-us |
| WEIGHT | 290kg (630lb) |
| LENGTH | Up to 4m (13ft) |
| HEIGHT | 1.5m (5ft) |
| MEANING OF NAME | 'Los Coloradis lizard', after the rock formation in which it was discovered |

### WHERE IN THE WORLD?

Only evidence found in the Los Colorados rocks of Argentina.

A plant-eating dinosaur with a small head, large eyes and a big body, Coloradisaurus may be an adult version of the *Mussaurus*, because they were both prosauropods from the same time and place but, unfortunately, evidence is scant.

**TEETH**
*Coloradisaurus* teeth had irregular edges suitable for tearing off and eating leaves.

**HIND LEGS**
It could probably stand upright on its hind legs, resting on its tail for stability, to reach high vegetation and maybe intimidate attackers.

### FOSSIL EVIDENCE

The remains of this quadrupedal herbivorous dinosaur were found in the same area as the nest of the baby dinosaur *Mussaurus*; some palaeontologists believe they are the same animal in juvenile and adult form. Although it walked on four legs, it was probably able to rise up on its rear legs. It could then use its front claws to reach food, and possibly swipe at predators. It was originally called *Colaradia*, but this was changed because the name had already been given to a moth.

### HOW BIG IS IT?

**TIMELINE** (millions of years ago)

| 540 | 505 | 438 | 408 | 360 | 280 | 248 | 208 | 146 | 65 | 1.8 to today |

LATE TRIASSIC

# Desmatosuchus

• **ORDER** • Aetosauria • **FAMILY** • Stagonolepididae • **GENUS & SPECIES** • *Desmatosuchus spurensis, D. smalli*

## VITAL STATISTICS

| | |
|---|---|
| Fossil Location | USA |
| Diet | Herbivorous |
| Pronunciation | des-mat-oh-SUE-kus |
| Weight | Unknown |
| Length | Up to 5m (17ft) |
| Height | 1.5m (5ft) |
| Meaning of name | 'Link crocodile' because it is similar to modern crocodiles in some ways |

With its shoulder spikes and armoured body, *Desmatosuchus* looked formidable. In fact, it was a gentle plant-eating reptile and its aggressive appearance was partly to deter predators.

### WHERE IN THE WORLD?

Remains found in Texas, USA, and named in 1920.

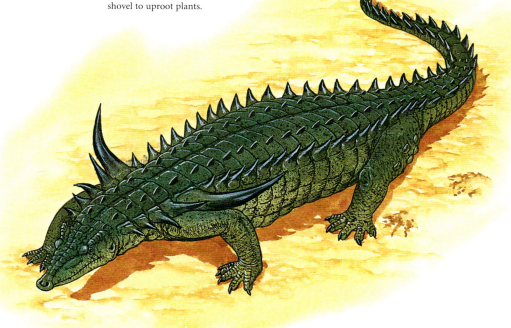

**SNOUT**
It had a pig-like head with a blunt, beaky snout probably used like a shovel to uproot plants.

**SPINES**
The longest of its many spines were those on the shoulders, which reached up to 45cm (18in) in length.

## FOSSIL EVIDENCE

*Desmatosuchus* looked like its archosaur relatives apart from the two rows of curved, serrated spikes that ran along its back. The largest jutted out from its shoulders and would have been valuable for putting off attackers. It was also defended by squares of body armour on its back, tail and belly. It probably could not have fought back with its weak, peg-like teeth, which were suitable only for munching plant matter that it dug up with its snout. The four legs were positioned beneath, rather than to the side of, the hip joint.

PREHISTORIC ANIMAL
LATE TRIASSIC

### HOW BIG IS IT?

## TIMELINE (millions of years ago)

| 540 | 505 | 438 | 408 | 360 | 280 | 248 | 208 | 146 | 65 | 1.8 to today |

83

LATE TRIASSIC

# Staurikosaurus

• **ORDER** • Saurischia • **FAMILY** • Staurikosauridae • **GENUS & SPECIES** • *Staurikosaurus pricei*

## VITAL STATISTICS

| | |
|---|---|
| Fossil Location | South America |
| Diet | Carnivorous |
| Pronunciation | STAW-rick-oh-SAWR-US |
| Weight | 28kg (62lb) |
| Length | 2m (7ft) |
| Height | 80cm (32in) |
| Meaning of name | 'Southern Cross lizard', after a constellation seen only in the southern hemisphere |

### FOSSIL EVIDENCE

Remains are scant, possibly because fossils rarely form in the forest environment that *Staurikosaurus* inhabited. However, because this was an early, primitive theropod, palaeontologists are confident they can picture the entire animal. It was small, with a large head on a slender neck. Long, muscular legs ended on five-toed feet, while the shorter forearms had four-digit hands. It balanced using its long, thin tail. A sliding joint in the lower jaw allowed it to flex back and forth to work prey back towards the throat, a feature that later theropods discarded.

DINOSAUR
LATE TRIASSIC

### WHERE IN THE WORLD?

The few remains were found in Brazil and Argentina.

This was one of the earliest dinosaurs, a speedy, carnivorous beast.

**Teeth**
The sharp, backwards-curving teeth set in a large head are thought by some to be big enough to allow *Staurikosaurus* to attack animals larger than itself.

**Legs**
Large hind legs probably made this animal the fastest on land for its time – a huge advantage for a meat-eater.

### HOW BIG IS IT?

**TIMELINE** (millions of years ago)

| 540 | 505 | 438 | 408 | 360 | 280 | 248 | 208 | 146 | 65 | 1.8 to today |

EARLY JURASSIC

# Abrictosaurus

• **ORDER** • Ornithischia • **FAMILY** • Heterodontosauridae • **GENUS & SPECIES** • *Abrictosaurus consors*

| VITAL STATISTICS | |
|---|---|
| FOSSIL LOCATION | South Africa |
| DIET | Herbivorous |
| PRONUNCIATION | ah-BRICK-tuh-SAWR-us |
| WEIGHT | 43kg (95lb) |
| LENGTH | 1.2m (4ft) |
| HEIGHT | 35cm (14in) |
| MEANING OF NAME | 'Wide awake lizard' to show it did not hibernate |

This animal is at the centre of much dispute. Did it hibernate or not? Does the lack of tusks show the remains are female? Is it really any different from a *Heterodontosaurus*?

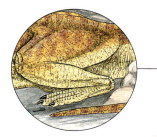

**LEGS**
*Abrictosaurus* was an ornithiscian (bird-hipped). It stood upright on its two hind legs, on which it could sprint away from predators that got too close.

**WHERE IN THE WORLD?**

Two individuals have been found, both in South Africa.

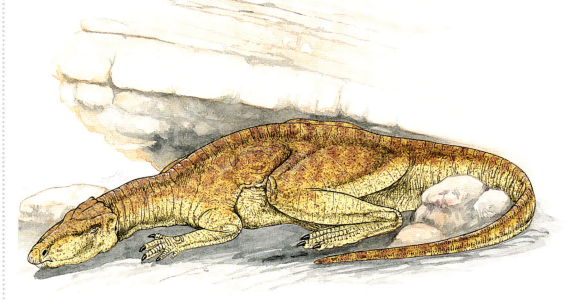

## FOSSIL EVIDENCE

This animal has been renamed as part of a palaeontological debate. Living in what is now South Africa, it had to survive a lack of plant food in the summer. It either roamed and browsed as best it could, or hibernated until the wet season produced sustenance. J.A. Hopson argued that its teeth show it was active all year round and gave it its 'wide awake' name. Debate also surrounds the lack of tusks (the fossil could be of a tusk-less female, or a juvenile) and it has been suggested the remains are actually of its close relative the *Heterodontosaurus*.

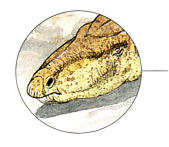

**TEETH**
Unlike other heterodontosaurids, it lacked canine-like teeth on it lower jaw. The cheek teeth are widely separated, suggesting a primitive early dinosaur.

**HOW BIG IS IT?**

DINOSAUR

EARLY JURASSIC

**TIMELINE (millions of years ago)**

| 540 | 505 | 438 | 408 | 360 | 280 | 248 | 208 | 146 | 65 | 1.8 to today |

EARLY JURASSIC

# Ammosaurus

• ORDER • Saurischia • FAMILY • Anchisauridae • GENUS & SPECIES • *Ammosaurus major*

| VITAL STATISTICS | |
|---|---|
| Fossil Location | USA |
| Diet | Herbivorous, possibly omnivorous |
| Pronunciation | AM-uh-SAWR-us |
| Weight | 290kg (640lb) |
| Length | 4m (13ft) |
| Height | 1.8m (6ft) |
| Meaning of name | 'Sand lizard' because it was found in sandstone |

### WHERE IN THE WORLD?

In sandstone rock layers of Connecticut and later in Arizona.

At only 4 metres (13ft) long, this was a small cousin of its fellow sauropodomorphs (some of which were huge). *Ammosaurus* was a plant-eater probably able to walk on four or two legs.

**STOMACH CONTENTS**
Fossilized stomach contents of prosauropods, such as *Ammosaurus*, included gastroliths (stones swallowed possibly to aid digestion) and remnants of a small reptile, suggesting they were omnivorous.

### FOSSIL EVIDENCE

In 1884, builders putting up Connecticut's South Manchester Bridge found the remnants of the back half of this animal. More bones were found when the bridge was demolished 85 years later, and four incomplete skeletons now exist. They reveal a long-necked, small-headed, apparently herbivorous dinosaur with a bulky body and a lengthy tail. It had big hands equipped with thumb claws that may have helped defend it against predators, but its main weapon was likely the flexibility offered by being both bipedal and quadrupedal. It is so like *Anchisaurus* that some palaeontologists believe they are the same animal.

**WALKING**
*Ammosaurus* was probably able to walk on four or two feet, giving it a range of feeding options and the ability to sprint.

### HOW BIG IS IT?

DINOSAUR

EARLY JURASSIC

**TIMELINE (millions of years ago)**

540 | 505 | 438 | 408 | 360 | 280 | 248 | 208 | 146 | 65 | 1.8 to today

EARLY JURASSIC

# Anchisaurus

• ORDER • Saurischia  **FAMILY** • Anchisauridae  **GENUS & SPECIES** • *Anchisaurus polyzelus*

| VITAL STATISTICS | |
|---|---|
| Fossil Location | USA |
| Diet | Herbivorous |
| Pronunciation | AN-key-SAWR-us |
| Weight | 30–70kg (66–154lb) |
| Length | 2–4m (7–13ft) |
| Height | 1m (40 in) |
| Meaning of name | 'Near lizard' to indicate its supposed place between ancestral and later dinosaurs |

So little was known about dinosaurs when *Anchisaurus* fossils were first found in 1818 that the bones were mistaken as being human in origin. However, by 1885 Anchisaurus was recognized as a dinosaur.

## WHERE IN THE WORLD?

Unearthed in Connecticut and Massachusetts, *Anchisaurus* was one of the first dinosaurs discovered in the USA.

## FOSSIL EVIDENCE

One of the smallest, most primitive, of the prosauropods, its powerful rear legs probably allowed *Anchisaurus* to stand upright to reach leaves, but it primarily walked on all fours. It was one of the first dinosaurs with a neck long enough to reach overhead leaves. *Anchisaurus*' jaw was lined with serrated teeth ideal for tearing and shredding plant matter. Gastroliths, small stones in its stomach, may have assisted with digestion by grinding up food. Anchisaurus is sometimes called *Yaleosaurus*, in reference to the Yale Peabody Museum where its fossil was stored.

**Claws**
Its claws grasped and tore at leaves as the dinosaur stood upright. While on all fours, the claws gripped the ground firmly.

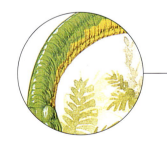

**Neck**
*Anchisaurus*' long neck made it front-heavy, rendering it more practical for the dinosaur to travel on all fours.

## HOW BIG IS IT?

DINOSAUR
EARLY JURASSIC

**TIMELINE (millions of years ago)**

| 540 | 505 | 438 | 408 | 360 | 280 | 248 | 208 | 146 | 65 | 1.8 to today |

EARLY JURASSIC

# Barapasaurus

• ORDER • Saurischia • FAMILY • Vulcanodontidae • GENUS & SPECIES • Barapasaurus tagorei

The impressive dimensions of one of the earliest-known sauropods, *Barapasaurus*, prove that even the earliest sauropods were enormous. Its long thigh bones earned it the name 'big-legged lizard'.

## VITAL STATISTICS

| | |
|---|---|
| FOSSIL LOCATION | India |
| DIET | Herbivorous |
| PRONUNCIATION | Buh-RAH-pah-SAWR-us |
| WEIGHT | Unknown |
| LENGTH | 18–20m (59–66ft) |
| HEIGHT | 6m (20ft) |
| MEANING OF NAME | 'Big-legged lizard' because its femur measured 1.7m (6ft) long |

### WHERE IN THE WORLD?

The limestone of India's Kota Formation provides the ideal material for preserving fossils of many prehistoric creatures.

### TEETH
Although no skull has been found, spoon-shaped teeth suitable for cropping leaves were discovered with the fossil.

### FOSSIL EVIDENCE

Fossils from eight or nine specimens of *Barapasaurus* were unearthed together in a single bonebed in India, suggesting that it lived in herds. Despite the discovery of many *Barapasaurus* fossils, not a single skull has been found. Teeth found with the fossil allow palaeontologists to approximate what its head may have looked like. To sustain its massive bulk, *Barapasaurus* probably survived by feeding throughout most of its waking hours. Its long neck and probably ability to stand on its hind legs made it easy for it to feast on leaves from high trees.

### SPINE
*Barapasaurus'* heavy vertebrae were almost solid, unlike the hollow, light-weight spines of later sauropods.

### HOW BIG IS IT?

DINOSAUR

EARLY JURASSIC

**TIMELINE (millions of years ago)**

| 540 | 505 | 438 | 408 | 360 | 280 | 248 | 208 | 146 | 65 | 1.8 to today |

EARLY JURASSIC

# Emausaurus

• **ORDER** • Ornithischia • **FAMILY** • Incertae cedis • **GENUS & SPECIES** • *Emausaurus ernsti*

| VITAL STATISTICS | |
|---|---|
| FOSSIL LOCATION | Germany |
| DIET | Herbivorous |
| PRONUNCIATION | EE-mau-SAWR-us |
| WEIGHT | 227kg (500lb) |
| LENGTH | 2m (7ft) |
| HEIGHT | 60cm (24in) |
| MEANING OF NAME | 'Ernst-Moritz-Arndt Universität lizard' after the university near the fossil location |

Incomplete fossil evidence renders *Emausaurus* something of a mystery. Debate about the dinosaur revolves around whether it is a primitive relative of the stegosaur or simply a small stegosaur.

**WHERE IN THE WORLD?**

Remains of ancient lava flows beneath Mecklenburg-Vorpommern in northern Germany yielded the fossil of *Emausaurus*.

**ARMOUR**
Rows of tough scales studded with bony scutes protected the body of *Emausaurus* from the bites of carnivorous predators.

## FOSSIL EVIDENCE

*Emausaurus* is classified as an ornithiscian dinosaur of the suborder Thyreophora, which means that it was armour-plated and walked on all fours. Only portions of the armour, skeleton and skull of *Emausaurus* have been uncovered, making it difficult to be certain about its habits. As a relative of *Scelidosaurus*, it walked on all four of its powerful limbs. Because its hind legs were longer than its front ones, it measured highest at it hips. Its leaf-shaped teeth and horned beak were ideal for nipping soft vegetation.

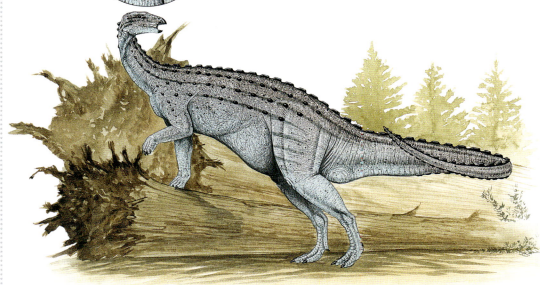

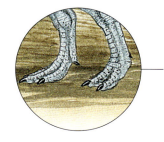

**FEET**
Broad feet were required to support this dinosaur's small but heavy body. They possibly also prevented it sinking into marshy ground.

**HOW BIG IS IT?**

DINOSAUR

EARLY JURASSIC

**TIMELINE** (millions of years ago)

| 540 | 505 | 438 | 408 | 360 | 280 | 248 | 208 | 146 | 65 | 1.8 to today |

EARLY JURASSIC

# Kotasaurus

• **ORDER** • Saurischia • **FAMILY** • Vulcanodontidae • **GENUS & SPECIES** • *Kotasaurus yamanpalliensis*

## VITAL STATISTICS

| | |
|---|---|
| Fossil Location | India |
| Diet | Herbivorous |
| Pronunciation | KOHT-ah-SAWR-us |
| Weight | Unknown |
| Length | 9m (30ft) |
| Height | 3m (10ft) |
| Meaning of name | 'Kota lizard' after India's Kota Formation, where the fossil was found |

*Kotasaurus* is believed to be the oldest-known sauropod because it shares traits of both sauropods and prosauropods. No other dinosaur has yet been discovered that possesses these transitional traits.

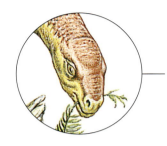

**BRAIN**
Its tiny brain rendered *Kotasaurus* dull-witted. Fortunately, high intelligence was not required in order to be a massive, lumbering herbivore.

### WHERE IN THE WORLD?

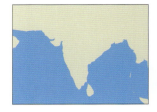

The Kota Formation of India is fossil-rich, yielding evidence of prehistoric fish, reptiles, lizards, plants and dinosaurs.

## FOSSIL EVIDENCE

*Kotasaurus* was among the first of the giant, long-necked dinosaurs. Its colossal size reflects characteristics of the sauropod family, but its bones are a mixture of features associated with both sauropods and prosauropods. With its long tail counter-balancing its long neck, *Kotasaurus* moved along slowly on thick legs. As a herbivore, speed was not necessary as it did not have to run after its lunch. Its little head contained a small brain and blunt teeth suitable for chewing leaves and soft plants.

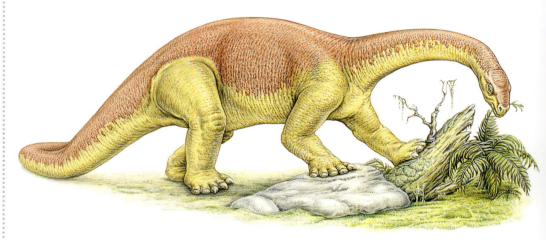

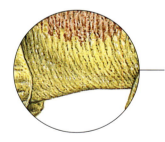

**GUT**
A large gut was necessary to contain and digest the massive amount of plant material consumed by *Kotasaurus*.

### HOW BIG IS IT?

DINOSAUR

EARLY JURASSIC

**TIMELINE (millions of years ago)**

| 540 | 505 | 438 | 408 | 360 | 280 | 248 | 208 | 146 | 65 | 1.8 to today |

EARLY JURASSIC

# Lufengosaurus

• **ORDER** • Saurischia • **FAMILY** • Plateosauridae • **GENUS & SPECIES** • *Lufengosaurus huenei*

## VITAL STATISTICS

| | |
|---|---|
| Fossil Location | China |
| Diet | Herbivorous |
| Pronunciation | LOO-FUHNG-oh-SAWR-us |
| Weight | Unknown |
| Length | 6m (20ft) |
| Height | 3m (10ft) |
| Meaning of name | 'Lu-feng lizard' after the Chinese rock formation where the fossil was discovered |

### WHERE IN THE WORLD?

The sedimentary Lufeng Formation of Yunnan Province, China contains a rich variety of fossils.

*Lufengosaurus* is both one of the oldest-known Chinese dinosaurs and the first complete dinosaur skeleton to be displayed in China. Discovery of its fossils in Yunnan Province indicates that prosauropods roamed worldwide.

## FOSSIL EVIDENCE

Typical of prosauropods, *Lufengosaurus*'s rear limbs were longer than its front limbs. Its long neck and probably ability to stand up on its rear legs allowed the dinosaur to feed on high foliage. The sharp, jagged teeth in its jaw led some to speculate that *Lufengosaurus* may have also eaten meat, but the idea was dismissed when gastroliths were discovered with its fossils. A similar type of sharp tooth has been observed in a herbivorous iguana, further discrediting the theory that *Lufengosaurus* was omnivorous.

**Head**
A small, flat head allowed *Lufengosaurus* to reach in and feed among tree limbs beyond the reach of other dinosaurs.

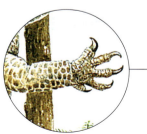

**Claws**
Its front limbs featured sharp claws with a prominent thumb claw suited for grasping and tearing vegetation or defending against predators.

**HOW BIG IS IT?**

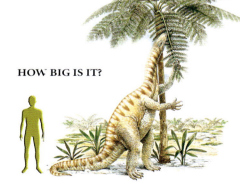

DINOSAUR
EARLY JURASSIC

**TIMELINE (millions of years ago)**

| 540 | 505 | 438 | 408 | 360 | 280 | 248 | 208 | 146 | 65 | 1.8 to today |

EARLY JURASSIC

# Lycorhinus

• **ORDER** • Ornithischia • **FAMILY** • Heterodontosauridae • **GENUS & SPECIES** • *Lycorhinus angustidens*

| VITAL STATISTICS | |
|---|---|
| Fossil Location | South Africa |
| Diet | Herbivorous |
| Pronunciation | LIE-koh-RINE-us |
| Weight | Unknown |
| Length | 1.2m (4ft) |
| Height | 40cm (16in) |
| Meaning of name | 'Wolf snout' in reference to the wolf-like teeth in its lower jaw |

### WHERE IN THE WORLD?

The Upper Elliot Formation spans South Africa and Lesotho. *Lycorhinus* fossils were recovered from its layers.

Originally known only from fragmentary remains, *Lycorhinus* was not recognized as a dinosaur for almost 40 years. Its jaw, which featured both blunt cheek teeth and sharp canines, was unusual for a dinosaur.

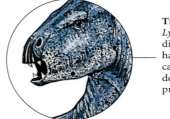

**Teeth**
*Lycorhinus* was a small dinosaur and may have used its sharp canine teeth as a defence against hungry predators.

### FOSSIL EVIDENCE

Like other herbivorous dinosaurs, *Lycorhinus* used its toothless beak to snip shoots and leaves. The first portion of the *Lycorhinus* fossil found was a partial jawbone and, because of its sharp canine teeth, it was misidentified as belonging to a primitive mammal-like reptile for years. Almost 40 years later, in 1962, *Heterodontosaurus* was discovered. *Lycorhinus* was recognized as its relative and finally claimed its rightful place among dinosaurs.

**Hind Legs**
In the hind legs of the *Lycorhinus*, the tibia was longer than the femur. This enabled it to run swiftly.

### HOW BIG IS IT?

| DINOSAUR |
|---|
| EARLY JURASSIC |

**TIMELINE (millions of years ago)**

| 540 | 505 | 438 | 408 | 360 | 280 | 248 | 208 | 146 | 65 | 1.8 to today |

# Megapnosaurus

EARLY JURASSIC

• **ORDER** • Saurischia **FAMILY** • Coelophysoidea
• **GENUS & SPECIES** • *Megapnosaurus rhodesiensis*, *M. kayentakatae*

### VITAL STATISTICS

| | |
|---|---|
| FOSSIL LOCATION | Africa, North America |
| DIET | Carnivorous |
| PRONUNCIATION | Sin-TAR-sus |
| WEIGHT | 30kg (66lb) |
| LENGTH | 3m (10ft) |
| HEIGHT | 80cm (32in) |
| MEANING OF NAME | 'Big dead lizard' |

This agile, light-weight dinosaur hunted small prey in packs. It was originally named *Syntarsus* until scientists discovered that the name was already assigned to a beetle.

### WHERE IN THE WORLD?

As well as in North America, almost 30 specimens of *Megapnosaurus* were discovered together in fossil beds in Zimbabwe.

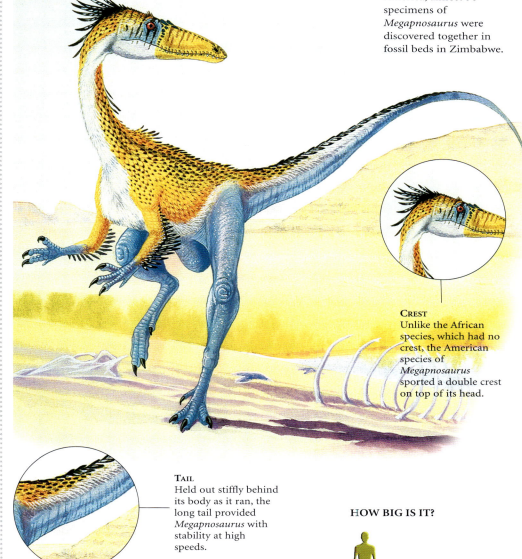

### FOSSIL EVIDENCE

Light, hollow bones and long, powerful legs made *Megapnosaurus* a swift bipedal runner. It may have hunted in packs, possibly pursuing small mammals and lizards. Discovery of *Megapnosaurus* in both Africa and North America suggests that it migrated across the continents when they were joined as Pangaea. Differences in the African and North American species reflect adaptations to the changes in environment as they migrated.

**CREST**
Unlike the African species, which had no crest, the American species of *Megapnosaurus* sported a double crest on top of its head.

**TAIL**
Held out stiffly behind its body as it ran, the long tail provided *Megapnosaurus* with stability at high speeds.

### HOW BIG IS IT?

DINOSAUR
EARLY JURASSIC

**TIMELINE (millions of years ago)**

| 540 | 505 | 438 | 408 | 360 | 280 | 248 | 208 | 146 | 65 | 1.8 to today |

EARLY JURASSIC

# Dilophosaurus

• ORDER • Saurischia • FAMILY • Ceratosauria • GENUS & SPECIES • *Dilophosaurus wetherelli*

| VITAL STATISTICS | |
|---|---|
| FOSSIL LOCATION | USA, China |
| DIET | Carnivorous |
| PRONUNCIATION | Die-LOF-oh-SAWR-us |
| WEIGHT | 300–450kg (650–1000lb) |
| LENGTH | 6–7m (20–23ft) |
| HEIGHT | 1.5m (5ft) at the hip |
| MEANING OF NAME | 'Double crested lizard' because of the two thin bone crests atop its head |

**WHERE IN THE WORLD?**

The Early Jurassic beds of northwestern Arizona's Monument Valley are composed of sandstone, siltstone and shale.

Endowed with a comparatively weak jaw, *Dilophosaurus* probably relied on its deadly claws to bring down its prey. Three *Dilophosaurus* skeletons were discovered together at one site, suggesting that they travelled around in packs.

**CRESTS**
Twin crescent-shaped crests on top of its head may have made *Dilophosaurus* appear attractive to females, but scared off potential predators.

**FOSSIL EVIDENCE**

Two men worked for three years clearing and then mounting the original *Dilophosaurus* skeleton. It was originally named *Megalosaurus* until a more complete fossil was discovered that showed distinctive double crests on its head. The crests meant this was a previously undiscovered theropod dinosaur. *Dilophosaurus* had powerful hind limbs and weighed as much as a small horse. It possessed sharp teeth, but because its jaw was not powerful, these teeth would have been better suited for plucking than stabbing. It probably killed with its hands and feet.

DINOSAUR

EARLY JURASSIC

**HOW BIG IS IT?**

**HANDS**
Its flexible hand featured five fingers (two were very small), which is the ancestral condition for theropods.

**DINOSAUR TIMELINE (millions of years ago)**

| 540 | 505 | 438 | 408 | 360 | 280 | 248 | 208 | 146 | 65 | 1.8 to today |

EARLY JURASSIC

# Scelidosaurus

• **ORDER** • Ornithischia • **FAMILY** • Incertae sedis • **GENUS & SPECIES** • *Scelidosaurus harrisonii*

## VITAL STATISTICS

| | |
|---|---|
| FOSSIL LOCATION | England, Western USA |
| DIET | Low-lying shrubs, ferns |
| PRONUNCIATION | SKEL-eye-doh-SAWR-us |
| WEIGHT | 200–250kg (440–550lb) |
| LENGTH | 3–4m (10–13ft) |
| HEIGHT | 1.2–1.8m (4–6ft) |
| MEANING OF NAME | 'Limb lizard' in reference to its powerful hind legs |

### WHERE IN THE WORLD?

The first *Scelidosaurus* fossil was discovered in 1861 amid layers of limestone and shale in Dorset, England.

*Scelidosaurus* was a primitive ornithischian, a slow, armoured dinosaur with powerful rear limbs that may have allowed it to rear up from its normal position on all fours to pluck foliage from plants.

**HEAD**
*Scelidosaurus*' small head, bony beak and leaf-shaped teeth were characteristic of herbivorous dinosaurs.

## FOSSIL EVIDENCE

*Scelidosaurus*' weak jaw only moved up and down, limiting its ability to chew. It likely swallowed gastroliths to help break down its meals of plants. In order to protect itself from larger, swifter predators, *Scelidosaurus* probably crouched down until its vulnerable belly was safe against the ground, leaving only its armoured back and sides exposed. Horizontal rows of bony plates lined its broad body, tail and limbs. Two sets of three-pointed osteoderms behind its ears protected its head.

### HOW BIG IS IT?

**OSTEODERMS**
Its body was protected by parallel rows of bony plates called osteoderms, which were hard knobs of bone under the skin that were covered with horny material.

DINOSAUR

EARLY JURASSIC

**TIMELINE** (millions of years ago)

| 540 | 505 | 438 | 408 | 360 | 280 | 248 | 208 | 146 | 65 | 1.8 to today |

EARLY JURASSIC

# Cryolophosaurus

## VITAL STATISTICS

| | |
|---|---|
| Fossil Location | Antarctica |
| Diet | Carnivorous |
| Pronunciation | cry-o-LOF-o-sawr-us |
| Weight | 525kg (1,160lb) |
| Length | 6m (20ft) |
| Height | 2m (7ft) |
| Meaning of name | 'Frozen crested lizard' because it was found on Antartica |

**FOSSIL EVIDENCE**
The fossils found include a partially crushed skull (cranium), a jaw bone (mandible), 30 vertebrae from the backbone, three hip bones (ilium, iscium and pubis), two leg bones (femur and fibula), plus ankle and foot bones (tibiotarsus and metatarsals). Many of the bones were still in their natural positions (articulated) and the damage to the skull was possibly caused by a glacier. They had a mix of primitive and advanced characteristics, making classification problematic. Once thought to be the earliest tetanuran, scientists now think it seems closer to the dilophosaurs. It is an important link in theropod evolution.

DINOSAUR

EARLY JURASSIC

**WHERE IN THE WORLD?**

The remains are from Mount Kirkpatrick, located 4000m (13,000ft) up on Antarctica, at 640km (400 miles) from the South Pole.

**MEANING OF NAME**
When *Cryolophosaurus* lived 195 million years ago, Antarctica was not the frozen land we know today, but a land of Jurassic forests. It was closer to the equator, and still attached to Africa. On coastal areas the temperature never dropped below freezing and a wide range of animals survived. This is the reason for the 'cold' in this dinosaur's name, not the conditions in which it was found.

**HOW BIG IS IT?**

Where is the toughest place to look for fossils? The answer has to be Antarctica, the frozen region around the South Pole where your eyelashes freeze together. You have to forget about the shovel and pick up the dynamite to do some investigative digging. Yet here, in 1991, palaeontologist Dr William Hammer and his team made a massive discovery: the first carnivorous dinosaur fossil found in Antarctica.

**SKULL**
The deep, narrow skull had small horns either side of the crest. Its sharp, backward-curving serrated teeth provided extra grip.

**TAIL**
The tail's weight balanced the dinosaur's body, which was strengthened by interlocking bones.

EARLY JURASSIC

• **ORDER** • Saurischia • **FAMILY** • Ceratosauria • **GENUS & SPECIES** • *Cryolophosaurus ellioti*

**SIZE**
At 6 metres (20ft) long and 2 metres (7ft) high, *Cryolophosaurus* was far smaller than later, more advanced carnivores, such as *Allosaurus*, which grew to twice this size.

**CREST**
The most notable feature of *Cryolophosaurus* is the crest running across its head. Many other dinosaurs sport crests, but theirs usually run lengthwise along the skull. This one rises and fans out over the eyes perpendicular to the midline of the skull. Its furrowed nature gives it a comb-like appearance. In fact, it is so reminiscent of Elvis Presley's lavish hairstyle that the dinosaur was informally known as 'Elvisaurus'. Rather like the singer's hair, the crest was for show, probably to attract a mate.

**TIMELINE (millions of years ago)**

| 540 | 505 | 438 | 408 | 360 | 280 | 248 | 208 | 146 | 65 | 1.8 to today |

EARLY-MID JURASSIC

# Yunnanosaurus

• ORDER • Saurischia • FAMILY • Yunnanosauridae • GENUS & SPECIES • *Yunnanosaurus huangi*

| VITAL STATISTICS | |
|---|---|
| Fossil Location | China |
| Diet | Herbivorous |
| Pronunciation | YOU-NAHN-oh-SAWR-us |
| Weight | 2000–3500kg (4409–7716lb) |
| Length | 7m (23ft) |
| Height | 2m (7ft) |
| Meaning of name | 'Yunnan lizard' in honour of China's Yunnan Province, where the fossil was unearthed |

### WHERE IN THE WORLD?

Southern China's Yunnan Province has been the site of significant fossil finds over the years.

Its self-sharpening teeth gave *Yunnanosaurus* an advantage over other prosauropods. Typically, herbivores' teeth wore down with use. Its unusual teeth suggest that *Yunnanosaurus* had a specialized diet.

**Teeth**
Sixty spoon-shaped, self-sharpening teeth lined *Yunnanosaurus*' jaws. As the dinosaur fed, its teeth ground against each other and whetted themselves.

**FOSSIL EVIDENCE**
Known from 20 skeletons and two skulls, *Yunnanosaurus* was a bulky prosauropod that travelled on all fours. Its elongated neck and probable ability to rise on its hind legs enabled it to graze treetops other species of the time could not reach.
*Yunnanosaurus* had the long neck, small head, heavy body, and five-toed feet with curved claws typical of prosauropods. Its long, substantial rear legs meant *Yunnanosaurus* measured highest at the hip when on all fours. This dinosaur was one of the last prosauropods.

DINOSAUR

EARLY-MID JURASSIC

### HOW BIG IS IT?

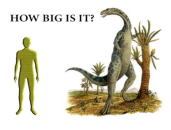

**Claws**
Its five-toed hands ended in hooked claws that allowed the dinosaur to clutch plants and defend itself from predators.

**TIMELINE (millions of years ago)**

| 540 | 505 | 438 | 408 | 360 | 280 | 248 | 208 | 146 | 65 | 1.8 to today |

JURASSIC

# Rhoetosaurus

• **ORDER** • Saurischia • **FAMILY** • Incertae cedis • **GENUS & SPECIES** • *Rhoetosaurus brownei*

| VITAL STATISTICS | |
|---|---|
| FOSSIL LOCATION | Australia |
| DIET | Herbivorous |
| PRONUNCIATION | REET-oh-SAWR-us |
| WEIGHT | Unknown |
| LENGTH | 12–17m (39–56ft) |
| HEIGHT | 5m (16ft) |
| MEANING OF NAME | 'Rhoetan lizard' after the giant Rhoetus of Greek mythology |

### WHERE IN THE WORLD?

One area of Queensland has yielded so many dinosaur bones that it's nicknamed the 'fossil triangle'.

*Rhoetosaurus* is one of Australia's largest sauropods and one of the world's earliest known sauropods. Although its exact weight is unknown, *Rhoetosaurus* is believed to have weighed as much as four elephants.

**SPINE**
Light-weight vertebrae with cartilage at the centre reduced the weight of *Rhoetosaurus*' spine and made it more flexible than solid bone.

### FOSSIL EVIDENCE

Discovered in stages, the first fragments of *Rhoetosaurus* were uncovered by a station manager named Arthur Browne near Roma, in Queensland, in 1924. When those fragments were confirmed to be vertebrae of a new species, an expedition set out to recover more of the *Rhoetosaurus* fossil. In 1926, parts of Rhoetosaurus' tail, neck, ribs and hind leg were found. More remains were discovered in 1975. The ponderous weight of this sauropod was supported by column-like legs featuring a femur that measured an impressive 1.5m (5ft).

**REAR FOOT**
Possibly designed for hauling *Rhoetosaurus*' bulk up slopes, the rear feet featured an enlarged claw on the first toe that dug into the ground for traction.

### HOW BIG IS IT?

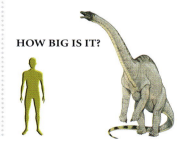

| DINOSAUR |
|---|
| JURASSIC |

**TIMELINE (millions of years ago)**

540 | 505 | 438 | 408 | 360 | 280 | 248 | 208 | 146 | 65 | 1.8 to today

99

MID JURASSIC

# Datousaurus

• ORDER • Saurischia • FAMILY • Incertae sedis • GENUS & SPECIES • *Datousaurus bashanensis*

| VITAL STATISTICS | |
|---|---|
| Fossil Location | China |
| Diet | Herbivorous |
| Pronunciation | DAH-toe-SAWR-us |
| Weight | 17.6 tonnes (16 tons) |
| Length | 14–15m (46–50ft) |
| Height | 5m (16ft) |
| Meaning of name | 'Chieftain lizard' after its large skull |

**WHERE IN THE WORLD?**

China's fossil-rich Dashanpu Formation contained the fossils of sauropods, stegosaurs, and theropods.

Although it was not the largest sauropod, *Datousaurus* was still immense enough to crush any carnivores that dared attack. It could strike a death blow with both its long neck and supple tail.

**Teeth**
Curved, spoon-shaped teeth allowed *Datousaurus* to scrape leaves off the highest branches in the lush, humid jungles prevalent in the Jurassic.

**FOSSIL EVIDENCE**
Often multiple sauropod skeletons are found at a single site, suggesting that they were social animals that lived in herds. However, the fossil remains of *Datousaurus* were found alone. This might mean this dinosaur was separated from its herd by predators or simply died of natural causes and was left behind. Elongated vertebrae gave *Datousaurus* a longer neck than other sauropods, enabling it to reach higher to the tops of trees when it fed. This meant it was not competing for the same food as its contemporaries, *Shunosaurus* and *Omeisaurus*.

**Tail**
Huge muscles attached to the tail vertebrae enabled *Datousaurus* to sweep it back and forth forcefully possibly to strike blows to predators.

**HOW BIG IS IT?**

| DINOSAUR |
| MID JURASSIC |

**TIMELINE (millions of years ago)**

| 540 | 505 | 438 | 408 | 360 | 280 | 248 | 208 | 146 | 65 | 1.8 to today |

MID JURASSIC

# Gasosaurus

• ORDER • Saurischia • FAMILY • Avetheropoda • GENUS & SPECIES • *Gasosaurus constructus*

## VITAL STATISTICS

| | |
|---|---|
| FOSSIL LOCATION | China |
| DIET | Carnivorous |
| PRONUNCIATION | GAS-oh-SAWR-us |
| WEIGHT | 150kg (331lb) |
| LENGTH | 3.5m (11ft) |
| HEIGHT | 1.3m (4ft) |
| MEANING OF NAME | 'Gas lizard' after the gas-mining company that discovered the fossil |

### WHERE IN THE WORLD?

China's Shaximiao Formation is a veritable dinosaur quarry. More than 8,000 bones have been recovered from the site.

The *Gasosaurus* fossil narrowly escaped being destroyed by bulldozers. Fragments of the dinosaur's bones were discovered when a Chinese natural gas company was clearing land to build a field facility.

**TEETH**
The knife-like teeth were used to puncture, slice and restrain prey. Theropods such as *Gasosaurus* possibly dealt lethal bites to victims' heads.

## FOSSIL EVIDENCE

Only one specimen of *Gasosaurus* has been discovered. The incomplete skeleton means that some of *Gasosaurus*' features are based on conjecture, and some speculation has arisen that it is the same species as *Kaijiangosaurus*. *Gasosaurus* was a bipedal dinosaur with short arms that each ended in three sharp claws, which it used to seize and rip the flesh of its victims. *Gasosaurus* possessed the typical theropod tail that was held out stiffly for balance as it ran. Its hollow bones lightened its frame and made it a speedy predator.

### HOW BIG IS IT?

**FEET**
Its three toes spread as *Gasosaurus* trod on hard ground and closed when it lifted its foot, similar to the step of modern birds.

DINOSAUR
MID JURASSIC

**TIMELINE (millions of years ago)**

| 540 | 505 | 438 | 408 | 360 | 280 | 248 | 208 | 146 | 65 | 1.8 to today |

MID JURASSIC

# Lapparentosaurus

• ORDER • Saurischia • FAMILY • Titanosauroformes • GENUS & SPECIES • *Lapparentosaurus madagascariensis*

| VITAL STATISTICS | |
|---|---|
| FOSSIL LOCATION | Madagascar |
| DIET | Herbivorous |
| PRONUNCIATION | Lap-pah-RENT-oh-SAWR-us |
| WEIGHT | Unknown |
| LENGTH | Unknown |
| HEIGHT | Unknown |
| MEANING OF NAME | 'Lapparent's lizard' in honour of French palaeontologist Albert de Lapparent |

## FOSSIL EVIDENCE

It took only about 12 years for *Lapparentosaurus* to reach its adult size. Dinosaurs that grew large fastest probably tended to have the best survival rate because large adults were less likely to be attacked by carnivores. Its nearly solid vertebrae made *Lapparentosaurus* heavier than other sauropods with hollow vertebrae. Its small, wide head contained little peg-like teeth for stripping leaves from high in trees or off ground-level plants. Because its neck resembles that of *Brachiosaurus*, some palaeontologists have speculated that *Lapparentosaurus* was a close relative.

DINOSAUR

MID JURASSIC

### WHERE IN THE WORLD?

Located off the southeast coast of Africa, Madagascar is the fourth-largest island in the world.

*Lapparentosaurus* belongs to a family of dinosaurs that includes the longest, tallest and heaviest dinosaurs known. Unlike later sauropods with hollow vertebrae that made them lighter and more flexible, *Lapparentosaurus* had almost solid vertebrae.

**FEET**
Its broad feet likely contained connective tissue that padded the bones and distributed *Lapparentosaurus'* colossal weight over a larger area.

**GUT**
*Lapparentosaurus* ate all day to sustain itself. An enormous gut churned the broad diet of plants necessary to power the tremendous creature.

### HOW BIG IS IT?

**TIMELINE (millions of years ago)**

| 540 | 505 | 438 | 408 | 360 | 280 | 248 | 208 | 146 | 65 | 1.8 to today |

MID JURASSIC

# Metriacanthosaurus

• ORDER • Saurischia • FAMILY • Sinraptoridae • GENUS & SPECIES • *Metriacanthosaurus parkeri*

When first discussed in 1923, *Metriacanthosaurus* was mistaken as a specimen of *Megalosaurus*. It took forty years to rectify the error and recognize *Metriacanthosaurus* as a separate and distinct dinosaur.

## VITAL STATISTICS

| | |
|---|---|
| FOSSIL LOCATION | England |
| DIET | Carnivorous |
| PRONUNCIATION | MET-ri-ah-CAN-thuh-SAWR-us |
| WEIGHT | 1.1 tonnes (1 ton) |
| LENGTH | 8m (26ft) |
| HEIGHT | 1.8m (6ft) at the hip |
| MEANING OF NAME | 'Moderate spined lizard' because its bony spine ridge was less pronounced than in other dinosaurs |

### WHERE IN THE WORLD?

The variety of species preserved there makes Dorset, England, a popular place for fossil-hunting.

**SPINES**
*Metriacanthosaurus* had 25-cm (10-in) spines along its backbone, which possibly supported a small skin sail, giving the dinosaur a slightly hump-backed appearance.

## FOSSIL EVIDENCE

Since *Metriacanthosaurus* is known from an incomplete fossil that lacks a skull, palaeontologists have disputed whether it belongs to the Sinraptoridae or another group within the Tetanurae. We do know that *Metriacanthosaurus* was a large, bipedal carnivore. The presence of spines along its backbone is interesting, as they were not tall enough to support a huge sail like those seen on *Dimetrodon* or *Spinosaurus*. There has been speculation that this species is related to the ancestor of spinosaurids.

DINOSAUR

MID JURASSIC

**HOW BIG IS IT?**

**TAIL**
Its tail was held stiff by a series of bony, interconnecting rods in the vertebrae. The rigid tail helped balance *Metria-canthosaurus*.

## TIMELINE (millions of years ago)

| 540 | 505 | 438 | 408 | 360 | 280 | 248 | 208 | 146 | 65 | 1.8 to today |

MID JURASSIC

# Omeisaurus

• ORDER • Saurischia • FAMILY • Unranked • GENUS & SPECIES • *Omeisaurus junghsiensis*

| VITAL STATISTICS | |
|---|---|
| Fossil Location | China |
| Diet | Herbivorous |
| Pronunciation | OH-may-SAWR-us |
| Weight | unknown |
| Length | 18–20m (59–66ft) |
| Height | 9m (30ft) |
| Meaning of name | 'Omei lizard' after China's Mount Omei near the fossil's location |

## WHERE IN THE WORLD?

Located in Sichuan, Mount Omei is one of the four Buddhist mountains of China that are considered sacred.

*Omeisaurus* was a typical sauropod with a herbivorous diet and four column-like legs. It did possess the unusual trait of having nostrils closer to the end of its snout than other sauropods.

**Neck**
*Omeisaurus* had more neck vertebrae than were found in most other sauropods. These vertebrae were also longer and stronger than most other sauropods'.

## FOSSIL EVIDENCE

Its amazingly long neck allowed *Omeisaurus* to feed on leaves at treetop level. However, it also meant that *Omeisaurus*' head was carried far above its heart. With at least three more neck vertebrae than the average sauropod, *Omeisaurus* probably required a heavy, powerful heart to pump blood the extended distance to its brain. Wide, muscular arteries in the neck carried blood at an extremely high pressure. Valves in the arteries prevented too much blood from rushing to *Omeisaurus*' brain when it bent its neck downwards.

DINOSAUR

MID JURASSIC

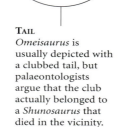

**Tail**
*Omeisaurus* is usually depicted with a clubbed tail, but palaeontologists argue that the club actually belonged to a *Shunosaurus* that died in the vicinity.

## HOW BIG IS IT?

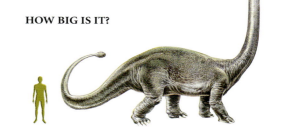

**TIMELINE (millions of years ago)**

| 540 | 505 | 438 | 408 | 360 | 280 | 248 | 208 | 146 | 65 | 1.8 to today |

MID JURASSIC

# Piatnitzkysaurus

• ORDER • Saurischia • FAMILY • Megalosauridae • GENUS & SPECIES • *Piatnitzkysaurus floresi*

## VITAL STATISTICS

| | |
|---|---|
| FOSSIL LOCATION | Argentina |
| DIET | Carnivorous |
| PRONUNCIATION | Pyat-NYIT-skee-SAWR-us |
| WEIGHT | Unknown |
| LENGTH | 4–6m (13–20ft) |
| HEIGHT | 2.1–2.5m (7–8ft) |
| MEANING OF NAME | 'Piatnitzky's lizard' in honour of Alejandro Piatnitzky, a Russian-born Argentine geologist |

*Piatnitzkysaurus* was a smaller, more primitive, version of its relative *Allosaurus*. This bipedal carnivore may have formed aggressive packs that hunted together, digging into the flesh of prey with their claw-tipped toes.

**TEETH**
This aggressive carnivore's jaw contained backwards-curving, serrated teeth designed to clamp onto and shred the flesh of struggling prey.

## WHERE IN THE WORLD?

This *Piatnitzkysaurus* specimen was discovered in the Canodon Asfalto Formation of Argentina's Chubut Province.

## FOSSIL EVIDENCE

*Piatnitzkysaurus*' remains were found near fossils of large sauropods. A pack of these theropods may have joined forces to hunt and attack the larger herbivorous dinosaurs. Once its teeth dug in, powerful muscles in its neck possibly allowed this creature to violently toss its head, shaking its prey and gouging chunks of meat from its body. Despite possessing lethal claws and teeth, *Piatnitzkysaurus* was outweighed by the bulky herbivores.

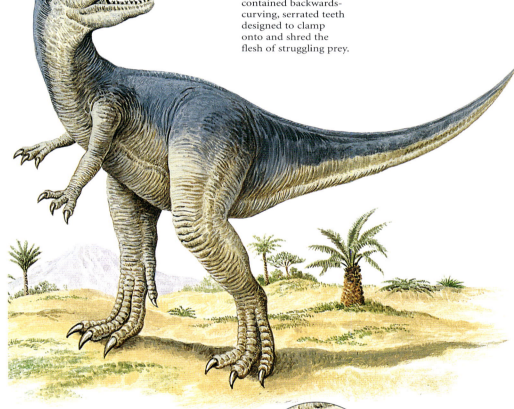

**NECK**
The powerful neck manoeuvred the dinosaur's massive head with ease. Hollows in the skull made it lighter and easy to move.

DINOSAUR

MID JURASSIC

### HOW BIG IS IT?

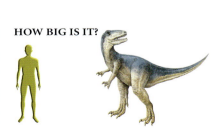

**TIMELINE** (millions of years ago)

| 540 | 505 | 438 | 408 | 360 | 280 | 248 | 208 | 146 | 65 | 1.8 to today |

MID JURASSIC

# Proceratosaurus

• ORDER • Saurischia • FAMILY • Coelurosauria • GENUS & SPECIES • *Proceratosaurus bradleyi*

| VITAL STATISTICS | |
|---|---|
| FOSSIL LOCATION | England |
| DIET | Carnivorous |
| PRONUNCIATION | Pro-ser-RAT-uh-SAWR-us |
| WEIGHT | 100kg (220lb) |
| LENGTH | Up to 3m (10ft) |
| HEIGHT | Unknown |
| MEANING OF NAME | 'Before *Ceratosaurus*' because of the belief it was Ceratosaurus' ancestor |

**WHERE IN THE WORLD?**

Discovered in Gloucestershire, England, the fossil of Proceratosaurus resides in London's Natural History Museum.

A victim of mistaken identity, *Proceratosaurus* was originally identified as an ancestor of *Ceratosaurus* because both bore small crests on their snouts. *Proceratosaurus* has since been reclassified as an early coelurosaur.

**TEETH**
Its varied teeth meant that *Proceratosaurus* could possibly shear the flesh of a variety of unlucky animals.

**FOSSIL EVIDENCE**
Known only from a fragmentary skull, *Proceratosaurus* possessed a small, bony crest on top of its snout. This reminded palaeontologists of *Ceratosaurus*' crest. Its function is still a mystery. Some believe the dinosaur's crest attracted females of the species. The jaw of *Proceratosaurus* was lined with smaller, more conical teeth in front than in the back. Having more than one type of tooth allowed *Proceratosaurus* to process a wider variety of foods.

DINOSAUR

MID JURASSIC

**HOW BIG IS IT?**

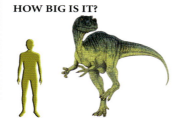

**LOWER LEG**
Typical of coelurosaurs, the lower leg bone is longer than the upper leg bone, making *Proceratosaurus* a nimble runner.

**TIMELINE (millions of years ago)**

| 540 | 505 | 438 | 408 | 360 | 280 | 248 | 208 | 146 | 65 | 1.8 to today |

MID JURASSIC

# Xiaosaurus

• ORDER • Ornithischia • FAMILY • Incertae cedis • GENUS & SPECIES • *Xiaosaurus dashanpensis*

## VITAL STATISTICS

| | |
|---|---|
| FOSSIL LOCATION | China |
| DIET | Herbivorous |
| PRONUNCIATION | Sheow-SAWR-us |
| WEIGHT | 7kg (15lb) |
| LENGTH | 1–1.5m (3–5ft) |
| HEIGHT | 30–50cm (12–20in) |
| MEANING OF NAME | 'Dawn lizard' because of its early occurrence in geological time |

### WHERE IN THE WORLD?

Southwestern China's Sichuan Basin, where *Xiaosaurus* was discovered, is surrounded on all sides by mountains.

Damaged, fragmentary fossils render it difficult to definitively describe or classify *Xiaosaurus*. However, many scientists believe that *Xiaosaurus* is the link between *Lesothosaurus* and *Hypsilophodon*, demonstrating how ornithiscian dinosaurs evolved.

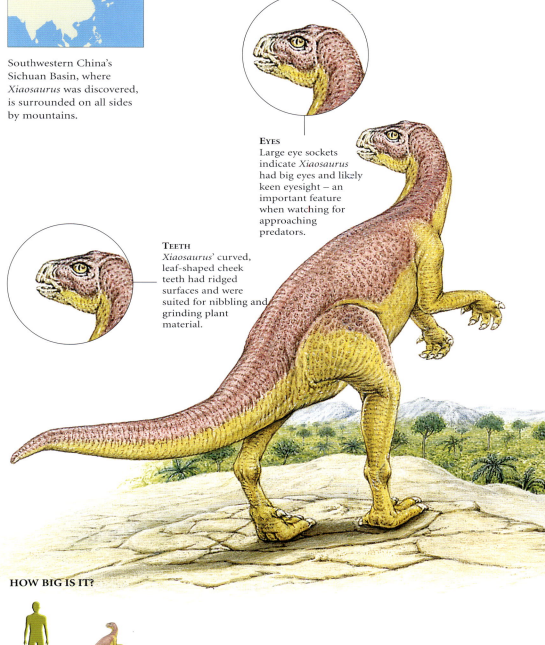

**EYES**
Large eye sockets indicate *Xiaosaurus* had big eyes and likely keen eyesight – an important feature when watching for approaching predators.

**TEETH**
*Xiaosaurus*' curved, leaf-shaped cheek teeth had ridged surfaces and were suited for nibbling and grinding plant material.

## FOSSIL EVIDENCE

*Xiaosaurus* was a bipedal herbivore that consumed low-lying plants. Its short snout had a beak-like mouth, which it used to snip foliage. Lightly built with long back legs, *Xiaosaurus* was an agile runner. As a small, vegetarian dinosaur, *Xiaosaurus*' main defence when threatened was probably fleeing. A strong tail balanced its body as it darted away from predators. Other survival tactics could have been travelling in packs and hiding in dense foliage. It displays characteristics of both early and late ornithiscian dinosaurs, so *Xiaosaurus* is considered to be a transitional animal.

DINOSAUR
MID JURASSIC

### HOW BIG IS IT?

### TIMELINE (millions of years ago)

| 540 | 505 | 438 | 408 | 360 | 280 | 248 | 208 | 146 | 65 | 1.8 to today |

MID JURASSIC

# Huayangosaurus

| VITAL STATISTICS | |
|---|---|
| Fossil Location | China |
| Diet | Herbivorous |
| Pronunciation | Hwah-YANG-oh-SAWR-us |
| Weight | 450kg (1000lb) |
| Length | 4m (13ft) |
| Height | 1.5m (5ft) |
| Meaning of name | 'Huayang lizard' after the region where it was found |

Stegosaurs have one of the most recognizable dinosaur profiles, because they all have two rows of vertical plates running along the top of their backbones. Stegosaurs also have a spiked tail. *Huayangosaurus* was an early, primitive version, far smaller than its younger relatives and with a shorter and higher skull. It lived in what is now China 20 million years before its famous cousin *Stegosaurus* roamed the lands of North America. Its descendants retained the spine plates and the small horizontal tail spikes, but did not have the much larger spikes that projected from its shoulders.

**BRAIN**
One myth is that stegosaurs had a 'second brain' in their spinal cord near the hips to control its hind limbs. In fact, this is an enlarged set of pelvic nerves common to many sauropsids.

**FOSSIL EVIDENCE**
Remains from 12 animals were discovered in 1982, a find that revealed the oldest known member of the stegosaurs. The way the plates on its back grow more like spikes towards the tail suggests that the plates evolved from spikes, particularly as the plates of later stegosaurs were less pointed. A *Huayangosaurus* relative may have been found in England: *Regnosaurus*. It has a similar lower jaw and may be a second member of the Huayangosauridae family.

**TEETH**
*Huayangosaurus* had 14 small teeth at the front of the upper jaw to crop low-lying plants. More derived stegosaurs had toothless beaks.

**SPIKES**
Unlike later stegosaurs, this creature had two large spikes that stuck out from its shoulders. Their purpose is unknown. They would certainly have given it a threatening appearance. They may also have had practical use as weapons against predators or nonspecific rivals or even displays for mates.

DINOSAUR

MID JURASSIC

**HOW BIG IS IT?**

# MID JURASSIC

• **ORDER** • Ornithischia • **FAMILY** • Huayangosauridae • **GENUS & SPECIES** • *Huayangosaurus taibaii*

### WHERE IN THE WORLD?

Found in a quarry in the Huayang, now known as Sichuan, province of China.

**FORELIMBS**
The forelimbs were three-quarters shorter than the hind limbs, perhaps to help it reach higher plants.

### DEFENCE

The double row of sharp, triangular plates running right down the arched back have obvious defensive benefits, making it difficult for predators to get a clean bite and threatening to injure them if they tried. Vein-like cavities in the plates suggest they could have been used to control body temperature, heating or cooling the blood, but they are far smaller than those of some later stegosaurs. They may also have helped attract a mate, and it is possible that they could change colour.

**TIMELINE (millions of years ago)**

| 540 | 505 | 438 | 408 | 360 | 280 | 248 | 208 | 146 | 65 | 1.8 to today |

MID JURASSIC

# Dimorphodon

• ORDER • Pterosauria • FAMILY • Dimorphodontidae • GENUS & SPECIES • *Dimorphodon macronyx*

A pterosaur from the early Jurassic, *Dimorphodon* swooped through the skies 180 million years ago. A flying reptile rather than a dinosaur, *Dimorphodon* was a fearsome predator with a massive head and impressive wingspan.

### VITAL STATISTICS

| | |
|---|---|
| Fossil Location | England |
| Diet | Carnivorous |
| Pronunciation | die-MORE-foe-don |
| Weight | Unknown |
| Length | 1m (3ft) with 1.4m (4.6ft) wingspan |
| Height | Unknown |
| Meaning of name | 'Two-form tooth' because of its two distinctive types of teeth |

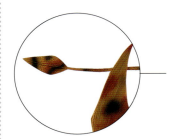

**Tail**
A diamond-shaped flap of skin that was likely at the end of *Dimorphodon's* tail aided in stabilizing the pterosaur during flight.

### WHERE IN THE WORLD?

Lyme Regis, England, where *Dimorphodon* was unearthed, is the home of Dinosaurland Fossil Museum.

### FOSSIL EVIDENCE

*Dimorphodon's* flexible neck had to be strong to support its oversized head. The large skull contained cavities in the bone that lessened the weight of the bulky head. *Dimorphodon* had a small brain. Although it was not one of the more intelligent prehistoric creatures, it survived by operating as a simple hunting machine. Perpetually in search of fish, squid or small reptiles, *Dimorphodon* waited to swoop down on a likely meal.

**HOW BIG IS IT?**

**Beak**
*Dimorphodon* is often depicted as having a puffin-like beak, which it used to snap up small animals.

PREHISTORIC ANIMAL

MID JURASSIC

**TIMELINE (millions of years ago)**

| 540 | 505 | 438 | 408 | 360 | 280 | 248 | 208 | 146 | 65 | 1.8 to today |

MID JURASSIC

# Megalosaurus

ORDER • Saurischia • FAMILY • Megalosauridae • GENUS & SPECIES • *Megalosaurus bucklandii*

## VITAL STATISTICS

| | |
|---|---|
| FOSSIL LOCATION | England, Wales, France, Portugal |
| DIET | Carnivorous |
| PRONUNCIATION | MEG-ah-lo-SAWR-us |
| WEIGHT | 1.2–1.8 metric tons (1.3–2 tons) |
| LENGTH | 7–9m (23–30ft) |
| HEIGHT | 3m (10ft) |
| MEANING OF NAME | 'Great lizard' because of its tremendous size |

### WHERE IN THE WORLD?

*Megalosaurus* fossils were first recovered from the limestone quarries and stonesfield slate of Oxfordshire in England.

The first dinosaur to be described, *Megalosaurus* was also the first dinosaur to be given a scientific name. This occurred before the word 'dinosaur' had even been coined.

## FOSSIL EVIDENCE

In 1676, part of a bone was discovered in an Oxfordshire quarry. It was first thought to be the thigh bone of a giant. More fossils were discovered in the early 1800s, but it was not until 1824 that scientists understood that the bones belonged to an enormous lizard-like creature. It was given the scientific name *Megalosaurus*, which means 'great lizard'. In 1842, the word 'dinosaur' was coined. Before scientists knew enough about dinosaurs to properly differentiate them, many dinosaur fossils were referred to as *Megalosaurus*.

DINOSAUR
MID JURASSIC

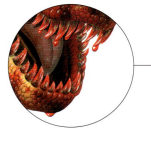

**JAWS**
Powerful jaws lined with curved, knife-like teeth could tear huge chunks of flesh from even the largest sauropods.

**LEGS**
*Megalosaurus* walked upright on two muscular legs. A swift runner, this aggressive theropod possibly relied on surprise while hunting, rushing at its prey.

### HOW BIG IS IT?

**TIMELINE (millions of years ago)**

| 540 | 505 | 438 | 408 | 360 | 280 | 248 | 208 | 146 | 65 | 1.8 to today |

111

MID JURASSIC

# Eustreptospondylus

| VITAL STATISTICS | |
|---|---|
| Fossil Location | England |
| Diet | Carnivorous |
| Pronunciation | You-STREP-to-SPON-die-lus |
| Weight | 200–250kg (440–550lb) |
| Length | 7–9m (23–30ft) |
| Height | 3–3.7m (10–12ft) |
| Meaning of name | 'True turned vertebrae' named for a form originally identified as a species of *Streptospondylus* which means 'reversed vertebra' for the shape of its vertebrae |

*Eustreptospondylus* was a typical theropod: a meat-eating monster that walked on two legs and was equipped to dispatch its victims with bloody efficiency. One intriguing question is why its remains were found in sediment laid down on an ocean bottom – 160 million years ago, southern England was made up of small islands in shallow seas. Was the carcass swept into the water, or did *Eustreptospondylus* sometimes plunge into the waves? It may have scavenged for dead bodies along beaches and estuaries, perhaps even wading in to grab fish and turtles. Or perhaps it learned to paddle with its back legs and swam between the islands.

**TEETH**
Sharp, serrated teeth lined the long jaw. A series of replacements were growing constantly.

**ARMS**
The arms were small and weak, but equipped with extremely sharp claws to rip victims apart.

**TAIL**
The tails of earlier theropods would sway from side to side as they walked, but *Eustreptospondylus* had shorter muscles between its tail and thigh, which would have made the tail much stiffer and less flexible.

## FOSSIL EVIDENCE

This is the best preserved theropod in Europe. It was first named as a *Megalosaurus* in 1841, but the mistake was corrected in 1964. The incomplete fossil is 5m (16ft) long, but the vertebrae show signs of not being fully developed, so it is thought to be a juvenile and to have been likely to grow another 2–4m (6–13ft). It has the powerful hind limbs, erect posture and small forelimbs characteristic of theropods. The skeleton is on display in England's Oxford University Museum.

DINOSAUR

MID JURASSIC

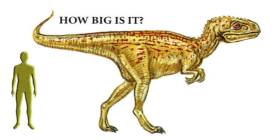

HOW BIG IS IT?

# MID JURASSIC

• **ORDER** • Saurischia • **FAMILY** • Megalosauridae • **GENUS & SPECIES** • *Eustreptospondylus oxoniensis*

## WHERE IN THE WORLD?

A specimen was found in a clay pit north of Oxford in the UK.

**NECK**
Powerful neck and back muscles allowed the animal to twist its head around while gripping its prey with its teeth.

**LEG BONES**
Thick-walled leg bones supported the bulky body and would have made sustained running difficult.

**EYE SOCKET**
The eyes were protected by bony ridges.

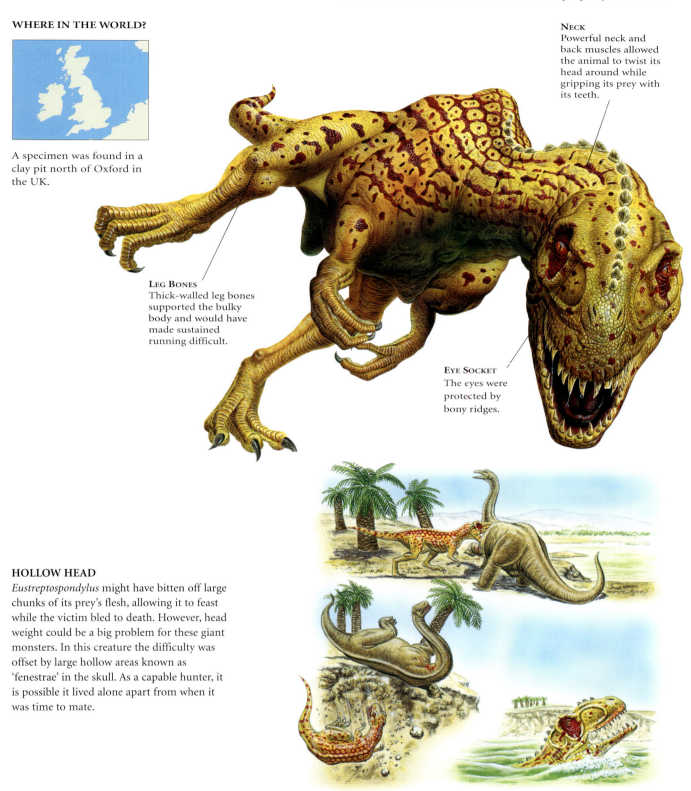

## HOLLOW HEAD

*Eustreptospondylus* might have bitten off large chunks of its prey's flesh, allowing it to feast while the victim bled to death. However, head weight could be a big problem for these giant monsters. In this creature the difficulty was offset by large hollow areas known as 'fenestrae' in the skull. As a capable hunter, it is possible it lived alone apart from when it was time to mate.

**TIMELINE (millions of years ago)**

| 540 | 505 | 438 | 408 | 360 | 280 | 248 | 208 | 146 | 65 | 1.8 to today |

MID JURASSIC

# Shunosaurus

| VITAL STATISTICS | |
|---|---|
| Fossil Location | China |
| Diet | Herbivorous |
| Pronunciation | SHOO-no-SAWR-us |
| Weight | 7.7 tonnes (7 tons) |
| Length | 10m (33ft) |
| Height | 4m (13ft) |
| Meaning of name | 'Shu lizard' from Shu, an old name for the Sichuan region of China where it was found |

**This dinosaur was able to give other animals – including us – a surprise. It was equipped with a bony club at the end of its tail with which it could deliver extremely painful blows to attackers. This club came as a surprise to palaeontologists who thought they knew *Shunosaurus*, too, because it was only identified in 1989, 10 years after the first specimen was discovered.**

**Nostrils**
The nostrils were positioned low down on the muzzle, an unusual feature compared to other sauropods.

**FOSSIL EVIDENCE**
More than 20 complete *Shunosaurus* skeletons have been found, and every one of its bones is known – a very unusual occurrence. The high number of finds suggests that this was a very common dinosaur in the mid-Jurassic. It was the dominant plant-eater, stretching high up to reach leaves with its long neck pretty much all the time to sustain its large body. The legs are relatively short, and indeed *Shunosaurus* would have been dwarfed by the giant sauropods of the late Jurassic.

DINOSAUR

MID JURASSIC

HOW BIG IS IT?

# MID JURASSIC

• ORDER • Saurischia • FAMILY • Cetiosauridae • GENUS & SPECIES • *Shunosaurus lii*

## WHERE IN THE WORLD?

All the specimens are from China's Sichuan Province, from the Dashanpu Formation, which has produced 8,000 pieces of bone so far.

*Shunosaurus* probably browsed on the horsetails, ferns and other plants of the Jurassic period. This diet was low in nutrients and difficult to digest. It must have spent most of its time feeding, probably as part of a herd. It was able to reach plants and leaves higher up than other herbivores of the time were able to reach.

### HEAD
The small head was equipped with slender, ladle-shaped teeth for cropping leaves from plants.

### LEGS
The legs were short and its bones were solid, unlike the lighter, hollow bones that allow later sauropods to grow even bigger.

### CLUBBED TAIL
Absent from the early finds was the bony club at the end of the tail, which is thought to be unique among dinosaurs. Enlarged vertebrae formed two spikes that created a club-like weapon that could have been swung with considerable force as it lashed its tail. It was a powerful weapon with the added benefit of surprise. The heavy tail would have balanced the weight of the long neck.

**TIMELINE** (millions of years ago)

| 540 | 505 | 438 | 408 | 360 | 280 | 248 | 208 | 146 | 65 | 1.8 to today |

MID JURASSIC
# Shunosaurus

MID JURASSIC

• ORDER • Saurischia • FAMILY • Cetiosauridae • GENUS & SPECIES • *Shunosaurus lii*

**THE DASHANPU FORMATION**

In 1972, a Chinese gas company was preparing to install supplies in the till-then inaccessible province of Sichuan in southwest China when its workers came across something quite unexpected – and extremely impressive: the fossils of a fair-sized dinosaur which, appropriately enough, was afterwards given the name *Gasosaurus*. As dinosaurs went, it was not particularly large at 4m(13ft) long and 1.3m (4.3ft) tall, but size was not its greatest significance. In exhuming *Gasosaurus*, the gas company had discovered the Dashanpu Formation, which later produced at least six sauropods including *Shunosaurus*, stegosaurs, theropods, a pterosaur, an amphibian and several ornithopods, as well as fish, turtles and crocodiles and marine reptiles. The fact that the Dashanpu Formation, which was up to 168 million years old, was once a well-stocked forest was proved by pieces of fossilized wood that were scattered among the dinosaur remains. Palaeontologists now believe that the area included a lake that received water from a large river nearby. In this context, the remains of *Shunosaurus* and other dinosaurs were probably swept along by the river into the lake and left there to build into the mass of prehistoric remains discovered after 1972.

MID- LATE JURASSIC

# Cetiosaurus

• ORDER • Saurischia • FAMILY • Cetiosauridae • GENUS & SPECIES • *Cetiosaurus oxoniensis*

### VITAL STATISTICS

| | |
|---|---|
| FOSSIL LOCATION | England |
| DIET | Herbivorous |
| PRONUNCIATION | SEE-tee-oh-SAWR-us |
| WEIGHT | 9.9 tonnes (9 tons) |
| LENGTH | 15–18m (49–60ft) |
| HEIGHT | 9m (16ft) |
| MEANING OF NAME | 'Whale lizard', because its vertebrae had a size and structure similar to a whale's |

### WHERE IN THE WORLD?

The northern coast of England's Isle of Wight has yielded a wide variety of fossils, including *Cetiosaurus*.

*Cetiosaurus* was the first sauropod to be discovered and described. *Cetiosaurus'* heavy vertebrae meant that it normally held its neck straight out from its body rather than raising its head.

**SPINE**
Its heavy vertebrae were massive and primitive, unlike the light-weight, hollowed-out bones of later sauropods. This made its long neck less flexible.

### FOSSIL EVIDENCE

Discovered in 1841, *Cetiosaurus* was not recognized as a dinosaur until 1869. At the time it was first studied, *Cetiosaurus* was the largest land animal known to scientists. Its fossil was initially mistaken for some sort of marine reptile. *Cetiosaurus* was actually a herbivore that ambled about on four enormous, pillar-like legs, stripping the countryside of foliage with its peg-like teeth.

**BACK LEG**
With a thigh bone measuring 1.8m (6ft), *Cetiosaurus* had enormous legs.

### HOW BIG IS IT?

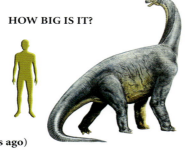

DINOSAUR
MID- LATE JURASSIC

**TIMELINE (millions of years ago)**

| 540 | 505 | 438 | 408 | 360 | 280 | 248 | 208 | 146 | 65 | 1.8 to today |

MID- LATE JURASSIC

# Cetiosauriscus

• ORDER • Saurischia • FAMILY • Diplodocidae • GENUS & SPECIES • *Cetiosauriscus stewartensi*

### VITAL STATISTICS

| | |
|---|---|
| Fossil Location | England |
| Diet | Herbivorous |
| Pronunciation | SEE-tee-oh-sawr-ISS-kus |
| Weight | Unknown |
| Length | 15m (49ft) |
| Height | 6m (20ft) |
| Meaning of name | 'Whale-lizard-like' because of its perceived similarity to *Cetiosaurus* |

### WHERE IN THE WORLD?

As the pounding sea erodes the cliffs of England's southern coastline, new fossils are slowly exposed, including those of *Cetiosauriscus*.

*Cetiosauriscus* turned out to be more closely related to *Diplodocus* than *Cetiosaurus* for which it was named. A lumbering sauropod, *Cetiosauriscus* swallowed leaves whole, possibly relying on gastroliths in its gizzard to pulverize the food.

**Tail**
*Cetiosauriscus* may have flicked its tapering tail to produce loud cracking noises when competing for female attention or frightening enemies.

**Teeth**
*Cetiosauriscus* cropped plants with the peg-like teeth in the front of its mouth. It swallowed plants whole without chewing them.

### FOSSIL EVIDENCE

*Cetiosauriscus* could not lift its head much higher than its shoulders, so it grazed on lower-lying vegetation. Its long neck poked into dense foliage or reached plants growing on ground too marshy to support its massive weight. From the base of the tail to its tip, the vertebrae of *Cetiosauriscus* got smaller, giving the tail a tapering appearance. A series of forked bones in the tail possibly protected the blood vessels if *Cetiosauriscus* was able to stand on its hind legs, using the tail as a prop.

### HOW BIG IS IT?

DINOSAUR

MID- LATE JURASSIC

**TIMELINE (millions of years ago)**

| 540 | 505 | 438 | 408 | 360 | 280 | 248 | 208 | 146 | 65 | 1.8 to today |

MID- LATE JURASSIC

# Lexovisaurus

| VITAL STATISTICS | |
|---|---|
| Fossil Location | Europe |
| Diet | Herbivorous |
| Pronunciation | lek-SOH-vi-SAWR-us |
| Weight | 2 tonnes (2.2 tons) |
| Length | 5m (16ft) |
| Height | 2.7m (9ft) |
| Meaning of name | 'Lexovii lizard' after the French tribe that once lived in the area the fossil was found in |

### WHERE IN THE WORLD?

The first remains were found near Lyons, France, and others are from England.

*Lexovisaurus* was an early stegosaur and close relative of the *Huayangosaurus*. It also resembles the African *Kentrosaurus*. It was a herbivore that possibly roamed in herds, browsing the plants of the Jurassic Period. It interests palaeontologists partly because of the contrast it shows with later, more developed stegosaurs. These dispensed with the shoulder spikes. Perhaps surprisingly, they also evolved with smaller heads and more pointed mouths, limiting their diet to a narrower range of plant food than this creature was able to eat.

**Spines**
It is possible the defensive back plates also helped the animal control its temperature by warming or cooling the blood.

### FOSSIL EVIDENCE

Three partial skeletons and 10 isolated bone remnants, mainly pieces of armour and limb bones, have allowed palaeontologists to reconstruct this early stegosaur. Some of the remains were later found to be from a different species, named *Loricatosaurus*. One difference evident from the two partial skeletons involved is that *Loricatosaurus* does not have shoulder spikes. The characteristic stegosaurid armoured plates were fairly short and narrow. There were 22 on the neck and back and 12 spines on the tail.

DINOSAUR

MID- LATE JURASSIC

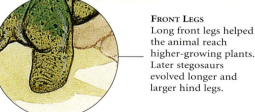

**Front Legs**
Long front legs helped the animal reach higher-growing plants. Later stegosaurs evolved longer and larger hind legs.

120

# MID-LATE JURASSIC

• **ORDER** • Ornithischia • **FAMILY** • Stegosauridae • **GENUS & SPECIES** • *Lexovisaurus durobrivensis*

## WELL ARMOURED

*Lexovisaurus* belied the gentle manner implied by it being a herbivore: it was a well-armoured animal perfectly capable of fighting back. In addition to the double plates along its back it had at least one, possibly two, sets of spikes on its tail. These would have struck a damaging blow if the tail was swung accurately.

## SHOULDER SPIKES

This animal had the largest pair of shoulder spikes of any stegosaur: a menacing 1.2m (4ft) long. Early reconstructions showed these projecting from the hips, but they are now thought to have been positioned at the shoulders. The confusion arises because the spikes were attached to the skin, not to the skeleton, and the skin was not fossilized so no record remains of its life position. The spikes would have offered valuable protection provided *Lexovisaurus* could turn quickly enough to place them in its attacker's way, using its powerful hind legs.

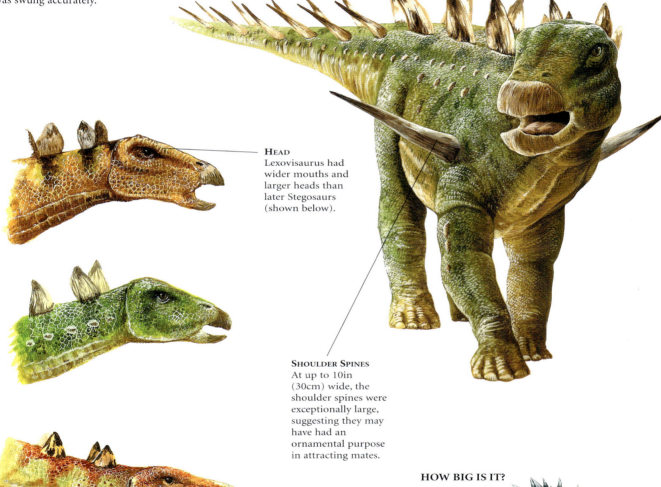

**HEAD**
Lexovisaurus had wider mouths and larger heads than later Stegosaurs (shown below).

**SHOULDER SPINES**
At up to 10in (30cm) wide, the shoulder spines were exceptionally large, suggesting they may have had an ornamental purpose in attracting mates.

**HOW BIG IS IT?**

**TIMELINE (millions of years ago)**

| 540 | 505 | 438 | 408 | 360 | 280 | 248 | 208 | 146 | 65 | 1.8 to today |

MID- LATE JURASSIC

# Liopleurodon

• ORDER • Plesiosauria • FAMILY • Pliosauridae • GENUS & SPECIES • *Liopleurodon ferox*

| VITAL STATISTICS | |
|---|---|
| Fossil Location | Europe |
| Diet | Carnivorous |
| Pronunciation | LIE-oh-PLOOR-oh-don |
| Weight | Unknown |
| Length | 7–10m (23–33ft) |
| Height | Unknown |
| Meaning of name | 'Smooth-sided tooth' because its teeth were smooth on one side |

### WHERE IN THE WORLD?

*Liopleurodon* patrolled the depths of Europe's Late Jurassic seas, striking at unwary prey from below.

This plesiosaur was the top killer of Europe's Jurassic seas. A predatory swimming reptile that lived at the same time as dinosaurs, *Liopleurodon* was one of the biggest plesiosaurs that ever lived.

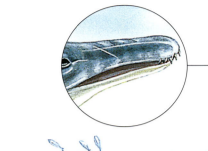

**JAWS**
The lethal, crushing jaws of *Liopleurodon* were filled with twice as many long, conical teeth as *Tyrannosaurus* had.

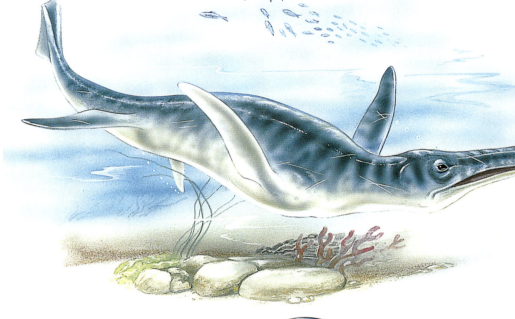

### FOSSIL EVIDENCE

*Liopleurodon* held the highest position on the marine food chain. This plesiosaur was ferocious, probably hunting the seas for fish, smaller plesiosaurs, ichthyosaurs and sharks. Its elongated skull measured 1.5m (5ft) long and anchored muscles that powered a bite possibly formidable enough to crush the bones of any animal. *Liopleurodon* may have swum with its jaws open, allowing water to enter the nostril openings in the roof of its mouth. As the water exited the nasal openings near its eyes, *Liopleurodon* could smell the scent of its prey in the water.

**FLIPPERS**
Four enormous flippers propelled *Liopleurodon* through the water, giving it the power to accelerate rapidly as it ambushed its next meal.

### HOW BIG IS IT?

PREHISTORIC ANIMAL
MID- LATE JURASSIC

**TIMELINE (millions of years ago)**

| 540 | 505 | 438 | 408 | 360 | 280 | 248 | 208 | 146 | 65 | 1.8 to today |

LATE JURASSIC

# Tuojiangosaurus

• **ORDER** • Ornithischia • **FAMILY** • Stegosauridae • **GENUS & SPECIES** • *Tuojiangosaurus multispinus*

## VITAL STATISTICS

| | |
|---|---|
| FOSSIL LOCATION | China |
| DIET | Herbivorous |
| PRONUNCIATION | TOO-oh-gee-ANG-oh-SAWR-us |
| WEIGHT | Unknown |
| LENGTH | 7m (23ft) |
| HEIGHT | 2m (7ft) |
| MEANING OF NAME | 'Tuo River lizard' in honour of the Chinese river where its fossils were first found |

### WHERE IN THE WORLD?

The River Tuo runs through China's Sichuan Province and empties into the River Yangtze.

The largest-known Asian stegosaur, *Tuojiangosaurus* dwelled among the flowering plants and subtropical forests of the Late Jurassic. The distinctive double row of plates along its spine was a typical feature of stegosaurs.

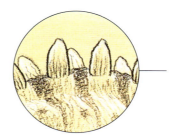

**PLATES**
Lining the neck, back and tail of *Tuojiangosaurus* were 17 pairs of pointed, vertical plates. These made it appear threatening to enemies.

## FOSSIL EVIDENCE

*Tuojiangosaurus*' small, narrow head held a tiny brain. It probably spent most of its waking hours cropping ferns, cycads, and other ground-level plants with its horny, toothless beak. The jaws of *Tuojiangosaurus* were lined with small, weak teeth. It shredded rather than chewed vegetation with its ridged, leaf-shaped cheek teeth. With hind legs longer than its front legs, *Tuojiangosaurus* could probably rear up on two legs to reach taller plants. A slow runner, *Tuojiangosaurus* relied on its spiked tail and plates to intimidate predators.

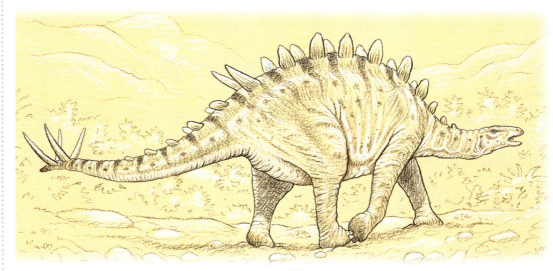

**TAIL**
Two sets of sharp spikes protruded dangerously from the end of *Tuojiangosaurus*' tail, making an effective defence against predators.

### HOW BIG IS IT?

DINOSAUR
LATE JURASSIC

**TIMELINE (millions of years ago)**

| 540 | 505 | 438 | 408 | 360 | 280 | 248 | 208 | 145 | 65 | 1.8 to today |

LATE JURASSIC

# Bothriospondylus

• **ORDER** • Saurischia • **FAMILY** • Brachiosauridae • **GENUS & SPECIES** • *Bothriospondylus robustus*

| VITAL STATISTICS | |
|---|---|
| Fossil Location | Madagascar, England, Tanzania |
| Diet | Herbivorous |
| Pronunciation | BAWTH-ree-oh-SPON-di-lus |
| Weight | Unknown |
| Length | 20m (66ft) |
| Height | 10.7m (35ft) |
| Meaning of name | 'Furrowed vertebrae' because of the shape of its vertebrae |

When *Bothriospondylus* fossils were first discovered, scientists had a difficult time envisioning such an enormous creature walking on land all the time without tiring. Experts originally thought it lived part of its life in water.

**WHERE IN THE WORLD?**

England, Madagascar and Tanzania.

## FOSSIL EVIDENCE

*Bothriospondylus* had longer front legs than rear legs, which helped support the weight of its incredibly long neck. Its elongated neck gave it the reach to graze the tops of trees, where its spoon-shaped teeth tore off leaves. Instead of chewing the leaves to a pulp, *Bothriospondylus* swallowed them in a shredded form. Gastroliths that it probably had in its gizzard did the work of grinding the plant material into a digestible form.

**TEETH**
Long, spoon-shaped teeth helped *Bothriospondylus* tear off coarse leaves from high up trees. It used these teeth to shred rather than chew.

**NOSTRILS**
Nostrils placed on top of its head allowed *Bothriospondylus* to breathe while eating without inhaling little pieces of plant material.

**HOW BIG IS IT?**

DINOSAUR
LATE JURASSIC

**TIMELINE (millions of years ago)**

| 540 | 505 | 438 | 408 | 360 | 280 | 248 | 208 | 146 | 65 | 1.8 to today |

LATE JURASSIC

# Camptosaurus

• **ORDER** • Ornithischia • **FAMILY** • Iguanodontidae • **GENUS & SPECIES** • *Camptosaurus dispar*

## VITAL STATISTICS

| | |
|---|---|
| Fossil Location | USA, England, Portugal |
| Diet | Herbivorous |
| Pronunciation | KAMP-toe-SAWR-us |
| Weight | Up to 1.1 tonnes (1 ton) |
| Length | 5–7m (16–23ft) |
| Height | 1m (40in) |
| Meaning of name | 'Flexible lizard' because of its supposedly unfused hip vertebrae |

### WHERE IN THE WORLD?

Reed's Quarry 13 near Como Bluff, Wyoming, has been the site of numerous fossil discoveries.

A primitive ancestor of *Iguanodon*, *Camptosaurus* may have walked upright on its hind legs, but fed on all fours. *Camptosaurus* appeared humpbacked while on all fours because of its shorter forelimbs.

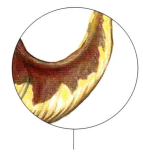

**Tail**
*Camptosaurus*' tail may have helped it make sudden turns when fleeing predators. A quick move of its tail at high speed may have helped change *Camptosaurus*' direction.

## FOSSIL EVIDENCE

*Camptosaurus* had a toothless, horny beak that nipped foliage from trees. The hundreds of leaf-like teeth that lined its jaws cut the vegetation into bits. Its strong hind legs made it an efficient sprinter, a necessary feature for survival with carnivorous predators such as *Allosaurus* lurking in search of a meal. *Camptosaurus* probably lived in migratory herds constantly in search of fresh supplies of plant material.

**Hooves**
Both *Camptosaurus*' front and rear limbs featured hooves rather than sharp claws. Its front hooves made it easier to walk when on all fours.

**HOW BIG IS IT?**

DINOSAUR
LATE JURASSIC

**TIMELINE (millions of years ago)**

| 540 | 505 | 438 | 408 | 360 | 280 | 248 | 208 | 146 | 65 | 1.8 to today |

LATE JURASSIC

# Chialingosaurus

• ORDER • Ornithischia • FAMILY • Stegosauridae • GENUS & SPECIES • *Chialingosaurus kuani*

| VITAL STATISTICS | |
|---|---|
| Fossil Location | China |
| Diet | Herbivorous |
| Pronunciation | CHEE-ah-LING-ah-SAWR-us |
| Weight | 150kg (331lb) |
| Length | 4m (13ft) |
| Height | Unknown |
| Meaning of name | 'Chialing lizard' in honour of China's River Chialing |

**WHERE IN THE WORLD?**

During the Jurassic, when *Chialingosaurus* roamed the Earth, China was part of a super-continent called Laurasia.

*Chialingosaurus*' brain was approximately the size of a golf ball. The formidable spikes protruding from its back provided this slow dinosaur with some protection against attackers.

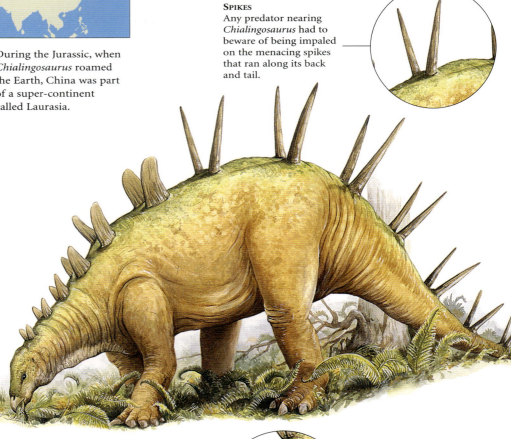

**SPIKES**
Any predator nearing *Chialingosaurus* had to beware of being impaled on the menacing spikes that ran along its back and tail.

**BRAIN**
The size of its tiny brain in comparison to its 150kg (331lb) body indicates that *Chialingosaurus* had very low intelligence.

**FOSSIL EVIDENCE**
*Chialingosaurus* rambled through the Jurassic landscape on all fours on feet that possessed hoof-like claws. Its hind legs were longer than its front ones, giving *Chialingosaurus* a frontwards slope. Its toothless beak, similar to a modern-day turtle's, snipped off plants. The small cheek teeth chopped up the plant material. Two rows of bony plates ran down its spine from its neck to its midsection. A double row of dangerous spikes ran from the middle of its back down its tail.

DINOSAUR
LATE JURASSIC

**HOW BIG IS IT?**

**TIMELINE (millions of years ago)**

| 540 | 505 | 438 | 408 | 360 | 280 | 248 | 208 | 146 | 65 | 1.8 to today |

LATE JURASSIC

# Coelurus

• ORDER • Saurischia • FAMILY • Coeluridae • GENUS & SPECIES • *Coelurus fragilis*

| VITAL STATISTICS | |
|---|---|
| FOSSIL LOCATION | Western USA |
| DIET | Carnivorous |
| PRONUNCIATION | See-LURE-us |
| WEIGHT | 20kg (44lb) |
| LENGTH | 2m (7ft) |
| HEIGHT | 1.8m (6ft) |
| MEANING OF NAME | 'Hollow tail' because of its hollow tail bones |

*Coelurus* was a bipedal carnivore with long legs and a light-weight skeleton that made it an agile runner. Its speed assisted when both making a kill and avoiding falling victim to larger predators.

**HANDS**
*Coelurus*' three-fingered hand ended in curved claws that tightly grasped and then punctured the struggling body of its victim.

**WHERE IN THE WORLD?**

The *Coelurus* fossil is stored at Yale University's Peabody Museum of Natural History in New Haven, Connecticut.

## FOSSIL EVIDENCE

*Coelurus*' hollow tail vertebrae lightened the weight of the appendage. With a long, light-weight tail that functioned as both a balance and a rudder, *Coelurus* could make sharp turns and change direction while running at top speed. Its long, slender legs were powerful, making *Coelurus* a nimble runner that could readily spring upon prey. This also made *Coelurus* adept at avoiding large, hungry carnivores. With its small, sharp teeth, *Coelurus* probably dined on lizards and small mammals that it may have leapt on and took by surprise.

**HOW BIG IS IT?**

**FOOT CLAWS**
*Coelurus* may have stood on its prey and slashed open its belly with its claws, gaining easy access to the victim's tender internal organs.

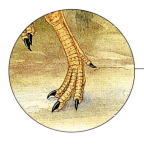

DINOSAUR
LATE JURASSIC

**TIMELINE (millions of years ago)**

| 540 | 505 | 438 | 408 | 360 | 280 | 248 | 208 | 146 | 65 | 1.8 to today |

LATE JURASSIC

# Dicraeosaurus

• ORDER • Saurischia • FAMILY • Dicraeosauridae • GENUS & SPECIES • *Dicraeosaurus hansemanni*

| VITAL STATISTICS | |
|---|---|
| FOSSIL LOCATION | Tanzania |
| DIET | Herbivorous |
| PRONUNCIATION | Die-CREE-oh-SAWR-us |
| WEIGHT | 16.5 tonnes (15 tons) |
| LENGTH | 13–20m (43–66ft) |
| HEIGHT | 6m (20ft) |
| MEANING OF NAME | 'Forked lizard' after the forked spines on its vertebrae |

### WHERE IN THE WORLD?

*Dicraeosaurus* bones were discovered in the Tendaguru beds of Tanzania, a fossil-rich site.

As a herbivore, *Dicraeosaurus* did not have many defences against predators. However, a sauropod of its size could rely on its size alone as a deterrent against attack.

**TEETH**
*Dicraeosaurus* had blunt teeth typical of sauropods. These teeth were suited for stripping the huge amount of vegetation it ate daily.

### FOSSIL EVIDENCE

*Dicraeosaurus* was a large sauropod that lived during the Late Jurassic. It coexisted with other herbivores that probably fed on different types of plants at varying heights, so they were not competing for the same vegetation. Once it had cleared an area of the plants it liked, *Dicraeosaurus* moved on to a new area. *Dicraeosaurus* may have travelled in herds, with the dominant adults leading the way, the young following closely behind, and the elderly bringing up the rear. The adults may have surrounded the young for protection when under attack.

DINOSAUR

LATE JURASSIC

### HOW BIG IS IT?

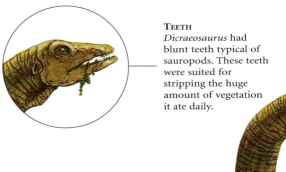

**LEGS**
Bulky *Dicraeosaurus* was not known for its speed. It probably plodded slowly on the four pillar-like legs that supported its hefty body.

**TIMELINE (millions of years ago)**

| 540 | 505 | 438 | 408 | 360 | 280 | 248 | 208 | 146 | 65 | 1.8 to today |

LATE JURASSIC

# Diplodocus

• ORDER • Saurischia • FAMILY • Diplodocidae • GENUS & SPECIES • *Diplodocus longus*

## VITAL STATISTICS

| | |
|---|---|
| FOSSIL LOCATION | Western USA |
| DIET | Herbivorous |
| PRONUNCIATION | Di-PLOD-oh-kus |
| WEIGHT | 11–22 tonnes (10–20 tons) |
| LENGTH | 27m (90ft) |
| HEIGHT | 5m (16ft) at the hips |
| MEANING OF NAME | 'Double-beamed' because of the structure of the bones in the underside of its tail |

**HOW BIG IS IT?**

**WHERE IN THE WORLD?**

Wyoming's Morrison Formation contains an abundance of fossils dating to the Jurassic rivers and floodplains.

*Diplodocus* is the longest dinosaur known from a complete skeleton. Its neck measured 8m (26ft) long and contained 15 vertebrae. Its tail stretched 14m (45ft) long and was composed of almost 80 vertebrae.

**BELLY RIBS**
A series of belly 'ribs', or gastralia, stretched across *Diplodocus*' underside and protected its internal organs. These 'ribs' were embedded in the muscles of the belly.

**TAIL**
*Diplodocus* held its whiplash tail above the ground rather than dragging it along the ground. The enormous tail counterbalanced its huge neck.

## FOSSIL EVIDENCE

Nostrils on top of its skull originally led scientists to believe that *Diplodocus* lived in water. They imagined that *Diplodocus* would submerge itself in water up to the top of its head and use the nostrils as a snorkelling device to breathe. The placement of the nostrils may have actually allowed *Diplodocus* to breathe while eating without inhaling tiny bits of stripped foliage. Like other sauropods, *Diplodocus* spent its days grazing on tender leaves. Fossilized skin impressions indicate *Diplodocus* had small spines along its back.

DINOSAUR
LATE JURASSIC

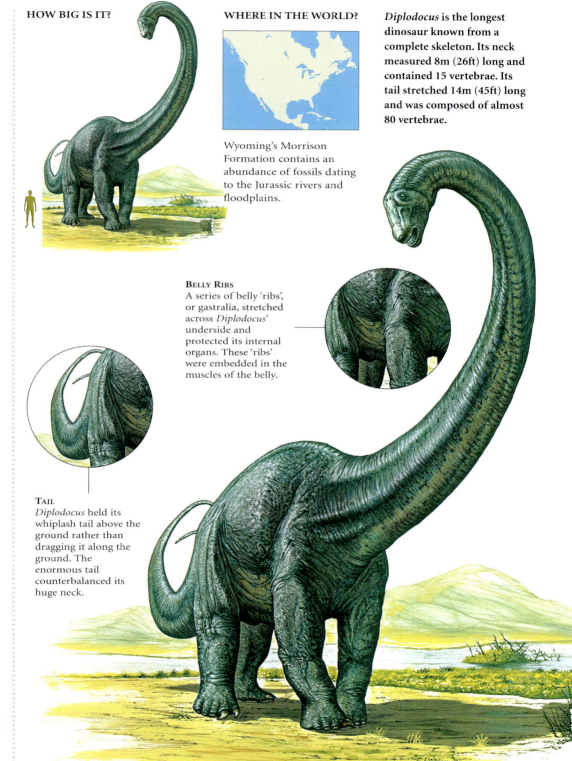

**TIMELINE (millions of years ago)**

| 540 | 505 | 438 | 408 | 360 | 280 | 248 | 208 | 146 | 65 | 1.8 to today |

LATE JURASSIC

# Dracopelta

• ORDER • Ornithischia • FAMILY • Incertae sedis • GENUS & SPECIES • *Dracopelta zbyszewskii*

*Dracopelta* was a primitive ankylosaur with five different types of armoured plates. *Dracopelta* may have crouched down and gripped the ground with its claws when under attack, leaving only its armoured back exposed.

## VITAL STATISTICS

| | |
|---|---|
| FOSSIL LOCATION | Portugal |
| DIET | Herbivorous |
| PRONUNCIATION | Drack-oh-PELL-tah |
| WEIGHT | Unknown |
| LENGTH | 2.1m (7ft) |
| HEIGHT | 80cm (32in) |
| MEANING OF NAME | 'Dragon shield' after its armour |

**EYELIDS**
*Dracopelta's* eyelids were supported by an internal bone. Such armoured eyelids probably protected *Dracopelta's* eyes from being gouged out by attacking predators.

### WHERE IN THE WORLD?

Portugal's Museu da Lourinhã houses a collection of fossils of dinosaurs discovered in the area.

### FOSSIL EVIDENCE

Armoured with scutes, *Dracopelta* was a tank-like creature that ambled slowly along on four strong, short legs. Its lack of speed and low intelligence meant it relied on its heavy armour for survival. The vulnerable belly lacked the spikes, knobs, scales and horn-studded plates that covered its back. When attacked, it may have pressed its stomach against the ground, clinging with its claws so it could not be flipped over. *Dracopelta's* pear-shaped head featured a toothless, horny beak for cropping plants.

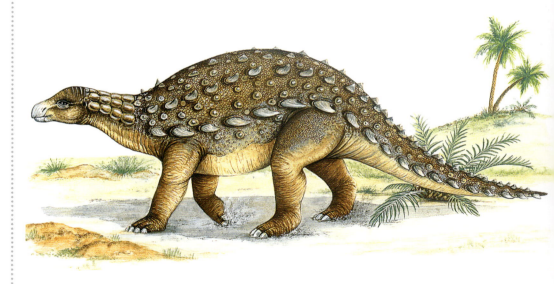

**HOW BIG IS IT?**

**SCUTES**
Attached to Dracopelta's skin were embedded bony plates called osteoderms. These osteoderms had a variety of shapes such as horns, spikes or knobs.

| DINOSAUR |
|---|
| LATE JURASSIC |

**TIMELINE (millions of years ago)**

| 540 | 505 | 438 | 408 | 360 | 280 | 248 | 208 | 146 | 65 | 1.8 to today |

LATE JURASSIC

# Elaphrosaurus

• ORDER • Saurischia • FAMILY • Neoceratosauria • GENUS & SPECIES • *Elaphrosaurus bambergi*

| VITAL STATISTICS | |
|---|---|
| FOSSIL LOCATION | Tanzania |
| DIET | Carnivorous |
| PRONUNCIATION | EL-ah-fro-SAWR-us |
| WEIGHT | 210kg (463lb) |
| LENGTH | 5–6.2m (16–20ft) |
| HEIGHT | 1.5m (5ft) at the hips |
| MEANING OF NAME | 'Light-weight lizard' because of its slender frame |

*Elaphrosaurus* was a lightly built, bipedal carnivore. This spry hunter probably sought out small mammals, lizards, insects and vulnerable young sauropods. An untended baby dryosaur would make the perfect meal for a marauding *Elaphrosaurus* pack.

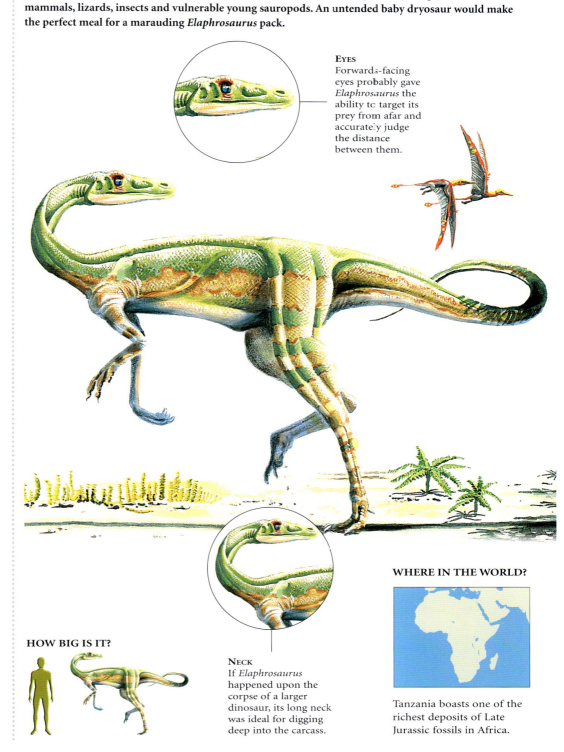

**EYES**
Forwards-facing eyes probably gave *Elaphrosaurus* the ability to target its prey from afar and accurately judge the distance between them.

**FOSSIL EVIDENCE**
Its long back legs and rigid tail made *Elaphrosaurus* a speedy and effective ambush predator. As a contemporary of giant sauropods, such as *Brachiosaurus* and *Dicraeosaurus*, *Elaphrosaurus* had to search for smaller meals probably in the form of little reptiles, amphibians and mammals. It may also have fed on the decaying flesh of large dinosaurs that had been killed by other predators. *Elaphrosaurus*' probable talent for sprinting aided it both in hunting and in escaping danger. It darted away with its tail stiffened for stability.

DINOSAUR

LATE JURASSIC

**HOW BIG IS IT?**

**NECK**
If *Elaphrosaurus* happened upon the corpse of a larger dinosaur, its long neck was ideal for digging deep into the carcass.

**WHERE IN THE WORLD?**

Tanzania boasts one of the richest deposits of Late Jurassic fossils in Africa.

**TIMELINE (millions of years ago)**

| 540 | 505 | 438 | 408 | 360 | 280 | 248 | 208 | 146 | 65 | 1.8 to today |

131

LATE JURASSIC

# Euhelopus

• ORDER • Saurischia • FAMILY • Euhelopodidae • GENUS & SPECIES • *Euhelopus zdanski*

| VITAL STATISTICS | |
|---|---|
| FOSSIL LOCATION | China |
| DIET | Herbivorous |
| PRONUNCIATION | You-HEL-oh-pus |
| WEIGHT | Unknown |
| LENGTH | 10–15m (33–49ft) |
| HEIGHT | 5m (16ft) |
| MEANING OF NAME | 'Good marsh foot', because its broad hind feet were presumed ideal for walking on marshy ground |

The nostrils atop *Euhelopus*' head are similar to those of swimming creatures. Scientists first thought that *Euhelopus* lived submerged underwater, poking only the top of its head out of water to breathe.

### WHERE IN THE WORLD?

China's Shandong Province, where *Euhelopus* was discovered, lies to the east of the majestic Taihang Mountains.

**NECK**
*Euhelopus*' long, somewhat flexible neck could stretch far between trees or over a lake's surface possibly to reach aquatic plants.

**FEET**
*Euhelopus* walked on four stout legs. Its broad feet distributed its weight over a wide area, preventing it from sinking into soft soil.

### FOSSIL EVIDENCE

The theory that *Euhelopus* spent any part of its life submerged in water has been disproved by careful examination of the dinosaur's anatomy. The water pressure against the chest of a creature the size of *Euhelopus* would have been too great for the animal to draw a breath. Euhelopus, like many other sauropods, could not raise its head much above the level of its shoulders.

**HOW BIG IS IT?**

DINOSAUR

LATE JURASSIC

**TIMELINE** (millions of years ago)

| 540 | 505 | 438 | 408 | 360 | 280 | 248 | 208 | 146 | 65 | 1.8 to today |

LATE JURASSIC

# Haplocanthosaurus

ORDER • Saurischia • FAMILY • Camarasauridae • GENUS & SPECIES • *Haplocanthosaurus priscus*

### VITAL STATISTICS

| | |
|---|---|
| FOSSIL LOCATION | Western USA |
| DIET | Herbivorous |
| PRONUNCIATION | Hap-lo-KAN-tho-SAWR-us |
| WEIGHT | 14.3 tonnes (13 tons) |
| LENGTH | 21.5m (71ft) |
| HEIGHT | 7m (23ft) |
| MEANING OF NAME | 'Single-spine lizard' because of the simplicity of its vertebrae |

The most primitive sauropod found in North America, *Haplocanthosaurus* had a shorter neck and tail than other sauropods. Slow-moving *Haplocanthosaurus* possibly migrated in herds, constantly in search of a fresh supply of plants.

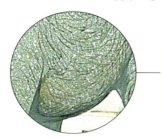

**GUT**
Because its probable diet of cycads and conifers was low in nutrients, *Haplocanthosaurus* had to consume massive amounts of plant material for energy.

### WHERE IN THE WORLD?

*Haplocanthosaurus* was first unearthed by a college student who was digging in Colorado's Morrison Formation.

### FOSSIL EVIDENCE

Fossilized tracks prove that sauropods like *Haplocanthosaurus* did not drag their tails along the ground. The lack of the drag mark the tail would have made indicates that it was held above the ground as a counterbalance for the dinosaur's long neck. A sauropod the size of *Haplocanthosaurus* could potentially leave a footprint in soft soil that was over 1m (3.2ft) long and as deep as 0.5m (18in). Its rear feet made larger tracks than its front feet.

**NECK**
The bones in *Haplocanthosaurus*' shoulder and neck were almost solid. This may have made it difficult for it to lift its head above its shoulders.

**HOW BIG IS IT?**

DINOSAUR
LATE JURASSIC

**TIMELINE (millions of years ago)**

| 540 | 505 | 438 | 408 | 360 | 280 | 248 | 208 | 145 | 65 | 1.8 to today |

LATE JURASSIC

# Othnielia

ORDER • Ornithischia • FAMILY • Euornithopoda • GENUS & SPECIES • *Othnielia rex*

| VITAL STATISTICS | |
|---|---|
| FOSSIL LOCATION | USA |
| DIET | Herbivorous |
| PRONUNCIATION | OTH-nee-EL-ee-ah |
| WEIGHT | Unknown |
| LENGTH | 1.4m (5ft) |
| HEIGHT | 60cm (24in) |
| MEANING OF NAME | 'For Othniel', in honour of American palaeontologist Othniel Marsh |

*Othnielia* was a contemporary of many giant, carnivorous dinosaurs that could easily make a meal of a snack-sized herbivore. Othnielia's small size suggest it was a runner, not a fighter.

### WHERE IN THE WORLD?

*Othnielia* bones were recovered by fossil hunters digging in the Morrison Formation in Colorado, Utah, and Wyoming.

**EYES**
Its large eyes faced forwards, probably giving *Othnielia* keen eyesight and excellent depth perception. This would have allowed it to spot and avoid predators.

### FOSSIL EVIDENCE

*Othnielia* was a small, bipedal herbivore. Its powerful legs were slim and had long shins, which made *Othnielia* a fast sprinter. Its best defence against becoming a meal for *Allosaurus* or *Ornitholestes* was dashing away, possibly briskly zigzagging by using its stiff tail to steer and stabilize its body. *Othnielia* used its horn-covered beak to snip off shrubs and low-growing plants. It probably also ate the occasional insect. Some scientists speculate *Othnielia* had cheek pouches to prevent food from spilling out of its mouth as it chewed.

DINOSAUR

LATE JURASSIC

### HOW BIG IS IT?

**TEETH**
Its enamelled, self-sharpening cheek teeth were shaped like little chisels. They honed themselves as they ground against each other.

**TIMELINE (millions of years ago)**

| 540 | 505 | 438 | 408 | 360 | 280 | 248 | 208 | 146 | 65 | 1.8 to today |

LATE JURASSIC

# Szechuanosaurus

ORDER • Saurischia • FAMILY • Tetanurae • GENUS & SPECIES • *Szechuanosaurus campi*

*Szechuanosaurus* resembled an undersized *Allosaurus*, but it was as well equipped for the kill as larger theropods. Its curved teeth, sharp claws and aggressive nature made *Szechuanosaurus* a fearsome hunter.

## VITAL STATISTICS

| | |
|---|---|
| FOSSIL LOCATION | China |
| DIET | Carnivorous |
| PRONUNCIATION | SESH-WAHN-uh-SAWR-us |
| WEIGHT | 100–150kg (220–331lb) |
| LENGTH | 6–8m (20–26ft) |
| HEIGHT | Unknown |
| MEANING OF NAME | 'Szechuan lizard', in honour of the Chinese province where its fossils were found |

**JAWS**
Its large head bore scissor-like jaws filled with pointed, curved teeth that were designed for tearing and eating flesh.

**HAND CLAWS**
*Szechuanosaurus'* grasping, three-fingered hands ended in claws, which it sank into its victim's flesh to rip open terrible wounds.

## FOSSIL EVIDENCE

*Szechuanosaurus* was endowed with the upright posture, sturdy body, thick neck and short arms characteristic of Jurassic theropods. Although it was not one of the fastest dinosaurs, *Szechuanosaurus* had robust legs and managed to run fast enough on its bird-like feet to be a feared hunter. It may have taken victims by surprise, leaping at them and sinking its claws into their flesh. *Szechuanosaurus'* foot claws tore muscle and tendon, but the deadliest trauma may have come as its mighty jaws bit through the neck of its prey.

DINOSAUR

LATE JURASSIC

### HOW BIG IS IT?

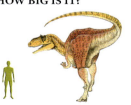

### WHERE IN THE WORLD?

East of the Himalayas lies the Sichuan basin where the fossils of *Szechuanosaurus* were discovered.

**TIMELINE (millions of years ago)**

| 540 | 505 | 438 | 408 | 360 | 280 | 248 | 208 | 146 | 65 | 1.8 to today |

LATE JURASSIC

# Ultrasauros

ORDER • Saurischia • FAMILY • Incertae sedis • GENUS & SPECIES • *Ultrasauros macintoshi*

| VITAL STATISTICS | |
|---|---|
| Fossil Location | Western USA |
| Diet | Herbivorous |
| Pronunciation | UL-trah-SAWR-ohs |
| Weight | 60.6–143.3 tonnes (55–130 tons) |
| Length | 25–30m (82–98ft) |
| Height | 15–16m (49–52ft) |
| Meaning of name | 'Ultra lizard' after its gigantic size |

If it managed not to fall victim to disease, an accident or a pack of hungry predators, gigantic *Ultrasauros* could have had a lifespan of up to 100 years.

### WHERE IN THE WORLD?

Colorado, where the majestic Rocky Mountains rise, was home to *Ultrasauros* during the late Jurassic period.

**Brain**
A dinosaur's intelligence is estimated by comparing its body size to its brain size. *Ultrasauros'* tiny head contained a small brain.

**Neck**
Like other sauropods *Ultrasauros* probably could not raise its head, much like a giraffe.

## FOSSIL EVIDENCE

*Ultrasauros* was so huge even a *Ceratosaurus* or *Allosaurus* might hesitate to attack it. A herd of these enormous sauropods would be too intimidating to approach. Moving slowly from tree to tree, *Ultrasauros* walked on four pillar-like legs. If these animals lived in herds, their approach must have caused the ground to tremble beneath their feet. Many scientists think that Ultrasauros may just be a large example of *Brachiosaurus*.

DINOSAUR
LATE JURASSIC

HOW BIG IS IT?

TIMELINE (millions of years ago)

| 540 | 505 | 438 | 408 | 360 | 280 | 248 | 208 | 146 | 65 | 1.8 to today |

136

LATE JURASSIC

# Yangchuanosaurus

**ORDER** • Saurischia  **FAMILY** • Sinraptoridae  **GENUS & SPECIES** • *Yangchuanosaurus shangyouensis*

### VITAL STATISTICS

| | |
|---|---|
| FOSSIL LOCATION | China |
| DIET | Carnivorous |
| PRONUNCIATION | YANG-choo-WAHN-oh-SAWR-us |
| WEIGHT | 2350kg (5180lb) |
| LENGTH | 6–10m (20–33ft) |
| HEIGHT | 4.6m (15ft) |
| MEANING OF NAME | 'Yang-chu'an lizard' after the area in China where it was found |

During its reign, *Yangchuanosaurus* was one of the largest predators living among stegosaurs and sauropods. It was classified as a carnosaur, a type of large, carnivorous theropod. *Yangchuanosaurus* probably hunted in packs.

**SKULL**
*Yangchuanosaurus'* large skull was not solid. It contained spaces, called fenestrae, which reduced the weight of its impressive head.

**BACK**
Some scientists think *Yangchuanosaurus* had a low crest on its spine, which would have given it a slightly hunchbacked appearance.

### FOSSIL EVIDENCE

*Yangchuanosaurus* is one of the most complete fossil dinosaur skeletons ever found in China. Only an arm, a foot and a piece of its tail were missing. The position in which *Yangchuanosaurus'* fossil was found shows that after it died, the ligaments in its spine contracted and pulled the body into a 'death pose'. Its powerful jaws contained serrated teeth that could rip flesh and splinter bones. If one of its teeth broke off while attacking prey, another one later grew in its place.

DINOSAUR
LATE JURASSIC

**HOW BIG IS IT?**

**WHERE IN THE WORLD?**

*Yangchuanosaurus* was uncovered during the construction of China's Shangyou Reservoir dam in Sichuan Province.

**TIMELINE (millions of years ago)**

| 540 | 505 | 438 | 408 | 360 | 280 | 248 | 208 | 146 | 65 | 1.8 to today |

LATE JURASSIC

# Dacentrurus

| VITAL STATISTICS | |
|---|---|
| Fossil Location | Europe |
| Diet | Herbivorous |
| Pronunciation | Dah-sen-TROO-rus |
| Weight | 1–2 tonnes (1–2 tons) |
| Length | Up to 10m (33ft) |
| Height | Unknown |
| Meaning of name | 'Very sharp tail' because of its long spikes |

### WHERE IN THE WORLD?

Remains found in England, France, Spain and the Louriñha site in Portugal.

Deciding what a dinosaur looked like from a jumbled and incomplete skeleton in which some of the bones may be of other animals is exceedingly difficult and the story can take many twists. This fact is illustrated by the *Dacentrurus*. It was first named *Omasaurus*, but it was then realized another dinosaur had already been given that name. Then the sizes attributed to it began to balloon. Sometimes listed as being 4.4m (15ft) long, it is now regarded as the largest of the stegosaurs at perhaps 10m (33ft) in length – rivalling its more famous relative *Stegosaurus*.

**TAIL SPIKES**
This was an ambling plant-eater that may have relied on its sharp tail spikes to protect it from faster predators.

### FOSSIL EVIDENCE

This was the first of the stegosaurs to be named, in 1875 in England. There were subsequent finds in France and Spain in the 1990s and, most usefully, in Portugal. Here five complete juvenile skeletons have emerged, as well as an egg that may have belonged to the species. The French specimens were destroyed during World War II when the Le Havre museum in which they were stored was bombed. The features of the specimens studied are so varied that some argue they are of different species.

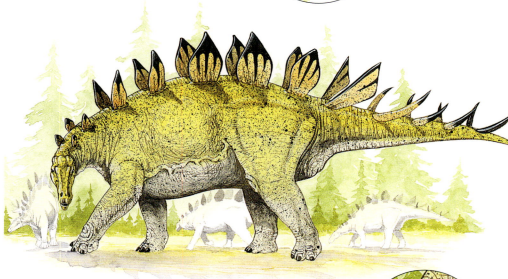

### HOW BIG IS IT?

**TOES**
*Dacentrurus* had three toes on each foot and four fingers on each hand, like other members of its family. The lack of claws emphasizes its reliance on its armour for defence.

DINOSAUR

LATE JURASSIC

138

# LATE JURASSIC

**ORDER** • Ornithischia • **FAMILY** • Stegosauridae • **GENUS & SPECIES** • *Dacentrurus armatus*

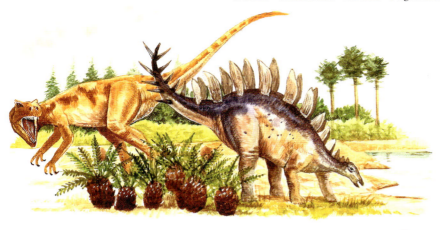

## BIGGER THAN FIRST THOUGHT

The largest pelvis discovered measures 1.5m (5ft) across, which is why some palaeontologists argue that the size of this beast has been significantly underestimated. An animal whose hips are that far apart is going to be very sizeable, especially when you consider that some femurs measure more than a metre (3ft). Part of the difficulty in agreeing on its size is that the more complete smaller specimens are juveniles.

**FORELIMBS AND VERTEBRAE**
The long forelimbs and primitive vertebrae are cited as further evidence that this is a very basal stegosaurid.

## BACK PLATES

The plates of bones on its back are more spike-like than triangular, indicating it is an early, primitive stegosaurid. The plates are positioned in two rows along the back, with a double row of long, sharp-edged spikes on the lower back and tail. The longest spike found is 45cm (18in) to the tip. This is a very different plate and spine configuration from the more derived *Stegosaurus*.

**TIMELINE (millions of years ago)**

| 540 | 505 | 438 | 408 | 360 | 280 | 248 | 208 | 146 | 65 | 1.8 to today |

LATE JURASSIC

# Ornitholestes

ORDER • Saurischia • FAMILY • Coelurosauria • GENUS & SPECIES • *Ornitholestes hermanni*

### WHERE IN THE WORLD?

Wyoming was located in the western part of the super-continent Laurasia during the late Jurassic.

*Ornitholestes* was a nimble, bipedal predator that sprinted after prey. As a small predator at a time when huge theropods ruled, *Ornitholestes* was constantly vigilant, watching for danger with its large eyes.

| VITAL STATISTICS | |
|---|---|
| Fossil Location | Western USA |
| Diet | Carnivorous |
| Pronunciation | Or-NITH-oh-LESS-teez |
| Weight | 11kg (24lb) |
| Length | 2m (7ft) |
| Height | 30cm (12in) |
| Meaning of name | 'Bird robber' because it was originally thought to eat birds |

**FOSSIL EVIDENCE**
The light-weight frame and slim, muscular legs of *Ornitholestes* made it quick and agile. Equipped with a balancing tail that aided in manoeuvring while running, *Ornitholestes* probably leapt at small mammals, lizards or unattended baby dinosaurs. Its narrow head had jaws lined with sharp, conical teeth. The largest were located at the front of its mouth for biting and gripping its prey. A broken nasal bone in the fossil's skull led many scientists to believe incorrectly that *Ornitholestes* had a crest on its snout.

**TAIL**
Its stiff tail served as both stabilizer and rudder, allowing it to make quick turns when charging after a victim.

**THUMB**
The third finger acted like a thumb, allowing *Ornitholestes* to maintain a death grip on its victims.

### HOW BIG IS IT?

DINOSAUR

LATE JURASSIC

**TIMELINE (millions of years ago)**

| 540 | 505 | 438 | 408 | 360 | 280 | 248 | 208 | 146 | 65 | 1.8 to today |

LATE JURASSIC

# Apatosaurus

ORDER • Saurischia • FAMILY • Diplodocidae • GENUS & SPECIES • *Apatosaurus ajax*

## VITAL STATISTICS

| | |
|---|---|
| FOSSIL LOCATION | Western USA, northwestern Mexico |
| DIET | Herbivorous |
| PRONUNCIATION | Uh-PAT-uh-SAWR-us |
| WEIGHT | 30.5 tonnes (30 tons) |
| LENGTH | 21–27m (70–90ft) |
| HEIGHT | 4.6m (15ft) at the hips |
| MEANING OF NAME | 'Deceptive lizard' because some details of its bones resembled those of other mosasaurs |

### WHERE IN THE WORLD?

Wyoming's Nine Mile and Bone Cabin quarries have yielded numerous dinosaur fossils.

Under the incorrect name of *Brontosaurus*, *Apatosaurus* became one of the most famous dinosaurs. However, the misunderstood *Apatosaurus* was saddled with the wrong name and wrong head for almost a century.

**REAR LEGS**
*Apatosaurus* may have reared up on its hind legs to reach high treetops, pressing its tail against the ground to help balance itself. Many scientists don't believe that sauropods were built to rear up.

## FOSSIL EVIDENCE

In 1879 Othniel Marsh described and named *Brontosaurus* at a time when he was involved in a heated competition called the 'bone wars' with his rival Edward Cope. Each man wanted to discover more new dinosaur species than the other. Marsh had a headless specimen of *Apatosaurus* that he was convinced was a new species. He gave it a skull based on *Camarasaurus*, which was thought to be a close relative and invented the *Brontosaurus*. *Apatosaurus* was not reunited with its proper skull until the 1970s. Although *Camarasaurus* and *Apatosaurus* were both sauropods, *Camarasaurus* had a larger skull.

DINOSAUR

LATE JURASSIC

**HOW BIG IS IT?**

**TAIL**
*Apatosaurus* had a whip-like tail that may have cracked loudly when lashed in anger.

## TIMELINE (millions of years ago)

| 540 | 505 | 438 | 408 | 360 | 280 | 248 | 208 | 143 | 65 | 1.8 to today |

LATE JURASSIC

# Archaeopteryx
ORDER • Archaeopterygiformes • FAMILY • Archaeopterygidae • GENUS & SPECIES • Archaeopteryx lithographica

| VITAL STATISTICS | |
|---|---|
| Fossil Location | Germany |
| Diet | Carnivorous |
| Pronunciation | Ark-ee-OP-ter-icks |
| Weight | 300–500g (11–18oz) |
| Length | 65cm (26in) |
| Height | 30cm (12in) |
| Meaning of name | 'Ancient wing' because it is the oldest known feathered animal |

**WHERE IN THE WORLD?**

The limestone of southern Germany beautifully preserved the impressions of *Archaeopteryx*'s feathers.

*Archaeopteryx* is the earliest known feathered animal. Because it displays the traits of both non-avian theropods and modern-day birds, some theories suggest it was a missing link between the two.

**WINGS**
The design of *Archaeopteryx*'s wing was better suited for gliding than flapping. Each of its wings had three fingers with claws.

**FOSSIL EVIDENCE**
*Archaeopteryx* displays characteristics of both birds and non-avian theropod dinosaurs. Its bird-like features include wings and feathers. In common with non-avian theropods are a wishbone, hollow bones, toothed jaws, a long bony tail, and clawed hands. *Archaeopteryx* may have used its claws to climb the trunks of trees. Debate continues concerning whether *Archaeopteryx* got airborne by dropping out of trees or by running along the ground first in search of small animals to eat.

**HOW BIG IS IT?**

**FEATHERS**
*Archaeopteryx*'s feathers insulated the primitive bird and regulated its body temperature. Like the feathers of modern birds, they also aided in flight.

DINOSAUR

LATE JURASSIC

TIMELINE (millions of years ago)

| 540 | 505 | 438 | 408 | 360 | 280 | 248 | 208 | 146 | 65 | 1.8 to today |

LATE JURASSIC

# Compsognathus

ORDER • Saurischia • FAMILY • Compsognathidae • GENUS & SPECIES • *Compsognathus longipes*

*Compsognathus* is one of the smallest non-avian dinosaurs known. Its fossil is so delicate and bird-like that it was used as an example in the argument that dinosaurs are the ancestors of modern-day birds.

## VITAL STATISTICS

| | |
|---|---|
| FOSSIL LOCATION | Germany, France |
| DIET | Carnivorous |
| PRONUNCIATION | KOMP-sog-NAY-thus |
| WEIGHT | 3kg (6.5lb) |
| LENGTH | 0.7–1.4m (2.3–4.6ft) |
| HEIGHT | 26cm (10in) at the hips |
| MEANING OF NAME | 'Delicate jaw' |

**FOSSIL EVIDENCE**

*Compsognathus* was a small, chicken-sized dinosaur. It bounded after small prey on its long, thin legs. The bones of fast-running lizards discovered in its stomach confirmed that *Compsognathus* was swift. *Compsognathus'* hands each had three claws that were ideal for gripping struggling victims. Its long tail acted as a counterbalance as the speedy dinosaur pursued prey or made fast turns when fleeing predators. Recent fossil discoveries show that *Compsognathus* was probably covered in fine, downy feathers, which insulated its body.

**TEETH**
*Compsognathus* had small, sharp teeth that were ideal for chomping down on lizards, insects, fish and small mammals.

**NECK**
Using its long, flexible neck, *Compsognathus* could look around and watch for danger or scan the area for prey.

**HOW BIG IS IT?**

DINOSAUR

LATE JURASSIC

**WHERE IN THE WORLD?**

A *Compsognathus* fossil was discovered embedded in limestone near Nice in southwestern France. It has also been found in Germany.

**TIMELINE (millions of years ago)**

| 540 | 505 | 438 | 408 | 360 | 280 | 248 | 208 | 146 | 65 | 1.8 to today |

LATE JURASSIC

# Allosaurus

| VITAL STATISTICS | |
|---|---|
| Fossil Location | USA and Europe |
| Diet | Carnivorous |
| Pronunciation | Al-oh-SAWR-us |
| Weight | 2.3 tonnes (2.5 tons) |
| Length | 8.5m (28ft) |
| Height | 4m (13ft) |
| Meaning of name | 'Strange lizard' because of its light vertebrae |

**WHERE IN THE WORLD?**

Most specimens are from the Morrison Formation in the USA, with others found in Portugal, and possibly Tanzania and Australia.

*Allosaurus* was a killing machine. Fast, powerful and able to attack anything it came across, it was the top predator for more than 10 million years. The wealth of specimens found has allowed palaeontologists to build a detailed picture of its anatomy and habits, but questions remain. How fast could it run? Although swift, it was top-heavy and risked serious injury if it fell onto its short front arms. Did it hunt in packs? Fossils have been found grouped together, but may have accumulated after death.

**FOSSIL EVIDENCE**

Thousands of bones have been found, mainly in the USA, together with footprints and possibly some eggs. Various species and ages of fossils reveal that Allosaurus grew into many different sizes. The measurements given are for a typical *Allosaurus fragilis*, the most common find, but the largest specimen is 9.7m (32ft) long. Larger claims may in fact belong to other species. *Allosaurus* had the typical large theropod features of a huge head, short S-shaped neck, reduced forelimbs and massive hind legs balanced by a long tail.

DINOSAUR

LATE JURASSIC

**HOW BIG IS IT?**

**CURVED CLAWS**

The front limbs had three sharp strongly curved claws capable of slicing through flesh, so at close quarters *Allosaurus* could possibly swipe at its prey and inflict mortal wounds. The innermost claw was positioned like a thumb in that it was slightly apart from the others. The claws could be up to 15cm (6in) long. It also used its claws to seize flesh while it was eating.

# LATE JURASSIC

**ORDER** • Saurischia • **FAMILY** • Allosauridae • **GENUS & SPECIES** • Several possible species within the genus *Allosaurus*

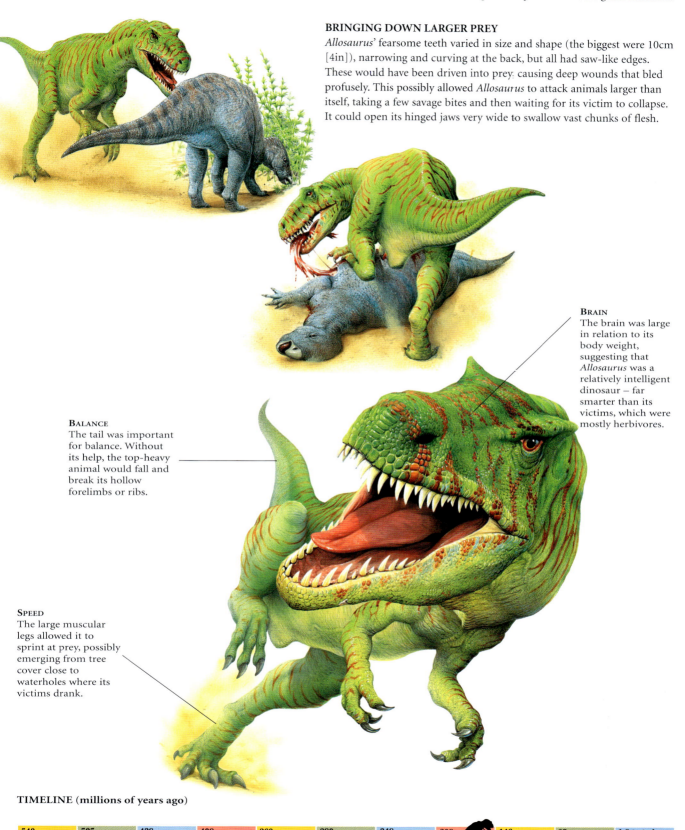

### BRINGING DOWN LARGER PREY
*Allosaurus'* fearsome teeth varied in size and shape (the biggest were 10cm [4in]), narrowing and curving at the back, but all had saw-like edges. These would have been driven into prey, causing deep wounds that bled profusely. This possibly allowed *Allosaurus* to attack animals larger than itself, taking a few savage bites and then waiting for its victim to collapse. It could open its hinged jaws very wide to swallow vast chunks of flesh.

**BRAIN**
The brain was large in relation to its body weight, suggesting that *Allosaurus* was a relatively intelligent dinosaur – far smarter than its victims, which were mostly herbivores.

**BALANCE**
The tail was important for balance. Without its help, the top-heavy animal would fall and break its hollow forelimbs or ribs.

**SPEED**
The large muscular legs allowed it to sprint at prey, possibly emerging from tree cover close to waterholes where its victims drank.

**TIMELINE** (millions of years ago)

| 540 | 505 | 438 | 408 | 360 | 280 | 248 | 208 | 146 | 65 | 1.8 to today |

LATE JURASSIC
# Allosaurus

# LATE JURASSIC

**ORDER** · Saurischia · **FAMILY** · Allosauridae · **GENUS & SPECIES** · Several possible species within the genus *Allosaurus*

## THE MORRISON FORMATION

Although the fossils of the theropod dinosaur Allosaurus have been found in many places across the world, the greatest concentration has come from the Late Jurassic sedimentary rock of the Morrison Formation. The Formation, which centres around Wyoming and Colorado in the USA, covers an area of 1.5 million square kilometres (600,000 square miles). It was named after the town of Morrison in Colorado. In 1877, geologist Arthur Lakes sent a fossilized vertebra specimen from the Formation to the palaeontologist Othniel Charles Marsh, who identified it and named it Allosaurus. But the first comprehensive find of Allosaurus remains did not occur until 1883, when an almost complete skeleton was discovered by Marshall P. Felch, a rancher living in Fremont County, Colorado. Since then, more than 60 Allosaurus fossils of the common species *A. fragilis* have been unearthed worldwide, most of them in the Morrison Formation. Several other dinosaurs have also been found, including *Ceratosaurus*, *Stegosaurus* and *Diplodocus*, while the megalosaur *Torvosaurus tanneri* was found in the area of the Formation comprising the Dry Mesa Quarry, Colorado, in 1971. *Torvosaurus*, one of North America's largest carnivores, was among the Quarry's first discoveries. It measured 11m (36ft) long and weighed in at 2000kg (4400lb).

LATE JURASSIC

# Brachiosaurus

| VITAL STATISTICS | |
|---|---|
| Fossil Location | USA, Europe, Africa |
| Diet | Herbivorous |
| Pronunciation | BRACK-ee-uh-SAWR-us |
| Weight | 32–37 tonnes (35–41 tons) |
| Length | 25m (82ft) |
| Height | 13m (42ft) |
| Meaning of name | 'Arm reptile' because of its long forelimbs |

This was once regarded as the giraffe dinosaur: a massive plant-eater that could stretch its long neck to reach the highest leaves. Many scientists now believe that they couldn't hold their necks vertically. Brachiosaurs are among the largest and heaviest land animals ever; their sheer size and leathery skin was probably sufficient protection from the Late Jurassic carnivores such as *Allosaurus*. The amount of fuel required to keep such a giant going has added to the discussion about whether some dinosaurs were warm-blooded. If it generated its own heat, *Brachiosaurus* would have needed to consume 200kg (440lb) of leaf matter a day. A cold-blooded animal would have required less.

**Teeth**
Brachiosaurus teeth were spatulate, like peg-shaped chisels, able to nip off fresh shoots at the top of trees. There were 26 each on the top and bottom jaws.

**HOW BIG IS IT?**

**FOSSIL EVIDENCE**
Estimates of the size and weight of *Brachiosaurus* vary wildly. The eyes and nostrils were set high on the dome. The large nasal openings near the snout suggest *Brachiosaurus* had a good sense of smell. It was not able to chew, and food was probably digested with the help of stones held in the gizzard, like chickens do today.

**Skull and Neck**
The skull had many hollows to restrict its weight. Lifting a solid skull on such a long neck would have been impossible.

**Forelegs**
*Brachiosaurus* takes its name from its extended forelegs. The thigh bone alone was 2m (6ft 6in) in length.

DINOSAUR

LATE JURASSIC

LATE JURASSIC

**ORDER** • Saurischia • **FAMILY** • Brachiosauridae • **GENUS & SPECIES** • *Brachisaurus altihorax, B. brancai*

## USES FOR A LONG NECK
The 14 vertebrae of the long, upright neck have hollow spaces in them – otherwise it would have been too heavy to lift. Early palaeontologists suggested the animal could have lived under water, using its neck to keep the nostrils sucking in air above the waves like a snorkel. While the water would have supported its huge frame, the pressure would have collapsed its lungs, and the soft mud would not have provided enough support for its narrow feet to stop the animal sinking. The neck may also have been used in battles for dominance between rival males.

## WHERE IN THE WORLD?

After the first discovery in western Colorado, USA, there have been finds in southern Europe and northern Africa.

## CARDIOVASCULAR POWER
An animal of this size needed an incredibly powerful heart to pump blood along the many metres of the neck to the brain. It would have had muscular blood vessels featuring many valves to prevent blood flowing backwards, and its blood pressure was probably three or four times as high as ours.

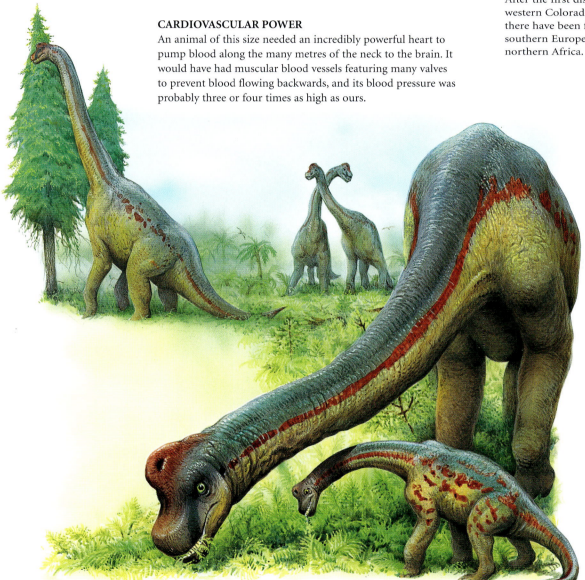

**TIMELINE (millions of years ago)**

| 540 | 505 | 438 | 408 | 360 | 280 | 248 | 208 | 146 | 65 | 1.8 to today |

LATE JURASSIC
# Brachiosaurus

LATE JURASSIC

**ORDER** • Saurischia **FAMILY** • Brachiosauridae **GENUS & SPECIES** • *Brachiosaurus altihorax, B. brancai*

### ONE OF THE LARGEST JURASSIC DINOSAURS

*Brachiosaurus*, one of the largest dinosaurs of the Jurassic Period, was a contemporary of *Stegosaurus*, *Dryosaurus*, *Apatosaurus* and *Diplodocus*, none of which was exactly small. In fact, *Apatosaurus* was slightly bigger than *Brachiosaurus*, while *Diplodocus* – up to 45m (147ft) long – was a massive 56 per cent bigger. Huge dinosaurs like these made gigantic demands on their environment and also needed a physique powerful enough to keep them alive and active. They lived on land rich in food – ferns, bennettites and horsetails flourished there – and also probably relied on the groves of cycads and ginkgos that grew in profusion in the forests. Yet *Brachiosaurus* alone probably required 182kg (401lb) of food every day. Some palaeontologists have suggested that they moved around in herds, so the amount of daily sustenance they took collectively from their environment would have been colossal. But their enormous size did, at least, give *Brachiosaurus* important advantages. Some palaeontologists have reckoned that *Brachiosaurus* and other large dinosaurs were gigantotherms. This meant that the ratio between their volume and their surface area enabled them to maintain a high body temperature. Also, with more of their body shielded from close contact with the outside, they lost less body heat to their environment.

LATE JURASSIC

# Ceratosaurus

| VITAL STATISTICS | |
|---|---|
| Fossil Location | USA, Africa and Europe |
| Diet | Carnivorous |
| Pronunciation | See-RAT-oh-SAWR-us |
| Weight | 1 tonne (1.1 tons) |
| Length | 6m (20ft) |
| Height | 2m (7ft) |
| Meaning of name | 'Horned lizard' because of its nose horn |

*Ceratosaurus* had large jaws, a distinctive horn on its nose and a strip of bony armour down its back. Smaller than its contemporary *Allosaurus*, this was a meat-eating beast capable of attacking creatures larger than itself.

**FOSSIL EVIDENCE**

Classifying *Ceratosaurus* is difficult because, although it resembles other bipedal predators such as *Allosaurus* and *T. rex*, these other theropods do not have a line of bony plates down their backs. Other features of *Ceratosaurus* are more avian. Teeth provide the only evidence of one species, *C. ingens*, but they suggest a much larger variant that would have been one of the biggest theropods. Teeth marks possibly from *Ceratosaurus* found on prey do not tell us whether it was a predator or scavenger.

**BACK AND SPINE**

In combat it always helps to look even bigger and fiercer, and the Ceratosaurus achieved this with a high arch on its back together with a line of bony plates along the spine. These would have provided protection and given the animal a particularly terrifying appearance. They may also have helped it radiate heat to cool down.

**HOW BIG IS IT?**

DINOSAUR

LATE JURASSIC

152

# LATE JURASSIC

**ORDER** • Saurischia • **FAMILY** • Ceratosauridae • **GENUS & SPECIES** • Several species within the genus *Ceratosaurus*

## SNOUT AND EYEBROW HORNS
The feature that gives *Ceratosaurus* its name is the large horn on the snout. Shaped like a blade, its purpose is a mystery. It may have been used in courtship to distinguish between the sexes; acted as a signal to encourage a potential mate; or threatened a rival. It also had a pair of hornlets like ridged eyebrows over the eyes. These horns grew as the animal aged.

### BROW HORNLETS
A pair of hornlets shaped like ridged eyebrows would have offered some protection to its large eyes in battle.

## WHERE IN THE WORLD?

Fossils have been found in Colorado and Utah in the USA, as well as in Portugal and Tanzania.

### FINGERS
Most large meat-eaters of the time had three fingers on the front hands, but *Ceratosaurus* had one extra – a primitive feature complicating classification.

### HIND LEGS
Powerful hind legs gave *Ceratosaurus* the advantage of speed over both victims and predators, allowing it to sprint or lunge rapidly to surprise and overpower an opponent.

**TIMELINE (millions of years ago)**

| 540 | 505 | 438 | 408 | 360 | 280 | 248 | 208 | 146 | 65 | 1.8 to today |

LATE JURASSIC
# Ceratosaurus

LATE JURASSIC

**ORDER** • Saurischia • **FAMILY** • Ceratosauridae • **GENUS & SPECIES** • Several species within the genus *Ceratosaurus*

## DINOSAUR QUARRIES

The Cleveland-Lloyd Dinosaur Quarry in central Utah is one of two sites in the United States where the fossils of *Ceratosaurus* have been discovered (the other is the Dry Mesa Quarry in Colorado). The Utah Quarry, where excavation and study work began in 1929, has proved to be one of the most prolific sources of dinosaur fossils in the world. The Quarry's fossil bed has yielded up the remains of some 70 creatures from 11 species and more than 12,000 dinosaur bones. These have been in the Quarry for around 147 million years and were revealed only after erosion wore away the surface mud and covering rock. The bones are mainly non-articulated finds, but when they are put together some may prove to be complete skeletons, like the *Stegosaurus* and *Allosaurus* already reassembled and put on display at the Quarry visitors' centre. In 1960, scientists and palaeontologists from the University of Utah embarked on a five-year project to excavate further fossils from the Quarry. The investigations were renewed in 2001, this time to conduct research into why the Quarry contains such copious quantities of dinosaur remains.

LATE JURASSIC

# Kentrosaurus

| VITAL STATISTICS | |
|---|---|
| Fossil Location | Africa |
| Diet | Herbivorous |
| Pronunciation | KEN-troh-SAWR-us |
| Weight | 400kg (880lb) |
| Length | 2.5–5m (8–16ft) |
| Height | 1m (3ft) at the hips |
| Meaning of name | 'Spiked lizard' because of its spiked back |

*Kentrosaurus* has helped us to understand how our world was shaped, because it is closely related to *Stegosaurus*. That might not seem particularly significant until you consider that the only specimens of *Kentrosaurus* are found in East Africa, while its cousin is from North America. This proves that what is now Tanzania was at one time joined to North America and thus contiguous with the Morrison Formation, a layer of sedimentary rocks in the western USA. About 150 million years ago, both areas were part of the super-continent Pangaea, and must have shared a similar climate in order to produce animals so alike. When the continents drifted apart, they took related fossilized remains with them.

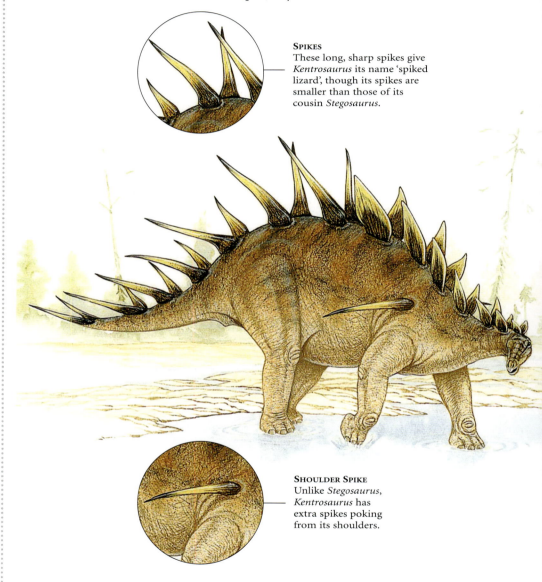

**SPIKES**
These long, sharp spikes give *Kentrosaurus* its name 'spiked lizard', though its spikes are smaller than those of its cousin *Stegosaurus*.

**FOSSIL EVIDENCE**
First discovered in 1909, *Kentrosaurus* is far smaller than its stegosaur cousin *Stegosaurus*, but there are similarities. They share the familiar stegosaur profile of plates and spines sticking out along the back, but those of *Kentrosaurus* are smaller and it has extra spikes poking from its shoulders. It also has a spiky tail and pairs of short triangular plates on its neck and shoulders. So many bones were found at the site that it is thought the animal lived in herds – the bones are estimated to have come from about 70 specimens.

**SHOULDER SPIKE**
Unlike *Stegosaurus*, *Kentrosaurus* has extra spikes poking from its shoulders.

DINOSAUR

LATE JURASSIC

**HOW BIG IS IT?**

**GROUND FEEDER**
Its low front legs allowed *Kentrosaurus* to bend low to reach plants on the ground. The plant matter was scooped up in its toothless beak. It would have spent most of its time eating just to sustain its large body.

# LATE JURASSIC

**ORDER** • Ornithischia • **FAMILY** • Stegosauridae • **GENUS & SPECIES** • *Kentrosaurus aethiopicus*

### POSITION OF SPIKES
*Kentrosaurus* gets its name from the spikes that stuck out sideways and backwards from its shoulders and above its spine. While they certainly would have given an aggressive look to this herbivorous dinosaur, their function can only be guessed at. They offered some sort of protection to the side of the body, but were not attached to the skeleton, so it is difficult to assess how strong they were or their exact positioning.

### WHERE IN THE WORLD?

The only remains are in Tanzania, East Africa, in the area of Tendaguru.

### Brain
The skull enclosed a tiny brain but highly developed olfactory bulbs, showing that *Kentrosaurus*, though probably dull-witted, likely had a very keen sense of smell.

### Hind Legs
The long hind legs suggest the animal was able to rear up to reach higher twigs and leaves, although it spent most of its time grazing.

### Spike Distribution
A double row of bony plates were embedded in its back, changing to sets of spikes in its rear half and down its long, stiff tail.

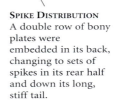

**TIMELINE (millions of years ago)**

540 | 505 | 438 | 408 | 360 | 280 | 248 | 208 | 146 | 65 | 1.8 to today

LATE JURASSIC

# Seismosaurus

| VITAL STATISTICS | |
|---|---|
| Fossil Location | USA |
| Diet | Herbivorous |
| Pronunciation | SIZE-moe-SAWR-us |
| Weight | Up to 50 tonnes (55tons) |
| Length | Up to 45m (148ft) |
| Height | Up to 13m (43ft) |
| Meaning of name | 'Earthquake reptile' because it could have made the ground shake |

This was one of the longest land animals that ever trod the Earth. It stomped through the Late Jurassic forests on short legs, swishing its enormous tail and stretching out its extended neck in a constant search for food. As so often in the fossil-studying community, arguments have raged on several points concerning *Seismosaurus*. One is whether it could have raised its neck up high as it is often depicted doing, given its weight. Another is whether it is a separate species from the extremely similar *Diplodocus* or just an even longer variant of that genus.

**Legs**
*Seismosaurus'* legs had to be like columns to carry its body. It may have struggled to remain stable on wetlands, preferring firm ground.

**Tail**
The tail was extremely long, made of about 80 small bones, so it was highly flexible and could possibly have been flung like a whip.

**FOSSIL EVIDENCE**
*Seismosaurus* remains were named in 1991 from a set of bones including various vertebrae and other spine parts, some ribs and part of the pelvis. With these fossils were many probable gastroliths: stones swallowed by the animal to possibly grind down the food in its gizzard. The reconstructed animal is longer than the *Diplodocus* it so resembles, but is assumed to have pursued the same tree-browsing lifestyle. It was protected by its whip-like tail and its sheer size: it was probably simply too big for an opponent to take on.

DINOSAUR

LATE JURASSIC

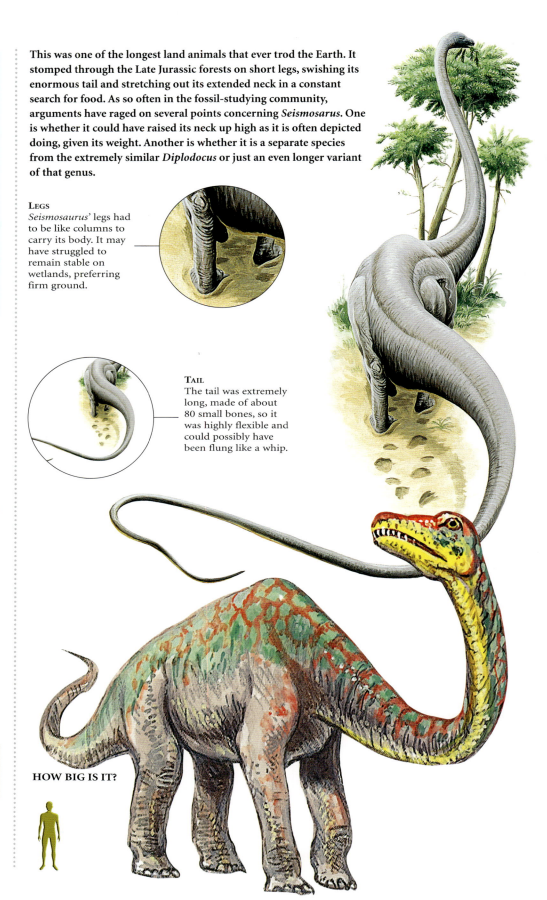

**HOW BIG IS IT?**

LATE JURASSIC

**ORDER** • Saurischia • **FAMILY** • Diplodocidae • **GENUS & SPECIES** • *Seismosaurus hallorum*

## WHERE IN THE WORLD?

The only location for remains is New Mexico, USA.

## TEETH

The teeth resemble long pegs with which the animal raked foliage from high branches. Study of wear patterns on the teeth suggests that one row would strip the foliage while the other row guided the direction of the mouth. Food was swallowed without chewing and was digested in the enormous gut probably with the help of gastroliths.

## NECK

The neck was held at a slight upwards angle from the ground. *Seismosaurus* would have found it difficult to force its unwieldy body through dense woodland, so probably pushed its neck in at the edges of forests. It may also have lowered its small head towards soft-leaved plants. Some scientists believe it could not have raised its head up very high because of the difficulty in holding up the weight and in keeping blood flowing.

**FOSSILS**
A fully grown adult collapses and dies. Soon, flying reptiles tear the meat from the body. The remaining tissue rots away, leaving bones that will be preserved for millions of years.

## TIMELINE (millions of years ago)

| 540 | 505 | 438 | 408 | 360 | 280 | 248 | 208 | 146 | 65 | 1.8 to today |

159

LATE JURASSIC

# Stegosaurus

| VITAL STATISTICS | |
|---|---|
| Fossil Location | USA, Europe, Asia, Africa |
| Diet | Herbivorous |
| Pronunciation | STEG-oh-SAWR-us |
| Weight | 3100kg (6800lb) |
| Length | 9m (30ft) |
| Height | 2.75m (9ft) at the hips, total height 4m (14ft) |
| Meaning of name | 'Roofed lizard' because its back plates were originally assumed to lie like roof tiles |

*Stegosaurus* is an iconic dinosaur because of the two walls of bony plates along its back that gave it a distinctive profile. It is a fascinating dinosaur because of its array of defence equipment, from the spiked tail to the armour-plated skin under its throat. This was necessary because *Stegosaurus* was a slow, heavy herbivore that needed to eat for most of the day, so it was likely to encounter one of the sharp-clawed aggressive Late Jurassic predators rather often. The purpose and mechanics of its back plates continue to fascinate palaeontologists.

**HEAD**
The head was tiny in relation to the body, and the sloping shape of the back suggests that the head was kept low most of the time.

**FOSSIL EVIDENCE**
A variety of specimens have been found, mainly in western USA, from the largest adult at 2.3 tonnes (2.6 tons) to juveniles the size of a big dog. Embedded in its back were a double row of 17 bony plates. Sometimes shown in a symmetrical arrangement, it is more likely they were in an alternating pattern: no matching identical plates have been found so far. Opinion is divided over whether it lived alone or in herds.

DINOSAUR

LATE JURASSIC

**STURDY LEGS**
The legs were sturdy and ended in flat feet. *Stegosaurus* may have been able to rise up on its hind legs, balancing on its tail.

**HOW BIG IS IT?**

**OSTEODERMS**
The soft throat was protected by bony plates known as osteoderms, which may also have covered part of the side of the animal.

LATE JURASSIC

• **ORDER** • Ornithischia • **FAMILY** • Stegosauridae • **GENUS & SPECIES** • Several species within the genus *Stegosaurus*

## BACK PLATES

At first, the back plates were first thought to overlap like roof tiles. Now they are known to have stood vertically. They probably would not have offered much protection. The plates contain tube-like tunnels fed by blood vessels. This raises the possibility that they served to warm or cool the blood (and thus the whole animal), and even that the plates could have changed colour when filled with blood, perhaps for mating purposes.

## WHERE IN THE WORLD?

Fossils are found in Western USA and Portugal.

## TAIL SPIKES

The four tail spikes were a powerful weapon. Once thought to stick up in the air, they are now believed to have sat horizontal. When the tail was swung the spikes would have been driven into an attacker's body. An *Allosaurus* fossil with a puncture wound in its back that perfectly matches the size of Stegosaurus' tail spike supports this idea.

**TIMELINE (millions of years ago)**

| 540 | 505 | 438 | 408 | 360 | 280 | 248 | 208 | 146 | 65 | 1.8 to today |

161

LATE JURASSIC
# Stegosaurus

LATE JURASSIC

• **ORDER** • Ornithischia • **FAMILY** • Stegosauridae • **GENUS & SPECIES** • Several species within the genus *Stegosaurus*

**A GREAT NEW FIND**

In January 2007, when fossils of the 150-million-year-old plated dinosaur *Stegosaurus* were discovered at Casal Novo, north of Lisbon, Portugal, they were hailed as a great new find. The fossils comprised a tooth and parts of the spinal column of *Stegosaurus ungulatus*, together with some leg bones. *Stegosaurus* had already been excavated in the United States in 1877, and was considered 'native' to the New World. The remains in Portugal were the first to be unearthed in Europe. This was not entirely surprising. During *Stegosaurus*' time, the Late Jurassic Period, North America and Europe were component areas of the supercontinent Pangaea, which formed around 250 million years ago and began to break apart around 70 million years later. Geophysicists have confirmed 'a very high probability' of a land corridor that joined Newfoundland, now in eastern Canada, with the Iberian landmass that now comprises Portugal and Spain. This corridor was likely to have been temporary, slipping below the water when sea levels rose and emerging again when they receded. In those circumstances, it would have been easy for *Stegosaurus* to cross from one future continent to the other.

LATE JURASSIC

# Ophthalmosaurus

• ORDER • Ichthyosauria • FAMILY • Ophthalmosauridae • GENUS & SPECIES • *Ophthalmosaurus discus*

*Ophthalmosaurus* was not a dinosaur, but an ichthyosaur, a prehistoric swimming reptile. Its huge, dolphin-shaped body cut through the warm Jurassic seas probably in search of fish and molluscs.

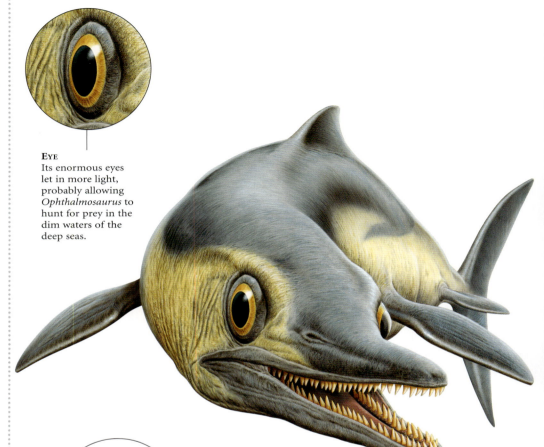

| VITAL STATISTICS | |
|---|---|
| Fossil Location | Europe, North America, Argentina |
| Diet | Fish, molluscs, squid |
| Pronunciation | Off-THAL-moh-SAWR-us |
| Weight | 3 tonnes (3 tons) |
| Length | Up to 6m (20ft) |
| Height | Unknown |
| Meaning of name | 'Eye lizard' after its huge eyes |

**EYE**
Its enormous eyes let in more light, probably allowing *Ophthalmosaurus* to hunt for prey in the dim waters of the deep seas.

## FOSSIL EVIDENCE

*Ophthalmosaurus'* most notable attributes were its oversized eyes. The larger the eye, the better the ichthyosaur was probably able to see in the murky depths of the ocean. *Ophthalmosaurus'* eyes contained bony rings. These rings may have prevented the eyes from changing shape, even under the greatest water pressure. Despite having few teeth, *Ophthalmosaurus'* mighty jaws readily snapped shut on smaller ocean creatures.

**FRONT FINS**
*Ophthalmosaurus'* front fins were larger and stronger than its rear ones. The front fins steered while the tail powered the ichthyosaur.

## HOW BIG IS IT?

**WHERE IN THE WORLD?**

Patagonia, in southern Argentina, has become one of the most popular places for fossil-hunting in the world.

PREHISTORIC ANIMAL

LATE JURASSIC

**TIMELINE** (millions of years ago)

| 540 | 505 | 438 | 408 | 360 | 280 | 248 | 208 | 146 | 65 | 1.8 to today |

LATE JURASSIC – EARLY CRETACEOUS

# Dryosaurus

• **ORDER** • Ornithischia • **FAMILY** • Dryosauridae • **GENUS & SPECIES** • *Dryosaurus altus*

## VITAL STATISTICS

| | |
|---|---|
| Fossil Location | Tanzania |
| Diet | Herbivorous |
| Pronunciation | DRY-oh-SAWR-us |
| Weight | Unknown |
| Length | 2.4m (8ft) |
| Height | Unknown |
| Meaning of name | 'Tree lizard' after its forest habitat |

*Dryosaurus* was a small, sleek dinosaur with a light-weight build. This bipedal herbivore utilized its speed and keen eyesight to avoid predators. Living in herds may have also helped keep *Dryosaurus* safe.

## WHERE IN THE WORLD?

Fossils representing *Dryosaurus* at all stages of its life were found in Tanzania's Tendaguru Formation.

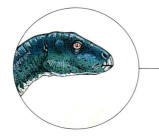

**Cheek**
*Dryosaurus*' cheeks may have contained pouches which probably prevented food from spilling out of its mouth as it ate.

## FOSSIL EVIDENCE

*Dryosaurus* had a horn-covered beak at the end of its snout for cropping plants: possibly conifers, gingkoes and cycads. Its self-sharpening cheek teeth broke down plant material into a size small enough to swallow. *Dryosaurus* had to use its large eyes to watch out for the predator *Allosaurus* to avoid becoming the carnivore's next meal. At the first sign of trouble, *Dryosaurus* would dart away on its long hind legs, using its stiff tail for balance.

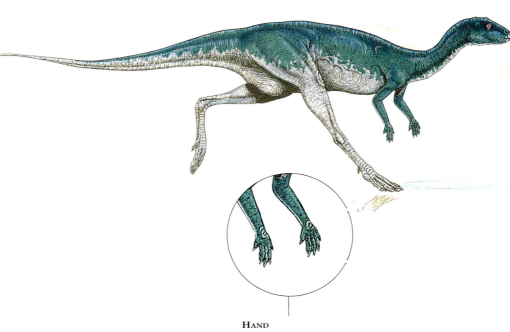

**Hand**
Each hand had five fingers with claws. *Dryosaurus* may have grasped leaves and stems in its claws to hold them steady as it nibbled.

## HOW BIG IS IT?

DINOSAUR
LATE JURASSIC

**TIMELINE (millions of years ago)**

| 540 | 505 | 438 | 408 | 360 | 280 | 248 | 208 | 146 | 65 | 1.8 to today |

LATE JURASSIC – EARLY CRETACEOUS

# Dsungaripterus

• ORDER • Pterosauria • FAMILY • Dsungaripteridae • GENUS & SPECIES • *Dsungaripterus weii*

| VITAL STATISTICS | |
|---|---|
| FOSSIL LOCATION | China |
| DIET | Crabs, fish, molluscs, plankton, insects |
| PRONUNCIATION | ZUNG-ah-RIP-tare-us |
| WEIGHT | 10kg (22lb) |
| LENGTH | 3m (10ft) wingspan |
| HEIGHT | Unknown |
| MEANING OF NAME | 'Junggar wing' in honour of China's Junggar Basin where its fossil was found |

*Dsungaripterus* was an Early Cretaceous pterosaur with a bony crest running along its snout. Its long, narrow jaws curved upwards towards the tip, giving them a tweezer-like appearance.

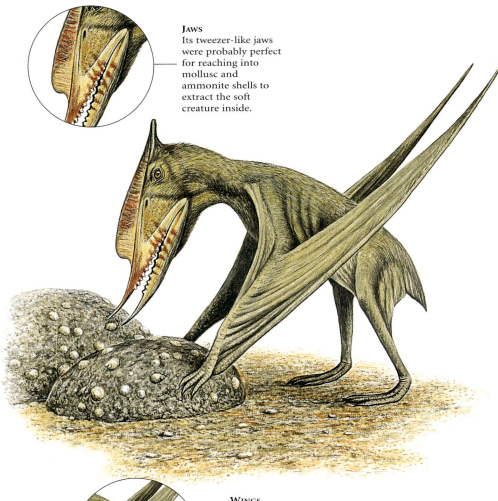

**JAWS**
Its tweezer-like jaws were probably perfect for reaching into mollusc and ammonite shells to extract the soft creature inside.

**FOSSIL EVIDENCE**
*Dsungaripterus'* pointed, narrow jaws were toothless in front. In the back of its jaw were blunt, tooth-like knobs it probably used to crush shellfish. It lived along beaches where it may have picked small marine animals from between rocks with its slender jaws. No one is certain if the strange crest on its snout functioned as a rudder or had some other purpose. Discovery of a large number of *Dsungaripterus* fossils together suggests that these pterosaurs may have lived in colonies where they cared for their young.

**WINGS**
The wings were covered by a thin but durable leathery membrane. *Dsungaripterus* flew long distances on these light-weight wings.

| PREHISTORIC ANIMAL |
|---|
| EARLY CRETACEOUS |

**HOW BIG IS IT?**

**WHERE IN THE WORLD?**

Northwestern China's Junggar Basin, where *Dsungaripterus* was discovered, is almost entirely surrounded by mountains.

**TIMELINE (millions of years ago)**

| 540 | 505 | 438 | 408 | 360 | 280 | 248 | 208 | 146 | 65 | 1.8 to today |

EARLY CRETACEOUS

# Gastonia

• **ORDER** • Ornithischia • **FAMILY** • Ankylosauridae • **GENUS & SPECIES** • *Gastonia burgei*

## VITAL STATISTICS

| | |
|---|---|
| FOSSIL LOCATION | Utah, USA |
| DIET | Herbivorous |
| PRONUNCIATION | gas-TOE-nee-ah |
| WEIGHT | Unknown |
| LENGTH | 6m (20ft) |
| HEIGHT | Unknown |
| MEANING OF NAME | 'Gaston's', because Rob Gaston found the first specimens |

### WHERE IN THE WORLD?

So far, *Gastonia* is known only from the Cedar Mountain Formation in Utah, USA. However, it is very similar to related forms from the Lower Cretaceous Period of southern England.

One look at *Gastonia* and the purpose for all those spikes and plates is evident: protection. But species recognition was another function.

**SACRAL SHIELD**
The upper part of Gastonia's pelvic area was protected by thick skin packed with osteoderms. These mostly tiny osteoderms fused together in life making a single bony shield.

**ARMOUR**
There was a huge amount of variation in the osteoderms in *Gastonia*'s skin. A good understanding of their anatomical position can help identify ones found loose on the ground.

### FOSSIL EVIDENCE

*Gastonia*, first named in 1998, is known from a wealth of disarticulated bonebed skeletal material from the Cedar Mountain Formation, which was deposited during the last half of the Early Cretaceous. Despite the difficulty in trying to tell how many osteoderms (bony skin-armour elements) an individual *Gastonia* had (due to disarticulation), its abundance makes it the most completely known polacanthine ankylosaur. It is also found with the large carnivorous dinosaur *Utahraptor* and many other dinosaurs. Even though dinosaurs are common and very diverse in these deposits, the study of this very rich fauna only really began in the early 1990s.

DINOSAUR

EARLY CRETACEOUS

### HOW BIG IS IT?

### AN EFFECTIVE DETERRENT

Polacanthines are a subgroup of armoured dinosaurs whose bodies were covered with spikes and plates, but whose tails lacked clubs. Instead, a dinosaur of this type had triangular, blade-like armour elements running down each side of its tail. This could have been used as an effective deterrent or weapon by swinging it side to side.

**TIMELINE (millions of years ago)**

| 540 | 505 | 438 | 408 | 360 | 280 | 248 | 208 | 146 | 65 | 1.8 to today |

EARLY CRETACEOUS

# Afrovenator

• **ORDER** • Saurischia • **FAMILY** • Megalosauridae • **GENUS & SPECIES** • *Afrovenator abakensis*

| VITAL STATISTICS | |
|---|---|
| Fossil Location | Niger |
| Diet | Carnivorous |
| Pronunciation | AF-roh-vee-NAY-tor |
| Weight | 500kg (1102lb) |
| Length | 8–9m (26–30ft) |
| Height | 2.5m (8ft) |
| Meaning of name | 'African hunter' |

**WHERE IN THE WORLD?**

The Agadez Region of Niger, former home of *Afrovenator*, is now part of the Sahara Desert.

*Afrovenator* was a relative of North America's *Allosaurus*. This link between African and North American dinosaurs indicates there could have been a land bridge between the two continents during the Early Cretaceous.

**Teeth**
A mouthful of sharp teeth, which each measured 5cm (2in) long, made the bite of an *Afrovenator* absolutely lethal.

**FOSSIL EVIDENCE**
An almost whole *Afrovenator* fossil was discovered in Niger, making it the most complete skeleton of a Cretaceous carnivore from Africa excavated so far. *Afrovenator* was a bipedal carnivore with deadly claws. To counterbalance its heavy front end, Afrovenator held its tail out stiffly behind it as it aggressively pursued prey. Its slender, graceful skeleton meant that *Afrovenator* was unusually agile for a theropod. This nimble dinosaur was also a vicious killer. Its remains were found with the fossil of a large sauropod that it may have attacked.

**Claws**
*Afrovenator*'s claws were curved like sickles. The two largest claws on each hand were probably used to gouge and disembowel prey.

**HOW BIG IS IT?**

| DINOSAUR |
| EARLY CRETACEOUS |

**TIMELINE** (millions of years ago)

540 | 505 | 438 | 408 | 360 | 280 | 248 | 208 | 146 | 65 | 1.8 to today

EARLY CRETACEOUS

# Atlascopcosaurus

• ORDER • Ornithischia • FAMILY • Euornithopoda • GENUS & SPECIES • *Atlascopcosaurus loadsi*

## VITAL STATISTICS

| | |
|---|---|
| FOSSIL LOCATION | Australia |
| DIET | Herbivorous |
| PRONUNCIATION | AT-las-KOP-ko-SAWR-us |
| WEIGHT | 125kg (275lb) |
| LENGTH | 2–3m (7–10ft) |
| HEIGHT | 1m (40in) |
| MEANING OF NAME | 'Atlas Copco lizard' in honour of the Atlas Copco Corporation that manufactured the equipment used to excavate the fossil |

*Atlascopcosaurus* is known only from fragments of its teeth and jaw, so its description is based on more complete remains of close relatives. What is known for certain is that its teeth displayed two different types of ridges.

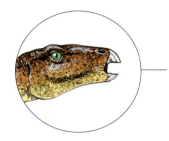

**TEETH**
Having two different sizes of ridges on its teeth may have allowed *Atlascopcosaurus* to feed on a wide variety of plants.

**EYES**
*Atlascopcosaurus* kept constant watch with its keen eyes to avoid attack. It scanned the horizon as it nibbled plants, always ready to run.

### WHERE IN THE WORLD?

Victoria is Australia's most densely populated state. When *Atlascopcosaurus* roamed, it was a lush floodplain.

## FOSSIL EVIDENCE

*Atlascopcosaurus* was a small plant-eating dinosaur that walked upright. Its build was slight, giving it the speed necessary to dodge the large predators that shared its territory. Tendons attached to its vertebrae held its tail out stiffly as it ran. The advantage of having a stiff tail was that it acted as a balance, allowing *Atlascopcosaurus* to execute fast turns when fleeing danger.

### HOW BIG IS IT?

DINOSAUR

EARLY CRETACEOUS

**TIMELINE (millions of years ago)**

| 540 | 505 | 438 | 408 | 360 | 280 | 248 | 208 | 146 | 65 | 1.8 to today |

EARLY CRETACEOUS

# Becklespinax

• ORDER • Saurischia • FAMILY • Tetanurae • GENUS & SPECIES • *Becklespinax altispinax*

*Becklespinax* was a large theropod dinosaur that hunted the Cretaceous landscapes in search of prey. It is known from three vertebrae bearing tall spines that were discovered in England in 1884.

| VITAL STATISTICS | |
|---|---|
| FOSSIL LOCATION | England |
| DIET | Carnivorous |
| PRONUNCIATION | BECK-el-SPY-nax |
| WEIGHT | 900kg (1984lb) |
| LENGTH | 5–8m (16–26ft) |
| HEIGHT | 3m (10ft) |
| MEANING OF NAME | 'Beckles' spine' in honour of Samuel Beckles who discovered the fossil |

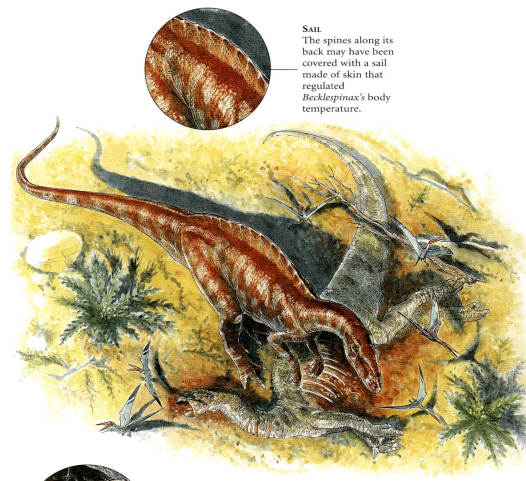

**SAIL**
The spines along its back may have been covered with a sail made of skin that regulated *Becklespinax's* body temperature.

## FOSSIL EVIDENCE

It proved so difficult to determine exactly what type of dinosaur *Becklespinax* was that it was not until 1991 that the theropod was assigned its current name. During the Cretaceous, *Becklespinax* probably stalked sauropods. It sank its knife-like claws into its victims, slicing open their hide so its sharp teeth could easily reach their muscles and internal organs. *Becklespinax's* strong jaws were able to tear away big hunks of meat that the dinosaur swallowed whole. When no live prey was available, *Becklespinax* may have scavenged for corpses.

**CLAWS**
*Becklespinax* may have dug its hand claws into the neck of its victim to steady itself while slashing away with its foot claws.

DINOSAUR

EARLY CRETACEOUS

**HOW BIG IS IT?**

**WHERE IN THE WORLD?**

The fossil of *Becklespinax* was found in the sandstone of the Hastings Beds of East Sussex.

**TIMELINE** (millions of years ago)

| 540 | 505 | 438 | 408 | 360 | 280 | 248 | 208 | 146 | 65 | 1.8 to today |

EARLY CRETACEOUS

# Chilantaisaurus

• **ORDER** • Saurischia • **FAMILY** • Spinosauridae • **GENUS & SPECIES** • *Chilantaisaurus tashuikouensis*

*Chilantaisaurus* was a large theropod dinosaur that roamed China during the Early Cretaceous. Although it was not the swiftest of runners, it was among the deadliest of predators when it ambushed other dinosaurs.

### VITAL STATISTICS

| | |
|---|---|
| Fossil Location | China |
| Diet | Carnivorous |
| Pronunciation | Chee-LAWN-tie-SAWR-us |
| Weight | 3630kg (8000lb) |
| Length | 6.1m (20ft) |
| Height | 2.7m (9ft) |
| Meaning of name | 'Chilantai lizard' after the name of the fossil's location in Inner Mongolia, China |

### FOSSIL EVIDENCE

When *Chilantaisaurus* was first described in 1964, it was assumed to be a carnosaur closely related to *Allosaurus*. Subsequent research suggested that it may be a primitive member of the spinosaur family or part of an offshoot group with affinities to both allosaurs and spinosaurs. More complete material will need to be found in order to get closer to an answer.

DINOSAUR

EARLY CRETACEOUS

**EYES**
Big, sharp-sighted eyes watched for the slightest hint of movement in the foliage, a signal to *Chilantaisaurus* that prey was nearby.

**ARMS**
Its long arms allowed *Chilantaisaurus* to hook its claws into the flesh of its victim while disembowelling the prey with its fierce teeth.

### WHERE IN THE WORLD?

Inner Mongolia in China.

**HOW BIG IS IT?**

**TIMELINE (millions of years ago)**

| 540 | 505 | 438 | 408 | 360 | 280 | 248 | 208 | 146 | 65 | 1.8 to today |

EARLY CRETACEOUS

# Fulgurotherium

• **ORDER** • Ornithischia • **FAMILY** • Hypsilophodontidae • **GENUS & SPECIES** • *Fulgurotherium australe*

## VITAL STATISTICS

| | |
|---|---|
| FOSSIL LOCATION | Australia |
| DIET | Herbivorous |
| PRONUNCIATION | FUL-gur-o-THEER-ee-um |
| WEIGHT | Unknown |
| LENGTH | 1.5m (5ft) |
| HEIGHT | Unknown |
| MEANING OF NAME | 'Lightning beast' because it was first found at Lightning Ridge |

*Fulgurotherium* was a small plant-eating dinosaur. It was a hypsilophodont, a type of primitive ornithopod. This group later contained the giant duck-billed hadrosaurs such as *Anatotitan* and *Corythosaurus*. Hypsilophodonts were diminutive bipedal herbivores with a global distribution.

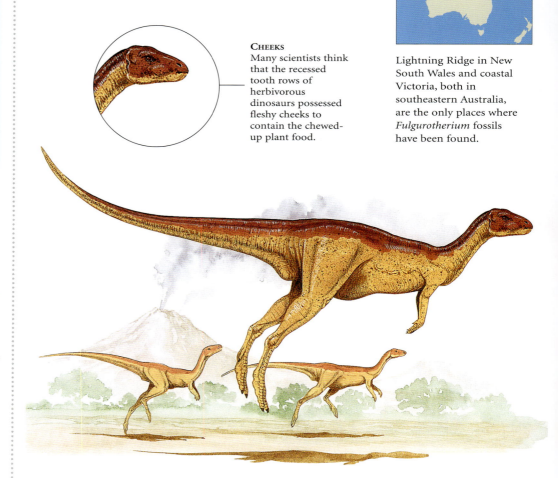

**CHEEKS**
Many scientists think that the recessed tooth rows of herbivorous dinosaurs possessed fleshy cheeks to contain the chewed-up plant food.

## WHERE IN THE WORLD?

Lightning Ridge in New South Wales and coastal Victoria, both in southeastern Australia, are the only places where *Fulgurotherium* fossils have been found.

## FOSSIL EVIDENCE

*Fulgurotherium* was an Early Cretaceous ornithopod dinosaur. The species *Fulgurotherium australe* was formalized by Friedrich von Huene in 1932, but more complete specimens need to be unearthed to determine whether the bones were from one or many kinds of small ornithopod. The first specimens came from Lightning Ridge in northern New South Wales, which is famous for its opal fields and yields some of the rarest and most beautiful fossils on Earth. The cavities left by dissolved bones buried in the ground at Lightning Ridge sometimes fill up with precious opal. Many of Australia's few Cretaceous mammals have also come from the opal fields.

**HOW BIG IS IT?**

**LEGS**
Like many early plant-eating dinosaurs, *Fulgurotherium* was an agile biped whose long hind limbs helped in quick escapes from predators.

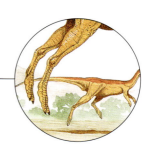

| DINOSAUR |
|---|
| EARLY CRETACEOUS |

**TIMELINE** (millions of years ago)

| 540 | 505 | 438 | 408 | 360 | 280 | 248 | 208 | 146 | 65 | 1.8 to today |

EARLY CRETACEOUS

# Gilmoreosaurus

• **ORDER** • Ornithischia • **FAMILY** • Hadrosauridae • **GENUS & SPECIES** • *Gilmoreosaurus mongoliensis*

## VITAL STATISTICS

| | |
|---|---|
| FOSSIL LOCATION | Asia |
| DIET | Herbivorous |
| PRONUNCIATION | GIL-mohr-o-SAWR-us |
| WEIGHT | 1400kg (3068lb) |
| LENGTH | 8m (26ft) |
| HEIGHT | Unknown |
| MEANING OF NAME | 'Gilmore's lizard' after Charles Gilmore, the American palaeontologist who originally misidentified it |

*Gilmoreosaurus* was a powerful hadrosaur, or duck-billed dinosaur. It walked on all fours but possibly rose onto its hind legs to run. It fed on plants, grinding them up with thousands of teeth.

**SKIN**
Like all hadrosaurs, *Gilmoreosaurus* probably had a thick leathery skin that was covered by large, oval tubercles.

### WHERE IN THE WORLD?

Only incomplete skeletons of *Gilmoreosaurus* have been found, all in Asia.

## FOSSIL EVIDENCE

The first fossilized remains of this duck-billed dinosaur were found in 1923. After 10 years it was named *Mandschurosaurus*, and then largely forgotten. Only in 1979 were the fossils studied closely, and reidentified. In 2003, analysis of fossilized vertebra showed the presence of several types of tumour. The cause is unknown, and it is striking that analysis of more than 10,000 fossil specimens from other dinosaurs showed that only *Gilmoreosaurus* and other hadrosaurs were affected.

DINOSAUR
EARLY CRETACEOUS

### HOW BIG IS IT?

**CLAWS**
The claw-like toes are similar to those of *Iguanodon*, suggesting that it may be an ancestor.

**TIMELINE (millions of years ago)**

| 540 | 505 | 438 | 408 | 360 | 280 | 248 | 208 | 146 | 65 | 1.8 to today |

173

EARLY CRETACEOUS

# Harpymimus

• **ORDER** • Saurischia • **FAMILY** • Harpymimidae • **GENUS & SPECIES** • *Harpymimus okladnikovi*

| VITAL STATISTICS | |
|---|---|
| Fossil Location | Mongolia |
| Diet | Omnivorous |
| Pronunciation | HAR-pi-MIME-us |
| Weight | 125kg (276lb) |
| Length | 2–3.5m (7–11ft) |
| Height | Unknown |
| Meaning of name | 'Harpy mimic' after the mystical bird-like Harpy, a monster with the head of a woman and the body of a bird |

**WHERE IN THE WORLD?**

*Harpymimus* was confined to Central Asia, though later ornithomimosaurs emigrated to Northern America.

With long-shinned legs, *Harpymimus* was a swift runner. It had to be, because probably its only defence from predators was to flee. A long tail, stiffening towards the end, provided balance.

**EYES**
With eyes set on the sides of its head, *Harpymimus* had panoramic vision – a good defence against predators.

**HANDS**
Its sharp, hook-like claws allowed *Harpymimus* to snatch small prey or to grasp branches.

**FOSSIL EVIDENCE**

*Harpymimus* is known from an incomplete specimen found in Mongolia, in part of the Shinkehudug Formation at Dundgovi Aimag. The skull is badly crushed, but it seems to have had a beak and small, blunt teeth in its lower jaw, about 10–11 on each side. These were likely used not for cutting but simply for catching and holding prey. Its small head (26cm/10in) sat on a neck about 60cm (24in) long.

**HOW BIG IS IT?**

DINOSAUR

EARLY CRETACEOUS

**TIMELINE (millions of years ago)**

| 540 | 505 | 438 | 408 | 360 | 280 | 248 | 208 | 146 | 65 | 1.8 to today |

EARLY CRETACEOUS

# Hylaeosaurus

• **ORDER** • Ornithischia • **FAMILY** • Anklyosauridae • **GENUS & SPECIES** • *Hylaeosaurus armatus*

*Hylaeosaurus* was seemingly not well equipped for attack, but spines and armour meant that it was vulnerable mainly if a predator flipped it over, exposing its underbelly.

| VITAL STATISTICS | |
|---|---|
| FOSSIL LOCATION | England and possibly France |
| DIET | Herbivorous |
| PRONUNCIATION | Hi-LEE-oh-SAWR-us |
| WEIGHT | 1 ton (1.1 tonnes) |
| LENGTH | 3–6m (10–20ft) |
| HEIGHT | Unknown |
| MEANING OF NAME | 'Forest lizard' named for the lower Cretaceous Wealden deposit at Tilgate Forest, Sussex, England |

**WHERE IN THE WORLD?**

*Hylaeosaurus* was certainly found in England, but a fossil from France may have been misidentified.

**MOUTH**
*Hylaeosaurus* used its bony beak to crop low-lying plants, and leaf-shaped cheek teeth for chopping.

**HEAD**
Its head was longer than it was wide, but since the best fossil is incomplete, little more is known.

## FOSSIL EVIDENCE

*Hylaeosaurus* was only the third dinosaur to be named (before the term 'dinosaur' had even been coined). The first fossil was discovered in 1832, and includes the front end of a skeleton without most of its head. Other fossils have since allowed us to conclude that it had three long spines on its shoulder, two on its hip and three rows of armour running down its back. The original specimen is displayed in the Natural History Museum in London, still embedded in the limestone in which it was discovered.

**HOW BIG IS IT?**

DINOSAUR

EARLY CRETACEOUS

**TIMELINE** (millions of years ago)

| 540 | 505 | 438 | 408 | 360 | 280 | 248 | 208 | 146 | 65 | 1.8 to today |

EARLY CRETACEOUS

# Leaellynasaura

• **ORDER** • Ornithischia • **FAMILY** • Hypsilophodontidae • **GENUS & SPECIES** • *Leaellynasaura amicagraphica*

| VITAL STATISTICS | |
|---|---|
| Fossil Location | Australia |
| Diet | Herbivorous |
| Pronunciation | Lee-EL-in-a-SAWR-a |
| Weight | Unknown |
| Length | 2–3m (7–10ft) |
| Height | Unknown |
| Meaning of name | 'Leaellyn lizard', after Leaellyn Rich, daughter of Thomas A. Rich and Patricia Vickers, the palaeontologists who discovered it |

**WHERE IN THE WORLD?**

*Leaellynasaura* lived in Australia, at a time when that landmass lay within the Antarctic Circle.

*Leaellynasaura* lived in the polar forests of the Antarctic, possibly feeding on ferns and horsetails. It may have laid its eggs in nests on the ground, and stood guard against predators until they hatched.

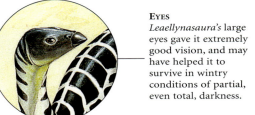

**Eyes**
*Leaellynasaura*'s large eyes gave it extremely good vision, and may have helped it to survive in wintry conditions of partial, even total, darkness.

**FOSSIL EVIDENCE**
The well-preserved head of *Leaellynasaura* was discovered in Dinosaur Cove, southeast Australia. Its skull had large eye sockets and two bumps further back, for its optic lobes. An endocast of its brain (a natural cast formed after the brain rotted away, to be replaced by sediment) shows an enlarged optic lobe area, suggesting that it had evolved to cope with the dark winters in the Antarctic Circle. No complete skeleton has yet been found, so assumptions about its body are based on its relatives, the hypsilophodonts.

**Body**
Unlike most other non-avian dinosaurs, *Leaellynasaura* may have been warm-blooded, allowing it to survive winter within the Antarctic Circle.

DINOSAUR

EARLY CRETACEOUS

**HOW BIG IS IT?**

**TIMELINE (millions of years ago)**

| 540 | 505 | 438 | 408 | 360 | 280 | 248 | 208 | 146 | 65 | 1.8 to today |

176

EARLY CRETACEOUS

# Muttaburrasaurus

• ORDER • Ornithischia • FAMILY • unknown • GENUS & SPECIES • *Muttaburrasaurus langdoni*

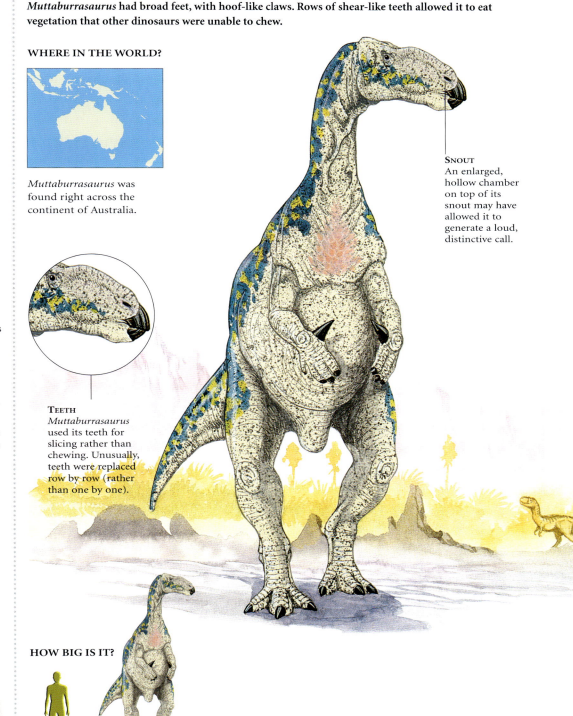

*Muttaburrasaurus* had broad feet, with hoof-like claws. Rows of shear-like teeth allowed it to eat vegetation that other dinosaurs were unable to chew.

### VITAL STATISTICS

| | |
|---|---|
| FOSSIL LOCATION | Australia |
| DIET | Herbivorous |
| PRONUNCIATION | Mutt-ah-BUHR-ah-SAWR-us |
| WEIGHT | 1–4 tons (1.1–4.4 tonnes) |
| LENGTH | 7m (23ft) |
| HEIGHT | Unknown |
| MEANING OF NAME | 'Muttaburra lizard' after the township of Muttaburra in Queensland, Australia, close to the fossil site |

### WHERE IN THE WORLD?

*Muttaburrasaurus* was found right across the continent of Australia.

### FOSSIL EVIDENCE

The fossil remains of this dinosaur had been trampled over by cattle for years, while local people had been picking up the fragments and taking them home. Only in 1963, when farmer Doug Langdon found an incomplete skeleton, was the importance of these fossils realized, and the locals were asked to return their souvenirs. Further remains have been found in Queensland and New South Wales, including the 'Dunluce Skull' found on Dunluce Station in 1987.

**SNOUT**
An enlarged, hollow chamber on top of its snout may have allowed it to generate a loud, distinctive call.

**TEETH**
*Muttaburrasaurus* used its teeth for slicing rather than chewing. Unusually, teeth were replaced row by row (rather than one by one).

### HOW BIG IS IT?

DINOSAUR
EARLY CRETACEOUS

### TIMELINE (millions of years ago)

| 540 | 505 | 438 | 408 | 360 | 280 | 248 | 208 | 146 | 65 | 1.8 to today |

EARLY CRETACEOUS

# Pelicanimimus

• ORDER • Saurischia • FAMILY • Ornithomimosauridae • GENUS & SPECIES • *Pelicanimimus polyodon*

| VITAL STATISTICS | |
|---|---|
| FOSSIL LOCATION | Spain |
| DIET | Carnivorous |
| PRONUNCIATION | PEL-uh-kan-uh-MEEM-us |
| WEIGHT | Up to 25kg (55lb) |
| LENGTH | 2m (7ft) |
| HEIGHT | Unknown |
| MEANING OF NAME | 'Pelican mimic' in reference to its long face and the pouch beneath its jaw |

### WHERE IN THE WORLD?

To date, *Pelicanimimus* has been found only in Spain.

Walking upright on long hind legs, *Pelicanimimus* may have waded into water in search of prey. These it caught in the three long fingers of its hook-like hands.

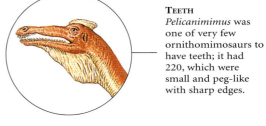

**TEETH**
*Pelicanimimus* was one of very few ornithomimosaurs to have teeth; it had 220, which were small and peg-like with sharp edges.

**SKIN**
Fossil evidence suggests that *Pelicanimimus* may have had smooth, naked skin, without scales, feathers or even hair.

### FOSSIL EVIDENCE

The Calizas de La Huérguina Formation in La Hoyas, Spain, has yielded many well-preserved fossils, and *Pelicanimimus* is no exception. Impressions of its soft tissue in the limestone reveal a crest on the back of its head, as well as a featherless pouch beneath its jaw, like the modern pelican. It may have used this pouch for storing fish. Impressions of the muscle tissue were also preserved, and in remarkable detail. Only one other fossil, from Brazil, has been similarly well preserved.

DINOSAUR

EARLY CRETACEOUS

### HOW BIG IS IT?

**TIMELINE** (millions of years ago)

| 540 | 505 | 438 | 408 | 360 | 280 | 248 | 208 | 146 | 65 | 1.8 to today |

EARLY CRETACEOUS

# Pelorosaurus

• ORDER • Saurischia • FAMILY • Brachiosauridae • GENUS & SPECIES • *Pelorosaurus conybeari*

## VITAL STATISTICS

| | |
|---|---|
| FOSSIL LOCATION | England, Portugal |
| DIET | Herbivorous |
| PRONUNCIATION | Pe-LOW-roh-SAWR-us |
| WEIGHT | Unknown |
| LENGTH | 15–24m (49–79ft) |
| HEIGHT | Unknown |
| MEANING OF NAME | 'Monstrous lizard' after its enormous vertebrae and limb bones |

### WHERE IN THE WORLD?

Fossil misidentification means that we can be certain only that *Pelorosaurus* lived in England and Portugal.

Against predators, *Pelorosaurus* needed only one defence: its vast bulk. Its remarkably long neck let it feed high in the foliage, grabbing leaves with its chisel-shaped teeth.

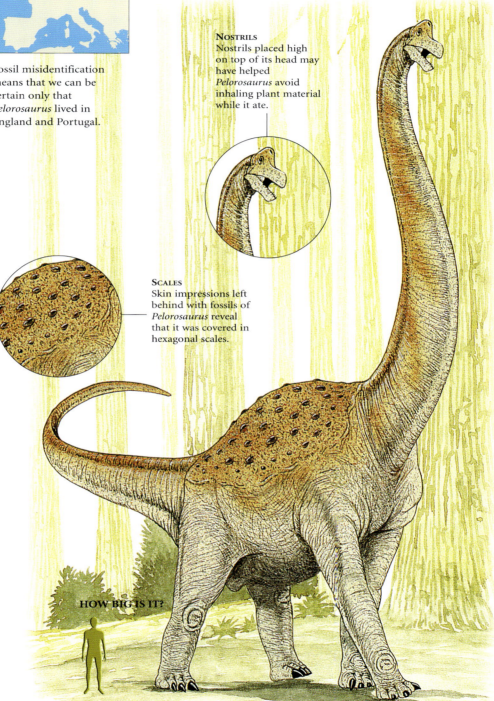

**NOSTRILS**
Nostrils placed high on top of its head may have helped *Pelorosaurus* avoid inhaling plant material while it ate.

**SCALES**
Skin impressions left behind with fossils of *Pelorosaurus* reveal that it was covered in hexagonal scales.

HOW BIG IS IT?

### FOSSIL EVIDENCE

*Pelorosaurus* was identified in the 1840s from several fossil specimens, some of which were later shown to belong to *Iguanodon*. Attempts to reclassify the dinosaur created further confusion, and the identification of later specimens found in Europe is now considered dubious. A specimen found in the Isle of Wight includes vertebrae and limb fragments.

DINOSAUR

EARLY CRETACEOUS

**TIMELINE (millions of years ago)**

| 540 | 505 | 438 | 408 | 360 | 280 | 248 | 208 | 146 | 65 | 1.8 to today |

EARLY CRETACEOUS

# Polacanthus

• ORDER • Ornithischia  FAMILY • Ankylosauridae  GENUS & SPECIES • *Polacanthus foxii*

| VITAL STATISTICS | |
|---|---|
| FOSSIL LOCATION | England |
| DIET | Herbivorous |
| PRONUNCIATION | Pole-ah-CAN-thus |
| WEIGHT | 1.1 tonnes (1 ton) |
| LENGTH | 4m (13ft) |
| HEIGHT | Unknown |
| MEANING OF NAME | 'Many prickles' |

*Polacanthus* was a squat creature advancing slowly on four legs in search of food. It may have lived in herds that travelled alongside *Iguanodon*.

**BODY**
Horn-covered plates along the top of its body, and spikes jutting from its shoulder, spine and tail, offered good defence against predators.

**WHERE IN THE WORLD?**

*Polacanthus* was found across what is now Western Europe

**FOSSIL EVIDENCE**
An incomplete skeleton of *Polacanthus*, missing its head, neck, forelimbs and anterior armour plates, was discovered on the Isle of Wight, England, in 1865. Found near the coast, erosion caused by the winds and waves exposed the fossil. It was initially assumed that *Polacanthus* did not have a club at the end of its tail, but later finds have brought this into question. The small number of samples means that little is known about several important anatomical features, including its head.

**SACRAL SHIELD**
Its sacral shield, a fused sheet of bone across its hips but not attached to them, was covered with tubercles.

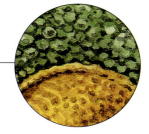

DINOSAUR

EARLY CRETACEOUS

**HOW BIG IS IT?**

**TIMELINE (millions of years ago)**

| 540 | 505 | 438 | 408 | 360 | 280 | 248 | 208 | 146 | 65 | 1.8 to today |

EARLY CRETACEOUS

# Silvisaurus

• **ORDER** • Ornithischia • **FAMILY** • Nodosauridae • **GENUS & SPECIES** • *Silvisaurus condrayi*

Relatively little is known about *Silvisaurus*. A primitive nodosaur, it had a pear-shaped head, with a horny beak to snip vegetation, cheek teeth and small, pointed teeth in its upper jaw.

## VITAL STATISTICS

| | |
|---|---|
| FOSSIL LOCATION | USA |
| DIET | Herbivorous |
| PRONUNCIATION | SILL-vah-SAWR-us |
| WEIGHT | Unknown |
| LENGTH | 2.5–4m (8–13ft) |
| HEIGHT | Unknown |
| MEANING OF NAME | 'Forest lizard' because of its presumed forest habitat |

### WHERE IN THE WORLD?

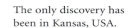

The only discovery has been in Kansas, USA.

**BONY SPINES**
It is possible that *Silvisaurus* was protected by bony spines on its shoulder and tail.

## FOSSIL EVIDENCE

To date, only the skull and sacrum of *Silvisaurus* have been found. This means that deductions about its size and lifestyle are based on comparison with similar dinosaurs. With a skull 33cm (13in) long and 25cm (10in) wide, *Silvisaurus* may have been up to 4m (13ft) long. The fossil was discovered in the Dakota Formation, which contains layers of sedimentary rock from several geological time periods but that has yielded very few dinosaur fossils. The few discoveries tend to have been found in Kansas.

**HEAD**
Balloon-like sinuses may have amplified sound, enabling *Silvisaurus* to generate a distinctive, echoing call.

### HOW BIG IS IT?

**TIMELINE (millions of years ago)**

| 540 | 505 | 438 | 408 | 360 | 280 | 248 | 208 | 146 | 65 | 1.8 to today |

EARLY CRETACEOUS

# Stenopelix

• ORDER • Ornithischia • FAMILY • Unknown • GENUS & SPECIES • *Stenopelix valdensis*

| VITAL STATISTICS | |
|---|---|
| FOSSIL LOCATION | Germany |
| DIET | Herbivorous |
| PRONUNCIATION | Ste-NOP-uh-liks |
| WEIGHT | Unknown |
| LENGTH | 1.5m (5ft) |
| HEIGHT | Unknown |
| MEANING OF NAME | 'Narrow pelvis' |

Little is known about *Stenopelix*. A small herbivore, it browsed on all fours while grazing, and may have relied on gastroliths (small stones) within its gizzard to break down tough plants.

**WHERE IN THE WORLD?**

The only fossil remains discovered to date locate *Stenopelix* in Germany.

**MOUTH**
*Stenopelix*'s skull is not known but probably had a parrot-like beak and blunt teeth that allowed it to snip plant material but not chew it.

**FOSSIL EVIDENCE**
*Stenopelix* is known only from incomplete remains, which means palaeontologists are uncertain about its identification. It was assumed to be a pachycelphalosaur, based on caudal ribs that were later shown to be sacral ribs, and a pubis that seemed to be missing but that was later shown to be part of the acetabulum (a perforation in the pelvis). The shape of part of its hip bone, the ischium, is also unlike any other pachycephalosaurs. Until further remains are discovered, assumptions about *Stenopelix* remain controversial.

**TAIL**
Its tail was likely stiffened by bony tendons, giving it balance and allowing it to rise onto its hind legs to escape predators.

**HOW BIG IS IT?**

DINOSAUR

EARLY CRETACEOUS

**TIMELINE (millions of years ago)**

| 540 | 505 | 438 | 408 | 360 | 280 | 248 | 208 | 146 | 65 | 1.8 to today |

EARLY CRETACEOUS

# Tapejara

• **ORDER** • Pterosauria • **FAMILY** • Tapejaridae • **GENUS & SPECIES** • *Tapejara wellnhoferi*

### VITAL STATISTICS

| | |
|---|---|
| FOSSIL LOCATION | Brazil |
| DIET | Unknown |
| PRONUNCIATION | Tah-pay-ZHAHR-a |
| WEIGHT | Up to 25kg (55lb) |
| LENGTH | 1.5m (5ft) wingspan |
| HEIGHT | Unknown |
| MEANING OF NAME | 'Old being' from the mythology of the Tupi Indians, indigenous to the area where it was discovered |

*Tapejara* was a pterosaur rather than a dinosaur and, as such, was part of the first group of vertebrates capable of powered flight. It was small, with a skull only 20cm (8in) long, and had a short tail.

### WHERE IN THE WORLD?

*Tapejara* was found in Brazil, near coastal areas.

### FOSSIL EVIDENCE

Like all pterosaurs, *Tapejara* had extremely light-weight bones. This means that its bones were likely to be crushed after death, so it was lucky that the first specimen was found preserved well and in three dimensions. *Tapejara's* strange skull was a surprise when first discovered. Without fossilized gut contents, its diet is a mystery. Since its discovery, several close relatives with similar skulls have been found. Its head crest may have functioned like a rudder.

**CREST**
Its head crest was possibly brightly coloured. It may have been used for display purposes.

**BEAK**
Although *Tapejara* probably used its beak to eat fruit and berries, some palaeontologists argue that it may have been used to catch fish.

**HOW BIG IS IT?**

PREHISTORIC ANIMAL

EARLY CRETACEOUS

**TIMELINE (millions of years ago)**

| 540 | 505 | 438 | 408 | 360 | 280 | 248 | 208 | 146 | 65 | 1.8 to today |

EARLY CRETACEOUS

# Tenontosaurus

• **ORDER** • Ornithischia • **FAMILY** • Tenontosauridae • **GENUS & SPECIES** • *Tenontosaurus tilletti*

| VITAL STATISTICS | |
|---|---|
| Fossil Location | Western USA |
| Diet | Herbivorous |
| Pronunciation | Ten-ONT-oh-SAWR-us |
| Weight | 1814kg (4000lb) |
| Length | 6.5m (21ft) |
| Height | Unknown |
| Meaning of name | 'Sinew lizard' after the strengthening tendons of its backbone |

*Tenontosaurus* walked (and ran) mainly on its back legs, but its front limbs were also very strong, with broad five-fingered hands. It probably used them when grazing among low-lying plant life.

**Teeth**
Strong, flat-topped cheek teeth let it chew tough plant material, as well, possibly, as the new flowering plants that had just evolved.

**FOSSIL EVIDENCE**
The skeletons of *Tenontosaurus* that have been discovered range from partial to complete, and from juveniles to adults. Some of the juveniles have been found in groups, one of which was accompanied by an adult. This suggests that young *Tenontosaurus* either gathered in groups, perhaps for protection against predators, or remained in family groups for some time after hatching. It seems that this herbivore likely fell prey to packs of *Deinonychus*, a small carnivore; partial fossils of *Deinonychus*, including broken teeth, have been found alongside it.

DINOSAUR

EARLY CRETACEOUS

**Tail**
*Tenontosaurus* had a tail that was two times longer than its body, kept rigid by a system of bony tendons.

**WHERE IN THE WORLD?**

*Tenontosaurus* was found in the USA, in Wyoming, Texas, Montana and Oklahoma.

**HOW BIG IS IT?**

**TIMELINE (millions of years ago)**

| 540 | 505 | 438 | 408 | 360 | 280 | 248 | 208 | 146 | 65 | 1.8 to today |

EARLY CRETACEOUS

# Tropeognathus

• ORDER • Pterosauria • FAMILY • Criorhynchidae • GENUS & SPECIES • *Tropeognathus mesembrinus*

## VITAL STATISTICS

| | |
|---|---|
| FOSSIL LOCATION | Brazil |
| DIET | Probably fish and cephalopods |
| PRONUNCIATION | TROP-ee-og-NATH-us |
| WEIGHT | 14kg (30lb) |
| LENGTH | 6.2m (20ft) wingspan |
| HEIGHT | Unknown |
| MEANING OF NAME | 'Keel jaw' because its jaw crest resembled a ship's keel |

## FOSSIL EVIDENCE

The Santana Formation in northeastern Brazil is one of the world's richest fossil-bearing deposits, yielding diverse specimens that are remarkably well preserved. Discovered in 1828, the formation continues to be studied, and still provides new discoveries. *Tropeognathus* was first identified in 1987, from remains that, curiously, are relatively poorly preserved. The result is that its identification remains controversial: should this keel-jawed pterosaur be assigned to its own genus (*Tropeognathus*) or should it be returned to its original identification, *Ornithocheirus*?

PREHISTORIC ANIMAL

EARLY CRETACEOUS

*Tropeognathus* is remarkable for the crests on its upper and lower jaws. Like a ship's keel, they may have helped it maintain a straight course as it skimmed through the water, fishing for prey.

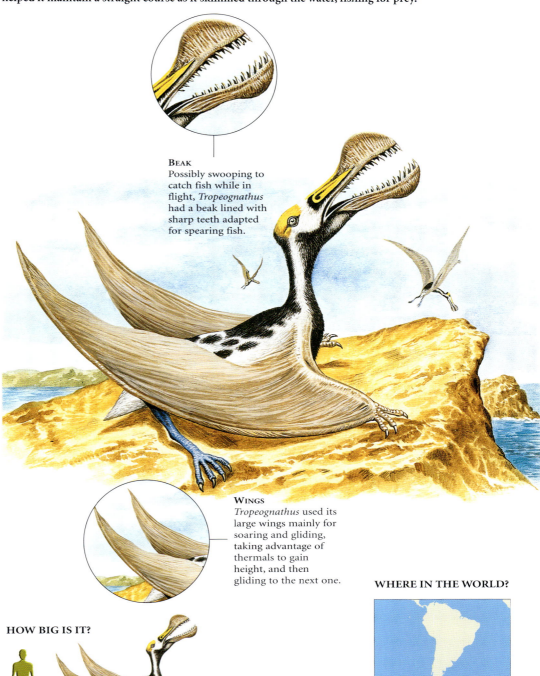

**BEAK**
Possibly swooping to catch fish while in flight, *Tropeognathus* had a beak lined with sharp teeth adapted for spearing fish.

**WINGS**
*Tropeognathus* used its large wings mainly for soaring and gliding, taking advantage of thermals to gain height, and then gliding to the next one.

**HOW BIG IS IT?**

**WHERE IN THE WORLD?**

*Tropeognathus* was found in Brazil.

**TIMELINE (millions of years ago)**

| 540 | 505 | 438 | 408 | 360 | 280 | 248 | 208 | 146 | 65 | 1.8 to today |

EARLY CRETACEOUS

# Wuerhosaurus

• **ORDER** • Ornithischia • **FAMILY** • Stegosauridae • **GENUS & SPECIES** • *Wuerhosaurus homheni*

Shorter than other stegosaurs, *Wuerhosaurus* probably grazed on low-lying plants. With forelimbs that were notably short, and hips over 1.2m (4ft) wide, it may have reared up on its hind legs to reach foliage.

## VITAL STATISTICS

| | |
|---|---|
| Fossil Location | China, Inner Mongolia |
| Diet | Herbivorous |
| Pronunciation | Woo-AYR-hoh-SAWR-us |
| Weight | 4.4 tonne (4 tons) |
| Length | 5–8.1m (16–27ft) |
| Height | Unknown |
| Meaning of name | 'Wuerho lizard' after the town of Wuerho, China, close to the fossil site |

### FOSSIL EVIDENCE

Until the discovery of *Wuerhosaurus*, it had been believed that the stegosaurs became extinct towards the end of the Jurassic period. A complete skeleton has yet to be found, and there is no fossil record of its hind legs. The shortness of the front limbs, however, suggests that *Wuerhosaurus* had an arched back, perhaps even more so than its relatives. Its armour plates are also unique, being small and rectangular rather than tall and triangular. For this reason, their purpose is debated.

**Plates**
*Wuerhosaurus* had bony rectangular plates on its back and tail, but its sides were probably unprotected.

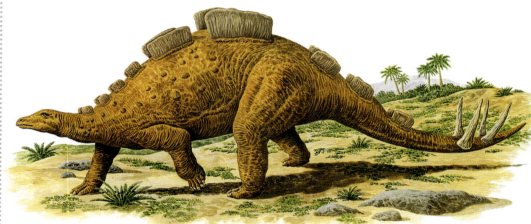

**Spikes**
At the tip of its tail, a thagomizer (an arrangement of four bony spikes) was probably a defence against predators.

**HOW BIG IS IT?**

DINOSAUR
EARLY CRETACEOUS

**WHERE IN THE WORLD?**

Found in central Asia, specifically Inner Mongolia, which is in western China.

**TIMELINE (millions of years ago)**

| 540 | 505 | 438 | 408 | 360 | 280 | 248 | 208 | 146 | 65 | 1.8 to today |

EARLY CRETACEOUS

# Yaverlandia

• ORDER • Saurischia • FAMILY • Unknown • GENUS & SPECIES • *Yaverlandia bitholus*

**The only fossil discovered suggests that *Yaverlandia* was a small creature. It probably walked upright, on its hind legs.**

## VITAL STATISTICS

| | |
|---|---|
| FOSSIL LOCATION | England |
| DIET | Herbivorous |
| PRONUNCIATION | Yah-ver-LAN-dee-ah |
| WEIGHT | Unknown |
| LENGTH | 90cm (35in) |
| HEIGHT | Unknown |
| MEANING OF NAME | 'From Yaverland Battery' after the site on the Isle of Wight where its skull was found |

## FOSSIL EVIDENCE

*Yaverlandia* was discovered in the 1930s but not formally named until 1971. It is known only from a partial skull, which means that its identification remains controversial. It was originally thought to belong to the family pachycephalosauridae, the thick-skulled dinosaurs, but this would make *Yaverlandia* the only pachycephalosaur to have been discovered outside China and North America. Furthermore, the twin domes on its skull are unique. Is it an early pachycephalosaur, from which later bone-headed dinosaurs developed? Or is it in fact a maniraptor, and thus related to birds? Current research identifies *Yaverlandia* as the latter.

DINOSAUR
EARLY CRETACEOUS

**SKULL**
The male *Yaverlandia* may have used its two-domed skull in shoving contests with other males to establish dominance or mating rights.

### WHERE IN THE WORLD?

To date, *Yaverlandia* has been found only on the Isle of Wight in England.

**HOW BIG IS IT?**

**FRONT LIMB**
*Yaverlandia* may have been a maniraptor, with long arms and three-fingered hands.

## TIMELINE (millions of years ago)

| 540 | 505 | 438 | 408 | 360 | 280 | 248 | 208 | 146 | 65 | 1.8 to today |

EARLY CRETACEOUS

# Minmi

## VITAL STATISTICS

| | |
|---|---|
| FOSSIL LOCATION | Australia |
| DIET | Herbivorous |
| PRONUNCIATION | MIN-mee |
| WEIGHT | Unknown |
| LENGTH | 3m (10ft) |
| HEIGHT | 0.9m (3ft) |
| MEANING OF NAME | 'Minmi' because it was discovered near Minmi Crossing, Australia |

This ankylosaur's long legs plus features in its back suggest it was comparatively fast for an armoured dinosaur.

**BODY ARMOUR**
A feature currently thought to be unique to ankylosaurs is *Minmi*'s bony armour, which protected the belly as well as the back, neck and tail.

### WHERE IN THE WORLD?

*Minmi* is found in Australia near *Minmi* Crossing in Queensland. Dinosaur fossils are very rare in Australia, but several specimens of *Minmi* are known.

**FOSSIL EVIDENCE**
*Minmi* is known from two good specimens, one of which is a nearly complete, articulated skeleton. Complete dinosaur fossils are extremely rare in Australia. There are also some fragments that may belong to this genus. The first specimen was unearthed in 1964 in Queensland's Bungil Formation and was described in 1980. *Minmi* was a primitive ankylosaur and does not fit into either of the two main groups in Ankylosauria: nodosaurids and ankylosaurids. It seems to have characteristics from both groups as well as some more akin to more primitive armoured dinosaurs.

**GUT CONTENTS**
One skeleton of *Minmi* is so well preserved it contains the remains of its last meal. These indicate that the plant pieces were probably mashed up without the benefit of gastroliths (stomach stones).

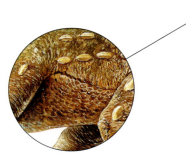

DINOSAUR

EARLY CRETACEOUS

### HOW BIG IS IT?

EARLY CRETACEOUS

• **ORDER** • Ornithischia  **FAMILY** • Unnamed  **GENUS & SPECIES** • *Minmi paravertebra*

## BONY PLATES
This small armoured dinosaur lacked a tail club and had triangular bony armour plates projecting back from the hips. With hind limbs longer than front ones and proportionally long legs, *Minmi* was shaped differently than most ankylosaurs. It also had a short neck, a wide skull and a very small brain.

## DID YOU KNOW?
*Minmi* had unique structures along its backbone called paravertebrae. Currently the function of these features is unknown, but they may have been tendons that became bony in life.

**TIMELINE (millions of years ago)**

| 540 | 505 | 438 | 408 | 360 | 280 | 248 | 208 | 146 | 65 | 1.8 to today |

EARLY CRETACEOUS

# Sauropelta

• ORDER • Ornithischia • FAMILY • Nodosauridae • GENUS & SPECIES • *Sauropelta edwardsorum*

| VITAL STATISTICS | |
|---|---|
| Fossil Location | USA |
| Diet | Herbivorous |
| Pronunciation | SAWR-oh-PEL-ta |
| Weight | 3 tonnes (3.3 tons) |
| Length | 7m (23ft) |
| Height | 1.8m (6ft) |
| Meaning of name | 'Shielded lizard' because of its horn-covered plates |

If an animal has to devote most of its time to eating plants it must have some protection from the carnivorous predators that will lurk near its food sources. Armour can be effective, but it is heavy and slows down movement, so a balance must be struck. *Sauropelta* illustrates the dilemma and the solution. It is a nodosaurid, a family of heavily built herbivorous dinosaurs that walked on four legs and had armoured plates on their skin. Under attack, they would crouch down to protect their soft underbelly. However, *Sauropelta* may have been less passive; it was equipped with large shoulder spikes.

**Hind Legs**
With hind legs considerably longer than the forelimbs, *Sauropelta* may have been able to stand up, but this would expose its unprotected underside.

## FOSSIL EVIDENCE

The top of the body is covered with hard, bony plates called osteoderms. These are similar to crocodile armour Those on the neck are long and pointed. This was common to the nodosaurid family that *Sauropelta* was part of, but it differs in that it had no tail club. It had a very long, thin tail that made up nearly half of its body length. This tail could have comprised as many as 50 vertebrae.

DINOSAUR

EARLY CRETACEOUS

**HOW BIG IS IT?**

**Legs**
The legs were longer than those of other nodosaurids, so *Sauropelta* might have been capable of a steady jog.

EARLY CRETACEOUS

## TANK-LIKE BUILD
*Sauropelta* was built like a tank. The upper body was shielded by rows of bony cones that alternated with small, bony studs. Large pointed spikes projected from the shoulders to protect the vulnerable neck. Behind this, triangular plates lined the tail on both sides. The only probable way a predator could give a decent bite at it would have been by flipping it over onto its armoured back and exposing the underbelly.

### GUT
The gut was huge because tough plant matter was likely digested by fermentation, which requires a big belly and produces a lot of gas.

## WHERE IN THE WORLD?

The only remains are in the American states of Wyoming and Montana, from the middle section of the Cloverly Formation.

## SKULL AND SNOUT
Viewed from above, the skull was triangular, tapering to the snout. The roof of the skull was flat, not domed like some other nodosaurids, whose heads are also bigger. At the snout was a tough, horny beak with which it broke off low-growing plants, perhaps ferns and horsetails.

**TIMELINE (millions of years ago)**

191

EARLY CRETACEOUS

# Zephyrosaurus

• **ORDER** • Ornithischia • **FAMILY** • Hypsilophodontidae • **GENUS & SPECIES** • *Zephyrosaurus schaffi*

| VITAL STATISTICS | |
|---|---|
| FOSSIL LOCATION | Western USA |
| DIET | Herbivorous |
| PRONUNCIATION | ZEF-i-ro-SAWR-us |
| WEIGHT | Unknown |
| LENGTH | 1.8m (6ft) |
| HEIGHT | Unknown |
| MEANING OF NAME | 'West wind lizard' after Zephyr, Greek god of the west wind, because it was found in the Western states of the USA |

Small and agile, *Zephyrosaurus* walked on its hind legs, possibly dropping onto all fours when feeding or resting. It was probably a vital food source for many larger predators.

**WHERE IN THE WORLD?**

Certainly found in the western states of the USA, *Zephyrosaurus* may also have been found in eastern states.

**JAWS**
The wear patterns on fossilized teeth suggest that *Zephyrosaurus* could move its jaws from side to side, not simply up and down.

## FOSSIL EVIDENCE

The small amount of fossil evidence (until recently only a partial skull and post-cranial fragments) means that *Zephyrosaurus* is difficult to classify and has been poorly studied. Distinguishing characteristics include a knob on the upper jaw and another on the cheekbone. New fossils under study include bones from all regions of the body, the remains of seven individuals. Features shared with other dinosaurs suggest *Zephyrosaurus* may have burrowed. Fossilized tracks, discovered in Maryland by an amateur palaeontologist, suggest a wider range for *Zephrosaurus*' family than previously thought.

**HIND LEGS**
With long hind legs, *Zephyrosaurus* was a swift runner, possibly moving in a zigzag direction to escape larger, less agile predators.

**HOW BIG IS IT?**

DINOSAUR

EARLY CRETACEOUS

**TIMELINE (millions of years ago)**

| 540 | 505 | 438 | 408 | 360 | 280 | 248 | 208 | 146 | 65 | 1.8 to today |

EARLY CRETACEOUS

# Giganotosaurus

• ORDER • Saurischia • FAMILY • Carcharodontosauridae • GENUS & SPECIES • Giganotosarus carolinii

## VITAL STATISTICS

| | |
|---|---|
| FOSSIL LOCATION | Argentina |
| DIET | Carnivorous |
| PRONUNCIATION | GI-gan-ot-oh-SAWR-us |
| WEIGHT | Up to 9.7 tonnes (8.8 tons) |
| LENGTH | 14m (46ft) |
| HEIGHT | Unknown |
| MEANING OF NAME | 'Giant lizard of the south' |

*Giganotosaurus* is one of the largest carnivores ever to have roamed the land, bigger even than *Tyrannosaurus rex*. In a skull 1.8m (6ft) long, its brain was the size and shape of a banana.

**FACE**
Ridges around its eyes obscured its forward vision, but large nostrils and an acute sense of smell enabled *Giganotosaurus* to find prey.

## WHERE IN THE WORLD?

Though fossils have been found only in Argentina, *Giganotosaurus* possibly lived all across South America.

**FORELIMBS**
Tiny arms seem to have been little used, though they may have helped *Giganotosaurus* to raise itself off the ground.

## FOSSIL EVIDENCE

The first specimen of *Giganotosaurus* was discovered only in 1993 by an amateur fossil hunter searching in the badlands near El Chocon, Argentina. About 70 per cent of the skeleton has been preserved, including the skull, pelvis, leg bones and most vertebrae. A second, larger specimen has since been found. It may have preyed on titanosaur, a giant herbivore, as suggested by remains found close to it. And it may have hunted in packs – as suggested by the discovery at one site of multiple fossils.

DINOSAUR

EARLY CRETACEOUS

**HOW BIG IS IT?**

**TIMELINE (millions of years ago)**

| 540 | 505 | 438 | 408 | 360 | 280 | 248 | 208 | 143 | 65 | 1.8 to today |

EARLY CRETACEOUS

# Hypsilophodon

• **ORDER** • Ornithischia • **FAMILY** • Hypsilophodontidae • **GENUS & SPECIES** • *Hypsilophodon foxii*

| VITAL STATISTICS | |
|---|---|
| Fossil Location | England, Portugal |
| Diet | Herbivorous |
| Pronunciation | Hip-sih-LO-fuh-don |
| Weight | 50kg (110lb) |
| Length | 2m (6ft 6in) |
| Height | Unknown |
| Meaning of name | 'Hypsilophus tooth', because its teeth resemble a type of iguana once called *Hypsilophus* whose name means "High crest" because of a ridge along its back |

*Hypsilophodon* was a relatively primitive dinosaur, barely changing between the Late Jurassic and the Early Cretaceous Periods. It had five digits on each 'hand' and teeth in the front of its jaw, features lost by its more derived contemporaries.

**WHERE IN THE WORLD?**

*Hypsilophodon* has been definitely located in the Isle of Wight in England and in Portugal.

**MOUTH**
Most of its teeth were set far back in its jaw, suggesting that *Hypsilophodon* had cheek pouches; which made chewing food easier.

**FOSSIL EVIDENCE**
Discovered in the early days of palaeontology, *Hypsilophodon* has been consistently misidentified. Found in 1849, it was mistaken for *Iguanodon*, a mistake corrected in 1870. The assumption, made in 1882, that it climbed trees was corrected only in 1974. In 1979, a thigh bone discovered in South Dakota was thought to be the first *Hypsilophodon* found outside Europe, but this has since been discounted. The discovery of thin mineralized plates seemed to point to armour, but is now believed simply to be an impression of cartilage.

DINOSAUR

EARLY CRETACEOUS

**HOW BIG IS IT?**

**HIND LEGS**
Long shins and short, powerful thighs made the light-weight *Hypsilophodon* a swift runner, able to outrun most predators.

**TIMELINE (millions of years ago)**

| 540 | 505 | 438 | 408 | 360 | 280 | 248 | 208 | 146 | 65 | 1.8 to today |

EARLY CRETACEOUS

# Kronosaurus

• **ORDER** • Plesiosauria • **FAMILY** • Pliosauridae • **GENUS & SPECIES** • *Kronosaurus queenslandicus*

*Kronosaurus* was a reptile that lived on the open seas, breathing air. It probably fed on fish, molluscs and other reptiles, including sharks. It is unknown whether it laid its eggs on land or had live birth at sea.

### VITAL STATISTICS

| | |
|---|---|
| FOSSIL LOCATION | Australia, Colombia |
| DIET | Carnivorous |
| PRONUNCIATION | KRON-oh-SAWR-us |
| WEIGHT | Up to 24 tonnes (22 tons) |
| LENGTH | 7–9 m (23–30ft) |
| HEIGHT | Unknown |
| MEANING OF NAME | 'Titan lizard', after the Titan Kronos of Greek mythology, a reference to its large size and appetite |

**FLIPPERS**
Four paddle-like flippers propelled *Kronosaurus* through the water at great speed. It may have enabled it to move on land like modern seals.

### WHERE IN THE WORLD?

*Kronosaurus* swam the shallow inland seas that covered Australia and Colombia.

### FOSSIL EVIDENCE

*Kronosaurus* is known from several specimens, one with a skull 3m (9ft) long (about one-third of its entire body length). Many specimens are poorly preserved, and it is unfortunate that the most famous, an incomplete skeleton displayed at Harvard University, was partially recovered from the surrounding rock by the use of dynamite. When it was reconstructed, large amounts of plaster of Paris were used, giving rise to its nickname Plasterosaurus. Today the reconstruction is believed to be about 4m (13ft) too long.

**MOUTH**
With muscle-packed jaws and front teeth up to 23cm (9in) long, *Kronosaurus* was a fierce predator.

**HOW BIG IS IT?**

PREHISTORIC ANIMAL

EARLY CRETACEOUS

**TIMELINE (millions of years ago)**

| 540 | 505 | 438 | 408 | 360 | 280 | 248 | 208 | 146 | 65 | 1.8 to today |

EARLY CRETACEOUS

# Ouranosaurus

• **ORDER** • Ornithischia • **FAMILY** • Unranked • **GENUS & SPECIES** • *Ouranosaurus nigeriensis*

*Ouranosaurus* was a dinosaur with few defences: its sail or hump added to its size, which may have been enough to deter predators. A plant-eater, it had a sharp beak well adapted for snipping plants.

| VITAL STATISTICS | |
|---|---|
| Fossil Location | Niger |
| Diet | Herbivorous |
| Pronunciation | Ooh-RAN-uh-SAWR-us |
| Weight | Up to 4.4 tonnes (4 tons) |
| Length | Up to 7m (23ft) |
| Height | Unknown |
| Meaning of name | 'Ourane lizard' from the Tuareg tribe of Niger's name for "monitor lizard" |

**Hump/Sail**
Along its back is a feature that may be a sail, for warming and cooling the blood, or a hump for storing food, as on a modern camel.

**WHERE IN THE WORLD?**

*Ouranosaurus* was found in Niger, at a time when the region was almost as hot as it is today.

**FOSSIL EVIDENCE**

*Ouranosaurus* is known from two complete skeletons found in 1966 in the Echkar Formation in the sands of the Sahara Desert. Originally thought to belong to *Iguanodon*, they were correctly identified ten years later. It is possible that both specimens are male. The skull has a pair of bumps on the nose, but their purpose is unclear, and it has been suggested that these are sexual features found only on males. The snout is much longer than *Iguanodon*, and seems to have been covered with a horny sheath.

**Hands**
A thumb claw made a good weapon, while its fifth digit was long and probably used to gather food.

**HOW BIG IS IT?**

DINOSAUR

EARLY CRETACEOUS

**TIMELINE (millions of years ago)**

| 540 | 505 | 438 | 408 | 360 | 280 | 248 | 208 | 146 | 65 | 1.8 to today |

196

# Psittacosaurus

EARLY CRETACEOUS
• **ORDER** • Ornithischia • **FAMILY** • Psittacosauridae
• **GENUS & SPECIES** • Several species within the genus *Psittacosaurus*

| VITAL STATISTICS | |
|---|---|
| Fossil Location | Mongolia, Russia, China, Thailand |
| Diet | Herbivorous |
| Pronunciation | SIT-uh-ko-SAWR-us |
| Weight | 50kg (110lb) |
| Length | 1.5m (5ft) |
| Height | Unknown |
| Meaning of name | 'Parrot lizard', for its parrot-like beak |

A herbivore, *Psittacosaurus* required huge amounts of food to meet its energy needs, so it probably spent most of its day eating. Its self-sharpening teeth were good for slicing tough vegetation.

### WHERE IN THE WORLD?

*Psittacosaurus* was a prolific dinosaur, found in Mongolia, Russia, China and Thailand.

**Body**
The skin of *Psittacosaurus* was covered in scales, some small and knob-like, others large and plate-like. This may have provided camouflage.

### FOSSIL EVIDENCE

There are more than 400 specimens of *Psittacosaurus*, ranging from hatchling to mature adult. These have enabled detailed study of its growth patterns. A nest containing one adult and 34 hatchlings was found in China in 2003, suggesting that parents protected their young. Discoveries of overlapping skeletons (presumably killed by natural disasters) suggest it gathered in herds, perhaps as a protection against predators. Some specimens contain gastroliths (small pebbles), suggesting that *Psittacosaurus* swallowed them to aid digestion just as modern birds swallow grit.

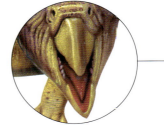

**Bill**
*Psittacosaurus* might have been able to pry open fruit and seeds with its tough, hook-like beak. This had a bone on its tip, a characteristic feature.

### HOW BIG IS IT?

## TIMELINE (millions of years ago)

| 540 | 505 | 438 | 408 | 360 | 280 | 248 | 208 | 146 | 65 | 1.8 to today |

EARLY CRETACEOUS

# Acrocanthosaurus

| VITAL STATISTICS | |
|---|---|
| Fossil Location | USA |
| Diet | Carnivorous |
| Pronunciation | AK-roh-CANthuh-SAWR-us |
| Weight | 2300kg (5000lb) |
| Length | Up to 12m (40ft) |
| Height | 5m (16ft) |
| Meaning of name | 'High spine lizard' due to the raised ridge on its spine |

**This was among the largest of the theropods and is similar to the slightly smaller *Allosaurus* in its refinement as a killing machine. The main difference is that *Acrocanthosaurus* has a low fin of muscled spines that runs down its back.**

**TEETH**
The massive jaw was equipped with 68 sharp, curved serrated teeth perfectly suited to the task of tearing up flesh.

### FOSSIL EVIDENCE

Some very large specimens, one with a skull nearly 1.3m (4.3ft) long, have been found. The long length of the femur bone suggests that *Acrocanthosaurus* probably could not run as quickly as some smaller dinosaurs. A fascinating find of a dinosaur trackway in Texas is thought to be of this animal. The sets of fossilized prints seem to show a pack of *Acrocanthosaurus* stalking a herd of sauropods, but the evidence is unclear. Large olfactory bulbs suggest that it could have hunted with its excellent sense of smell as well as its good eyesight.

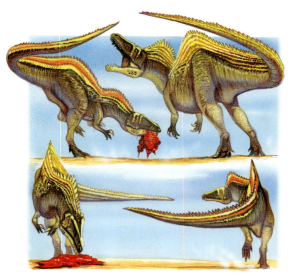

### RESTRICTED MOVEMENT

Close research into an entire fossilized arm reveals that the limb did not move very freely or with a wide range of movement. For example, it could not reach its own neck. This suggests that when this dinosaur hunted, it led with its mouth, using its claws to grip and slash the already damaged victim wriggling in its jaws.

| DINOSAUR |
| EARLY CRETACEOUS |

EARLY CRETACEOUS

• ORDER • Saurischia • FAMILY • Allosauridae or Carcharodontosauridae • GENUS & SPECIES • *Acrocanthosaurus atokensis*

## SPIKES
*Acrocanthosaurus* had a set of elongated vertebral prongs running from the neck to the tail. Some over the back are 43cm (17in) high and their height reduces towards the tail. It seems that these were attached to powerful muscles that formed a thick fleshy ridge along its body rather like a fin. This could well have been brightly coloured and used for signalling, fat storage or temperature control. It is far smaller than the skin sail of *Spinosaurus*, another large theropod.

## WHERE IN THE WORLD?

Remains are mainly in the southern US states of Oklahoma and Texas, with some possibly east, in Maryland.

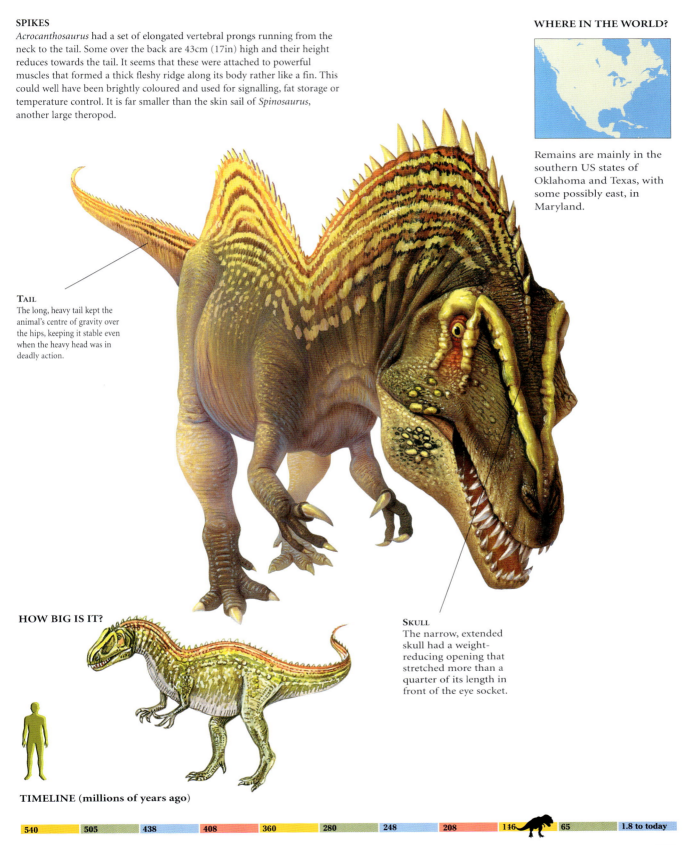

### Tail
The long, heavy tail kept the animal's centre of gravity over the hips, keeping it stable even when the heavy head was in deadly action.

### Skull
The narrow, extended skull had a weight-reducing opening that stretched more than a quarter of its length in front of the eye socket.

## HOW BIG IS IT?

**TIMELINE (millions of years ago)**

| 540 | 505 | 438 | 408 | 360 | 280 | 248 | 208 | 146 | 65 | 1.8 to today |

EARLY CRETACEOUS

# Acrocanthosaurus

200

EARLY CRETACEOUS

• **ORDER** • Saurischia • **FAMILY** • Allosauridae or Carcharodontosauridae • **GENUS & SPECIES** • *Acrocanthosaurus atokensis*

**THE GRAND UNVEILING**

The unveiling of the massive skeleton of *Acrocanthosaurus atokensis* on 8 September 1996 was an important event for the Black Hills Museum of Natural History in Hill City, in the US state of South Dakota. It was the culmination of years of painstaking work to convert a find that had been buried in the bed of an ancient stream for some 120 million years into an exhibit worthy of its extraordinary aura. As Terry Wentz, one of the people who worked on the model, put it: 'Even those of us involved in the preparation of this skeleton are awed by the sense of power we feel in (its) presence.' This was not surprising, for *Acrocanthosaurus*, though slightly smaller, was a no less terrifying relative of the *Tyrannosaurus rex*. It took three years, between 1983 and 1986, for two amateur fossil collectors, Cephis Hall and Sidney Love, to excavate the huge skeleton of *Acrocanthosaurus* from a private site in McCurtain County in southeast Oklahoma. From there, the skeleton was taken to the Black Hills Institute in South Dakota, where the bones were carefully cleaned by experts and the skeleton restored. It was considered too valuable to risk by putting it on display, so a cast replica was made.

EARLY CRETACEOUS

# Amargasaurus

| VITAL STATISTICS | |
|---|---|
| Fossil Location | Argentina |
| Diet | Herbivorous |
| Pronunciation | Uh-MARG-uh-SAWR-us |
| Weight | 5000kg (5.5 tons) |
| Length | 10m (33ft) |
| Height | 4m (13ft) |
| Meaning of name | 'La Amarga lizard', from the canyon where it was found |

**Many mysteries remain to be solved in our quest to understand the world of the dinosaurs, and one is the intriguing question of what *Amargasaurus* actually looked like. It was a medium-sized sauropod with the usual features of a bulky body, a long tail and neck and a small head. The enigma lies in the two rows of long spikes running down its back. Were they connected by a thin membrane of skin to form an elaborate double sail, or did they support some sort of fleshy ridge or a coloured frill? And what were they for?**

**Head**
The head was small, and the nostrils were positioned on top of the skull above the eyes.

**Neck**
The animal would probably not have been too agile because of its bulk, and the movement of its neck would have been restricted by its neural spines.

**FOSSIL EVIDENCE**
An almost complete skeleton from a single individual was reconstructed and named in 1991. The notable feature was the parallel row of tall spines. They are longest at the neck, the tallest reaching 50cm (20in), and decrease in height as they reach the hips. Neural spines are not unusual in Sauropods, but these are the most elaborate by far.

DINOSAUR

EARLY CRETACEOUS

EARLY CRETACEOUS

• **ORDER** • Saurischia • **FAMILY** • Dicraeosauridae • **GENUS & SPECIES** • *Amargasaurus cazaui*

## A BIGGER APPEARANCE

The most obvious explanation for the spines is that they made *Amargasaurus* look bigger. This could have helped to deter predators and to attract mates during courtship rituals. As a defence measure it was of limited value because the spines were quite fragile and would break rather than offer a serious threat to an attacker. As skin-covered sails, they could have been used for thermo-regulation, controlling the temperature of the blood, and for display. As a coloured mane or frill, it may have changed colour to signal to other creatures.

### WHERE IN THE WORLD?

The only remains are from the La Armaga canyon in Patagonia in the west of Argentina.

**FEET**
*Amargasaurus* walked on four broad feet, each with five toes, one of them a sharp claw.

**HOW BIG IS IT?**

**TIMELINE (millions of years ago)**

| 540 | 505 | 438 | 408 | 360 | 280 | 248 | 208 | 146 | 65 | 1.8 to today |

203

EARLY CRETACEOUS
# Amargasaurus

EARLY CRETACEOUS

• **ORDER** • Saurischia • **FAMILY** • Dicraeosauridae • **GENUS & SPECIES** • *Amargasaurus cazaui*

**LA AMARGA CREEK DISCOVERY**

Models of dinosaurs displayed in museums are not purely artistic creations but the result of a great deal of research. A case in point is a model of *Amargasaurus* discovered at La Amarga Creek, Neuquén Province, in Argentina in 1991. The model went on show in the foyer of the Melbourne Museum in Victoria, Australia, which opened in 2000. The display shows the the entire skeleton of *Amargasaurus cazaui*, complete with elongated spines along its neck and back. The *Amargasaurus* skeleton found in Argentina was in fact far from complete: what was found comprised a portion of the skull, some of the vertebrae with spines, part of the pelvis, and bones from the dinosaur's limbs. With so much missing, the model had to be constructed by using casts of what was there, and filling in the missing parts by using more complete skeletons from other dinosaurs closely related to *Amargasaurus*. One of the relatives used was the enormous herbivore *Diplodocus*, which had similar, though smaller, spines on its back. To ensure that the *Amargasaurus* looked as lifelike as possible, some of the bones had to be remade and special research into the dinosaur's posture was undertaken.

EARLY CRETACEOUS

# Baryonyx

| VITAL STATISTICS | |
|---|---|
| Fossil Location | Europe, Africa |
| Diet | Piscivorous |
| Pronunciation | Bare-ee-ON-iks |
| Weight | 1700kg (1750lb) |
| Length | 8.5m (28ft) |
| Height | 3m (10ft) |
| Meaning of name | 'Heavy claw' because of its massive claws |

In 1983 William Walker spotted a huge claw sticking out of a clay pit near Dorking in Surrey, England. As an amateur fossil hunter he knew the 30cm (1ft) specimen was something special. Eventually the site yielded an almost complete skeleton of a new species of dinosaur, which was named after its finder. *Baryonyx walkeri* is unusual because it is one of few suspected non-avian dinosaurs that ate fish, possibly sitting on a riverbank and swiping them out of the water with a massive claw, in a similar manner to the salmon-catching bears of today.

**Arms**
The long, strong arms suggest that *Baryonyx* could walk on all fours, and if so, would make it the only theropod known to do so.

**FOSSIL EVIDENCE**
About 70 per cent of the skeleton was recovered including, crucially, the skull that is so vital in building a detailed picture of the animal's appearance and lifestyle. Once *Baryonyx* was classified, various assorted remains originally attributed to the very similar *Suchosaurus* were assigned to it. The skeleton may not have been fully grown, so its adult measurements could have been larger than those given here, possibly as long as 12m (37ft) and as heavy as 3600kg (7930lb).

DINOSAUR

EARLY CRETACEOUS

**HOW BIG IS IT?**

# EARLY CRETACEOUS

• **ORDER** • Saurischia • **FAMILY** • Spinosauridae • **GENUS & SPECIES** • *Baryonyx walkeri*

## CROCODILE CHARACTERISTICS

The skull and long, flat jaw have crocodilian characteristics. The skull had twice as many teeth as many of its relatives did: 64 in the lower jaw, 32 larger ones in the upper jaw. The notch near the snout is a crocodilian feature probably designed to stop prey escaping. There was a small crest on the top of the head.

## WHERE IN THE WORLD?

Fossils have been discovered in England, Portugal and West Africa.

**NECK**
The neck was long, straight and less flexible than the S-shaped necks of most theropods.

**LAST SUPPER**
A fossilized meal in the stomach contained fish scales and bones, plus remnants of an *Iguanodon*, so it also ate non-piscivorous meat.

## FORELIMB

Each forelimb ended with a curved claw, like a scimitar, 30cm (12in) long. This thumb-claw was possibly used to spear and scoop fish from the water, just like grizzly bears of North America. It may have sat at the edge of the edge of the water ready to leap with its powerful legs. This enormous blade could also have been useful for picking at the remains of dead animals because it appears likely that *Baryonyx* also scavenged for food on the plains and deltas of what is now northern Europe.

**TIMELINE (millions of years ago)**

| 540 | 505 | 438 | 408 | 360 | 280 | 248 | 208 | 146 | 65 | 1.8 to today |

EARLY CRETACEOUS

# Deinonychus

| VITAL STATISTICS | |
|---|---|
| Fossil Location | USA |
| Diet | Carnivorous |
| Pronunciation | Die-NON-ih-kus |
| Weight | 73kg (160lb) |
| Length | 3.4m (11ft) |
| Height | 1.2m (4ft) |
| Meaning of name | 'Terrible claw' after its sickle-like weapons |

*Deinonychus* was a small but deadly theropod whose discovery showed that dinosaurs were not all slow-witted and prompted the debate about whether dinosaurs were warm- or cold-blooded. Palaeontological opinion is inconsistent about whether it hunted in packs, how fast it could run, how it used its enlarged toe claw, and even the question of whether it is a bird-like dinosaur or a dinosaur-like bird. It is agreed that this was an intelligent predator with a large brain for its size and that it was a formidable hunter.

**FOSSIL EVIDENCE**

*Deinonychus* fossils were first uncovered in 1931, but it was not described until 1969 because the site yielded other specimens of more interest originally. In the meantime, more finds came to light. The discovery of several *Deinonychus* near a larger dinosaur suggests it may have hunted in packs, but this is disputed. After all, the herd may have been scavenging a corpse or the skeletons could have washed together in a river. Previously unnoticed fossilized eggshells found very close to a skeleton have led to the suggestion that *Deinonychus* sat over its eggs, incubating them with body heat as modern birds do.

DINOSAUR

EARLY CRETACEOUS

**HOW BIG IS IT?**

# EARLY CRETACEOUS

• **ORDER** • Saurischia • **FAMILY** • Dromaeosauridae • **GENUS & SPECIES** • *Deinonychus antirrhopus*

## SICKLE-SHAPED TOE CLAW

*Deinonychus* is named after its sickle-shaped toe claw, the second of the four digits on its feet. This was like a 13-cm (5-in) blade that was kept upright when the animal was running (to prevent wearing it blunt) and was used to stab or slash its victims. The extent of curvature in the claw varies between specimens. It may have changed according to the gender or age of the animal.

### TEETH
The sharp teeth point backwards, making them suitable for biting into and holding prey.

### WHERE IN THE WORLD?

Nine incomplete skeletons have been found in mid-western USA in the states of Montana, Oklahoma, Wyoming, Utah and possibly Maryland.

### FEATHERS
Although no fossil evidence has been found, some believe it had feather-like coverings on all or part of its body, for insulation and display.

### LEGS
*Deinonychus* is thought to have been able to run quickly on its slender legs, possibly sprinting at 40kmh (25mph).

### BONY TENDONS

Opinion is divided on whether the tail was stiff or flexible. Some argue that bony extensions growing from the tail vertebrae would have gripped the neighbouring vertebrae, making the tail solid. Others claim, based on an articulated tail of a close relative, that the tail was flexible, allowing it to curve in an S-shape and that it would have swished from side to side to balance the animal as it ran.

**TIMELINE (millions of years ago)**

| 540 | 505 | 438 | 408 | 360 | 280 | 248 | 208 | 146 | 65 | 1.8 to today |

209

EARLY CRETACEOUS

# Deinonychus

# EARLY CRETACEOUS

• **ORDER** • Saurischia • **FAMILY** • Dromaeosauridae • **GENUS & SPECIES** • *Deinonychus antirrhopus*

**HUNTING TACTICS**

Some palaeontologists believe that *Deinonychus* not only hunted prey in gangs, but used several deadly techniques when attacking dinosaurs much larger than themselves. Despite their comparatively small size, *Deinonychus* was a terrifyingly efficient killing machine, with all the equipment needed for slaughtering prey. Any victim confronted with a whole gang of them probably stood little chance of survival. Some have suggested that this is what may have happened to the 8m (27ft)-long, 2000kg (1.9 ton) herbivore *Tenontosaurus* savaged by a number of *Deinonychus* more than a hundred million years ago, in territory now situated in the US state of Montana. *Deinonychus*' toe claw was capable of goring and even disembowelling its victims.

EARLY CRETACEOUS

# Iguanodon

| VITAL STATISTICS | |
|---|---|
| Fossil Location | Europe, Africa, USA |
| Diet | Herbivorous |
| Pronunciation | Ig-WAHN-oh-don |
| Weight | 3 tonnes (3.5 tons) |
| Length | 10m (33ft) |
| Height | 2.7m (9ft) at the hips, 5m (16ft) when upright |
| Meaning of name | 'Iguana tooth' because its teeth resembled those of this reptile |

*Iguanodon* was discovered so early on in the history of fossil-hunting (found 1822, named 1825) that the term 'dinosaur' did not exist, and it was named after the iguana reptile its tooth resembled. It is now known to have been one of the most successful herbivores, spreading worldwide during the 10-million-year or so lifespan of the species. It thrived because it had more advanced eating apparatus than other plant-eaters, and was protected both by sheer size, and possibly its threatening thumb spike. In early reconstructions, this spike was placed on the nose so it looked like a rhinoceros.

### WHERE IN THE WORLD?

The first tooth was found in England, and many skeletons have been found in Belgium. It is also found in Germany, northern Africa and western USA.

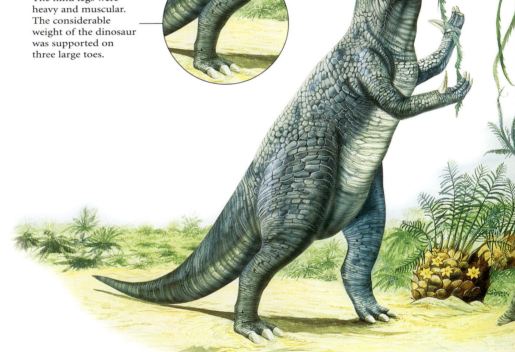

**Hind Legs**
The hind legs were heavy and muscular. The considerable weight of the dinosaur was supported on three large toes.

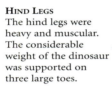

## FOSSIL EVIDENCE

The first find, in England, was of a single tooth. The 1878 discovery of 24 almost complete skeletons revealed *Iguanodon* in its entirety. First said to walk on all fours, then to be bipedal, it is now known to have been able to do both. The finding of many specimens together has fostered the view that *Iguanodon* was a social animal that moved as a herd.

DINOSAUR

EARLY CRETACEOUS

**HOW BIG IS IT?**

### TOOTHLESS SNOUT

Efficient eating was a major factor in the survival of this dinosaur. There were no teeth in the front of the snout, where food was shovelled into a bony beak. However, there were parallel rows of broad teeth in the back of the jaw to break food down, and the animal could flex its jaw outwards to chew.

EARLY CRETACEOUS

• **ORDER** • Ornithischia • **FAMILY** • Iguanodontidae • **GENUS & SPECIES** • *Iguanodon bernissartensis, I. anglicus*

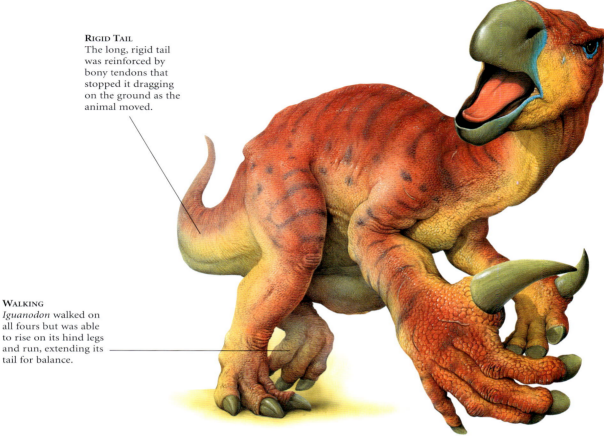

**RIGID TAIL**
The long, rigid tail was reinforced by bony tendons that stopped it dragging on the ground as the animal moved.

**WALKING**
*Iguanodon* walked on all fours but was able to rise on its hind legs and run, extending its tail for balance.

**THUMB SPIKE**
Originally mistaken for a nose horn, the conical thumb spike sat perpendicular to the other four fingers. It ranged in size from 5–15cm (2–6in) and its main function is not clear. It could have been used for defence against predators, such as *Aristosuchus* and *Baryonyx*. However, it would also have been valuable in gathering leaves, shoots and small branches. The fifth or little finger was dextrous like a thumb and was able to grasp small items.

**TIMELINE (millions of years ago)**

| 540 | 505 | 438 | 408 | 360 | 280 | 248 | 208 | 146 | 65 | 1.8 to today |

EARLY CRETACEOUS

# Iguanodon

EARLY CRETACEOUS

• **ORDER** • Ornithischia • **FAMILY** • Iguanodontidae • **GENUS & SPECIES** • *Iguanodon bernissartensis, I. anglicus*

**VICTORIAN SCULPTURES**

Benjamin Waterhouse Hawkins (1807–1894) was a Victorian sculptor and natural history artist. He was fascinated by dinosaurs, which excited enormous public curiosity at a time of new scientific revelations. Hawkins' chance to express his interest came after he was appointed Assistant Superintendent of the Great Exhibition of 1851, which was held at London's Crystal Palace. Hawkins was commissioned to produce 33 life-size models in concrete of extinct dinosaurs, to be sited in the park adjacent to the palace. He created as his centrepiece a model of an *Iguanodon*; it was so big that he was able to accommodate twenty guests inside it for a dinner held on New Year's Eve, 1853. Hawkins' work – and showmanship – attracted American patrons, and in 1868 he travelled to New York to produce models of dinosaurs for display in the city's Central Park in what was to be called the Paleozoic Museum. He also completed a model of a *Hadrosaurus* skeleton that went on display at the Academy of Natural Sciences in Philadelphia. While in the United States, Hawkins reconstructed several dinosaur skeletons for the Smithsonian Institution in Washington, D.C., and produced a series of dinosaur paintings for the Centennial Exhibition held in Philadelphia in 1876. Pictured is an *Iguanodon* reconstruction at The Natural History Museum in London.

215

# EARLY CRETACEOUS

## Probactrosaurus

• **ORDER** • Ornithischia • **FAMILY** • unranked • **GENUS & SPECIES** • Probactrosaurus gobiensis

| VITAL STATISTICS | |
|---|---|
| FOSSIL LOCATION | China |
| DIET | Herbivorous |
| PRONUNCIATION | Pro-back-tro-SORE-us |
| WEIGHT | 1.016kg (1 ton) |
| LENGTH | 6m (19.6ft) |
| HEIGHT | 3m (9.8ft) |
| MEANING OF NAME | 'Before *Bactrosaurus*', because it was originally thought to be ancestral to that dinosaur, whose name, in turn, means 'Club lizard' for the high massive club-shaped spines on its vertebrae |

*Probactrosaurus* was first described in 1966 by the Russian palaeontologist Anatoly Rozhdestvensky. A member of the hadrosauroid family, *Probactrosaurus* was originally thought to be a predecessor of the duck-billed dinosaurs.

### WHERE IN THE WORLD?

Fossils have been found at Nei Mongol Zizhiqu in the Gobi Desert in China.

### FOSSIL EVIDENCE

*Probactrosaurus* had a slender snout and double rows of compressed cheek teeth. Its lower jaw was elongated. *Probactrosaurus* has also been described as a large iguanodontid, although it had columns of replacement teeth, whereas *Iguanodon*, another herbivore, replaced its teeth one at a time. A skull of *Probactrosaurus gobiensis* has been placed on display at the Palaezoological Museum of China. This skull is of the plain iguanodont type, without crests or knobs.

**FIRST APPEARANCE**
According to the fossil record, *Probactrosaurus* first appeared around 121 million years ago, and survived until 99 million years ago.

**DUCK-BILLED**
As its name suggests, *Probactrosaurus* was originally thought to be a predecessor of the duck-billed *Bactrosaurus*, which was among the earliest known hadrosaurs.

### HOW BIG IS IT?

DINOSAUR
EARLY CRETACEOUS

**TIMELINE (millions of years ago)**

| 540 | 505 | 438 | 408 | 360 | 280 | 248 | 208 | 146 | 65 | 1.8 to today |

EARLY CRETACEOUS

# Pterodaustro

• **ORDER** • Pterosauria • **FAMILY** • Ctenochasmatidae • **GENUS & SPECIES** • *Pterodaustro guinazui*

## VITAL STATISTICS

| | |
|---|---|
| FOSSIL LOCATION | Argentina |
| DIET | Possibly omnivorous |
| PRONUNCIATION | ter-ah-DAWS-tro |
| WEIGHT | Unknown |
| LENGTH | 132cm (52in) |
| HEIGHT | Unknown |
| MEANING OF NAME | 'Wing from the South' |

One of the most unusual pterosaurs, *Pterodaustro* possessed a mouth full of long, thin, flexible teeth that were probably used as a strainer or filter.

## WHERE IN THE WORLD?

This unusual pterosaur is known only from San Luis Province in Argentina.

## FOSSIL EVIDENCE

The exposures in which this pterosaur's fossils were first found in 1970 still produce great specimens of this animal, either as isolated bones, disarticulated body parts or complete skeletons. The rocks the fossils come from are thinly bedded lake sediments that split cleanly to reveal the preserved remains. The beautiful fossils from this deposit also include complete fish called semionotids, which have heavy, enamel scales. When first exposed, the fossils are often bluish-green for a short time before they dry out. They are extremely fragile and hardeners need to be applied quickly to protect the specimens.

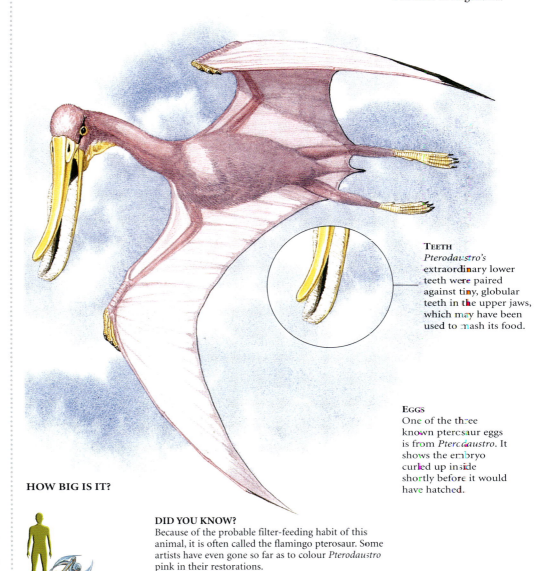

**TEETH**
*Pterodaustro's* extraordinary lower teeth were paired against tiny, globular teeth in the upper jaws, which may have been used to mash its food.

**EGGS**
One of the three known pterosaur eggs is from *Pterodaustro*. It shows the embryo curled up inside shortly before it would have hatched.

**HOW BIG IS IT?**

PREHISTORIC ANIMAL

EARLY CRETACEOUS

**DID YOU KNOW?**
Because of the probable filter-feeding habit of this animal, it is often called the flamingo pterosaur. Some artists have even gone so far as to colour *Pterodaustro* pink in their restorations.

**TIMELINE (millions of years ago)**

| 540 | 505 | 438 | 408 | 360 | 280 | 248 | 208 | 146 | 65 | 1.8 to today |

EARLY CRETACEOUS

# Utahraptor

| VITAL STATISTICS | |
|---|---|
| Fossil Location | USA |
| Diet | Carnivorous |
| Pronunciation | YOU-ta-RAP-tor |
| Weight | 700kg (150lb) |
| Length | 6.5m (21ft) |
| Height | 2m (6.6ft) |
| Meaning of name | 'Utah thief' after the site where it was discovered |

**Meeting a *Utahraptor* was bad news for the herbivores of the Early Cretaceous Period. Possibly one of the most intelligent dinosaurs of its time, it was fast and agile and equipped with a pair of formidable toe claws.**

**Skull**
The skull was 45cm (18in) long and featured large eyes as well as a large brain.

## FOSSIL EVIDENCE

The well-preserved and almost complete skeleton was discovered at a quarry near Moab in 1991 and named in 1993. *Utahraptor* was a dromaeosaurid, a group of light, quick-moving and quick-witted dinosaurs with large brains, keen senses, and a murderous claw on each foot. Fossil finds of several of these dromaeosaurids around giant plant-eaters suggest they were able to hunt in packs, disabling their victim with numerous stabs and cuts. They were bipedal and may have been at least partly covered with feathers.

DINOSAUR

EARLY CRETACEOUS

## CONFIDENT PREDATOR
The second toe had a sickle-like claw that was 23–38cm (9–15in) long. Enlarged joints held it upright while the animal moved, helping it stay sharp.

# EARLY CRETACEOUS

• **ORDER** • Saurischia • **FAMILY** • Dromaeosauridae • **GENUS & SPECIES** • *Utahraptor ostrommaysorum*

## FINE BODY RODS
The long, thick tail was stiffened by a sheath of fine body rods so it would have been almost rigid. It probably worked like an acrobat's pole to help *Utahraptor* keep stable while running and enhanced its manoeuvrability.

## WHERE IN THE WORLD?

The remains were found at a quarry in Grand County, Utah, USA. Some possible fragments have been found in South America.

### Teeth
The serrated teeth were 5cm (2in) long and set in a powerful jaw that could crush victims or cut through flesh. New teeth grew to replace any that broke.

### Claws
Each hand held three broad, flattened talon-like claws that could pierce flesh to provide a strong grip.

## HOW BIG IS IT?

## TIMELINE (millions of years ago)

| 540 | 505 | 438 | 408 | 360 | 280 | 248 | 208 | 116 | 65 | 1.8 to today |

219

EARLY CRETACEOUS

# Utahraptor

# EARLY CRETACEOUS

• **ORDER** • Saurischia • **FAMILY** • Dromaeosauridae • **GENUS & SPECIES** • *Utahraptor ostrommaysorum*

## FOOTPRINTS TELL A STORY

Dinosaur remains are not only fossils. Dinosaur footprints – or 'trackways', as they are also termed – have also a great deal to tell. Footprints possibly left behind by *Utahraptor ostrommaysorum* have been discovered in a coal mine in Utah, some of them impressed singly, others in quantity, showing that many *Utahraptor* walked this way millions of years ago. In one trackway, footprints were discovered showing where 23 *Utahraptor* moved across what was in prehistoric times a peat bog, but now lies deep beneath a mountain. One of the footprints was left by a *Utahraptor* 3m (10ft) tall, 6m (20ft) long and weighing more than 680kg (1500lb). Another *Utahraptor* left an impression of its large heel in the ancient peat. Further on, though, the trail of heel prints became fainter until they stopped marking the peat altogether: from this, palaeontologists have theorized that the *Utahraptor* was running towards its prey, with its heel lifting away from the peat as it neared its victim. A further set of *Utahraptor* footprints showed that each of the 14 steps the dinosaur took was exactly 109cm (3ft 6in) apart.

EARLY CRETACEOUS

# Suchomimus

• ORDER • Saurischia • FAMILY • Spinosauridae • GENUS & SPECIES • *Suchomimus tenerensis*

*Suchomimus* was a big, powerful predator that may have preyed on fish in the lush swamps of what is now the Sahara, wading in to trap them in its claws or jaws.

| VITAL STATISTICS | |
|---|---|
| FOSSIL LOCATION | Africa |
| DIET | Carnivorous |
| PRONUNCIATION | Soo-koe-MY-mus |
| WEIGHT | 6 tonnes (6 tons) |
| LENGTH | 12m (40ft) |
| HEIGHT | 4m (13ft) |
| MEANING OF NAME | 'Crocodile mimic' because of its crocodile-like mouth |

**SPINES**
Tail spines along the backbone may have supported a fleshy fin used for display and possibly to help it warm up and cool down.

**WHERE IN THE WORLD?**

Found in 1997 in the Sahara, near the Tenere Desert of Niger, East Africa.

**FOSSIL EVIDENCE**
Although it had a crocodilian mouth and nostrils on top of the snout, the rest of *Suchomimus* was more like *T. rex*: big and powerful with a long, strong tail. The forelimbs were equipped with a huge curved thumb. It was very similar to *Baryonyx*, apart from the tail spines on its backbone.

**TEETH**
*Suchomimus* had about 100 teeth. They were conical and, unlike most other toothed theropods, they were not serrated.

**HOW BIG IS IT?**

DINOSAUR

EARLY CRETACEOUS

**TIMELINE** (millions of years ago)

540 | 505 | 438 | 408 | 360 | 280 | 248 | 208 | 146 | 65 | 1.8 to today

MID CRETACEOUS

# Argentinosaurus

• ORDER • Saurischia • FAMILY • unranked • GENUS & SPECIES • *Argentinosaurus huinculensis*

This may have been the largest dinosaur that existed, but only a few fossils have been found so far. They reveal a giant at the limit of how big animals can become and survive.

| VITAL STATISTICS | |
|---|---|
| FOSSIL LOCATION | South America |
| DIET | Herbivorous |
| PRONUNCIATION | Ahr-JEN-oh-SAWR-us |
| WEIGHT | 80 tonnes (88 tons) |
| LENGTH | 35m (115ft) |
| HEIGHT | 21.4m (70ft) |
| MEANING OF NAME | 'Argentinian lizard' after the country where it was found |

**WHERE IN THE WORLD?**

Found in 1987 in the Rio Limay Formation in Neuquén Province, Argentina.

**SKIN**
As a titanosaur, *Argentinosaurus*' skin would have likely been armoured with a mosaic of osteoderms, or bony studs.

**FOSSIL EVIDENCE**
From the few bones found, including a 1.5m (5ft) shin bone, it is clear this was a massive, long-necked plant eater. One of the vertebrae is 1.3m (4ft) long and has a diameter of 1.7m (5ft 6in). Part of it is wing-shaped, allowing it to contain the powerful muscles needed to hold up the animal. The tail was not proportionately as long as a *Diplodocus*'. *Argentinosaurus* is thought to have roamed in herds across the wide floodplains of South America that could accommodate its size – and it may have been even bigger than the figures supplied here.

**HOW BIG IS IT?**

**HEAD LEVEL**
Despite early depictions, *Argentinosaurus* probably couldn't raise its head much above shoulder height because the blood pressure required would have burst its veins.

DINOSAUR
MID CRETACEOUS

**TIMELINE (millions of years ago)**

| 540 | 505 | 438 | 408 | 360 | 280 | 248 | 208 | 146 | 65 | 1.8 to today |

MID CRETACEOUS

# Carcharodontosaurus

| VITAL STATISTICS | |
|---|---|
| Fossil Location | Africa |
| Diet | Carnivorous |
| Pronunciation | Car-car-owe-dont-owe-SORE-us |
| Weight | 2900kg (1315lb) |
| Length | 13.5m (44ft) |
| Height | 3.65m (12ft) |
| Meaning of name | 'Carcharodon tooth lizard' because its teeth resemble those of Carcharodon, the great white shark, whose name, in turn, means 'jagged toothed' |

*Carcharodontosaurus* arrived on the palaeontological scene in 1927, when a skull and a few bones were found in the Sahara Desert of North Africa. It was at first named *Megalosaurus saharicus*, which meant 'huge Sahara lizard' but was renamed by the German palaeontologist Ernst Stromer in 1931. Like the ill-fated remains of *Spinosaurus*, the *Carcharodontosaurus* finds were destroyed when the Munich museum housing them was bombed by the British during World War II. Fortunately, the renowned American palaeontologist Paul Sereno and his team made another, even larger find in North Africa in 1996.

## FOSSIL EVIDENCE

*Carcharodontosaurus* is among the most gigantic meat-eating dinosaurs ever discovered. Its jaws were huge and contained teeth that were up to 3cm (8in) long. *Carcharodontosaurus saharicus*' skull measured 1.75m (5ft 6in) in length; the skull of *C. iguidensis* was longer, at 1.95m (6ft 4in). Probably the only small-sized features this predator possessed were its short arms, but even they were dangerous weapons, carrying three-fingered, sharp-clawed hands. In addition, *Carcharodontosaurus* was possibly a fast runner, able to home in on its prey with terrifying speed.

DINOSAUR

MID CRETACEOUS

### HOW BIG IS IT?

### SMALL BRAIN

*Carcharodontosaurus* might have had a huge skull, but its brain was very small – smaller than the brain of *Tyrannosaurus rex*, with which it is often compared. Palaeontologists could study *Carcharodontosaurus*' brain by means of an endocast, or endocranial cast, which is a cast made from the impression made by the brain on the inside of the braincase. Such a cast enables palaeontologists to study dinosaur brains without damaging the original skull.

MID CRETACEOUS

• **ORDER** • Saurischia • **FAMILY** • Carcharodontosauridae • **GENUS & SPECIES** • *Carcharodontosaurus saharicus, C. iguidensis*

## TEETH DISCOVERY
The destruction of the Munich museum in a bombing raid in 1944 left palaeontologists with a puzzle that took more than half a century to solve, until the American palaeontologist Paul Sereno matched the teeth from his 1996 Moroccan find with those described by Ernst Stromer in 1915. This proved that Stromer's finds and Sereno's discoveries belonged to the same predator.

### WHERE IN THE WORLD?
Fossils have been found in Morocco and Niger.

**TEETH**
Its long, jagged teeth were each serrated like steak-knives and it had flesh-tearing claws on its hands and feet.

**TAIL**
A thick, heavy tail looked as though it could kill or seriously injure with a whiplash blow of enormous power.

**CLAWS**
Fearsome claws designed for tearing flesh were located on both its hands and feet.

**LEGS**
Strong, muscular legs enabled *Carcharodontosaurus* to move quickly.

**TIMELINE (millions of years ago)**

| 540 | 505 | 438 | 408 | 360 | 280 | 248 | 208 | 146 | 65 | 1.8 to today |

MID CRETACEOUS

# Carcharodontosaurus

MID CRETACEOUS

• **ORDER** • Saurischia • **FAMILY** • Carcharodontosauridae • **GENUS & SPECIES** • *Carcharodontosaurus saharicus, C. iguidensis*

**HUGE CARNIVORE**

The most exciting event in palaeontology is the discovery of a formerly unknown species. This is what happened in 1997, when the remains of a huge meat-eating dinosaur, *Carcharodontosaurus*, were found in the Republic of Niger in northwest Africa. The announcement of the new species, named *Carcharodontosaurus iguidensis*, was not made until December 2007, but it presented palaeontologists with what was claimed to be one of the biggest carnivores that ever lived. *C. Iguidensis* was reckoned to measure around 13.5m (44ft) long, with a skull some 1.75m (5ft 8in) in length and had a mouth packed with teeth the size of bananas. The find comprised several pieces of skull including parts of the snout, the lower jaw and the braincase, in addition to part of the great dinosaur's neck. *Carcharodontosaurus iguidensis* lived around 95 million years ago, when levels of the Earth's seas and oceans were the highest ever known. In addition, the climate was the warmest it had ever been. As a result, Niger and Morocco, where the previous *Carcharodontosaurus* finds had been made in 1996, were separated from each other, allowing the Niger dinosaurs to evolve in different ways from the Moroccan ones.

MID CRETACEOUS

# Spinosaurus

| VITAL STATISTICS | |
|---|---|
| Fossil Location | Egypt, Morocco |
| Diet | Carnivorous |
| Pronunciation | Spy-know-SORE-us |
| Weight | 9000kg (8.85 tons) |
| Length | 18m (59ft) |
| Height | 5.6m (18ft 4in) including 'sail' |
| Meaning of name | 'Spine lizard' for the enormously elongated vertical prongs on its vertebrae |

The carnivorous dinosaur *Spinosaurus* lived in North Africa about 100 million years ago. It first became known from a description of remains found in the Egyptian desert in 1912 by the German palaeontologist Ernst Stromer. These remains were destroyed during World War II, but at least not all of Stromer's work was lost. In 1915, after three years' work, he had recorded a detailed description of his Egyptian finds. Further progress on *Spinosaurus* had to wait another 80 years, until another Spinosaurus find was made in Morocco by the Canadian palaeontologist Dale Russell.

**FOSSIL EVIDENCE**

At present, only the skull and backbone of *Spinosaurus* have been described in any great detail. The spines of the dinosaur's backbone were up to 11 times as tall as the vertebrae they grew from. It is thought that these spines were a good deal stronger and larger than the thin rods of pelycosaurs like *Dimetrodon*. More information about the skull emerged in 2005 when it was revealed that it was one of the longest known in a carnivorous dinosaur: the skull was estimated to measure 1.75m (5ft 9in) in length with a slender snout.

DINOSAUR

MID CRETACEOUS

**HOW BIG IS IT?**

**SPINOSAURUS SURROUNDINGS**

The surroundings in which *Spinosaurus* lived covered most of present-day North Africa before the River Nile system in Egypt had formed. The Egyptian *Spinosaurus* probably lived along a shore that featured channels, streams and tidal flats. It shared mangrove swamps with other large carnivorous predators, such as *Bahariasaurus*, *Carcharodontosaurus* and the 10m (33ft)-long prehistoric crocodile, *Stomatosuchus*.

# MID CRETACEOUS

• **ORDER** • Saurischia • **FAMILY** • Spinosauridae • **GENUS & SPECIES** • *Spinosaurus aegyptiacus*

**SPINES**
*Spinosaurus*' name comes from the spines that extended from its vertebrae possibly to form a large sail 1.8m (6ft) in length from its back.

### WHERE IN THE WORLD?

Fossils have been found in the Valley of the Golden Mummies at Bahariya Oasis in the Western Desert of Egypt, and in Morocco.

**TEETH**
The long, conical teeth lining *Spinosaurus*' jaws were very unlike most theropod teeth which were serrated and more flattened. *Spinosaurus*' jaws were ideal for catching and eating the fish on which it lived.

### DISCOVERY OF BONES

In 2000, Assistant Professor Josh Smith of Washington University in St Louis, Missouri, discovered photographs of the *Spinosaurus* bones found by Ernst Stromer in 1912. Until then, it had been thought that after his fossils and other records were destroyed in 1944 the only remnants of Stromer's find were his drawings. The discovery of the photographs made it possible to study Stromer's discoveries more closely than before, and to confirm his findings.

**TIMELINE (millions of years ago)**

| 540 | 505 | 438 | 408 | 360 | 280 | 248 | 208 | 146 | 65 | 1.8 to today |

LATE CRETACEOUS

# Abelisaurus

• ORDER • Saurischia • FAMILY • Abelisauridae • GENUS & SPECIES • *Abelisaurus comahuensis*

This was a primitive theropod, a two-legged meat-eater, which was of interest because it shows such creatures most probably evolved separately in the Southern as well as the Northern Hemisphere.

| VITAL STATISTICS | |
|---|---|
| FOSSIL LOCATION | Argentina |
| DIET | Carnivorous |
| PRONUNCIATION | Ah-beli-i-SAWR-us |
| WEIGHT | 1.4 tonnes (1.4 tons) |
| LENGTH | 6.5–7.9m (21–26ft) |
| HEIGHT | 2m (6ft) at the hips |
| MEANING OF NAME | 'Abel's lizard' after Roberto Abel, who discovered it |

**TEETH**
The teeth are smaller than those of a *T. Rex* and quite heavy.

**WHERE IN THE WORLD?**

Found in the Anacleto Formation in the Rio Negro Province of Argentina.

**FOSSIL EVIDENCE**
Only part of a skull was found in 1985. Reconstructions are based on the structures of other bipedal carnivores with a similarly large head, so it is assumed to have had slender legs, short front limbs and a long tail balancing the weight of the skull. The major difference between *Abelisaurus* and the tyrannosaurids of the Northern Hemisphere is the large gap in front of the eyes. It was one of South America's fiercest predators, using its speed to pounce on the slower plant-eaters and tearing at their flesh with its numerous teeth.

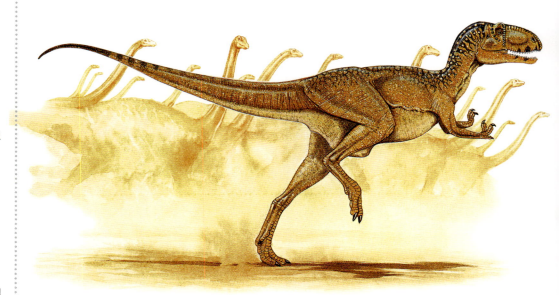

**SKULL**
The 85cm (33in) skull features large window-like openings (fenestrae) to reduce its weight.

**HOW BIG IS IT?**

DINOSAUR

LATE CRETACEOUS

**TIMELINE (millions of years ago)**

540 | 505 | 438 | 408 | 360 | 280 | 248 | 208 | 146 | 65 | 1.8 to today

LATE CRETACEOUS

# Adasaurus

• **ORDER** • Saurischia • **FAMILY** • Dromaeosauridae • **GENUS & SPECIES** • *Adasaurus mongoliensis*

| VITAL STATISTICS | |
|---|---|
| FOSSIL LOCATION | Mongolia |
| DIET | Carnivorous |
| PRONUNCIATION | ADD-ah-SAWR-us |
| WEIGHT | 15kg (33lb) |
| LENGTH | 2m (7ft) |
| HEIGHT | 70cm (28in) |
| MEANING OF NAME | 'Ada lizard' after a Mongolian mythological evil spirit |

Named after a female Mongolian evil spirit that was able to take many forms, the arrival of *Adasaurus* spelt trouble for Late Cretaceous lizards and small mammals.

### WHERE IN THE WORLD?

Both specimens are from Mongolia's Gobi Desert region.

**FEATHERS**
As a dromaeosaurid, *Adasaurus* is likely to have been at least partially covered in feathers on its body and tail.

### FOSSIL EVIDENCE

Two specimens have been found, neither complete. They are enough to establish this was a coelurosaur, a theropod closely related to birds that some analysts suggest was more like a true bird than a non-avian dinosaur. About the size of a large dog, it was intelligent, quick and agile, able to out-think and outrun smaller prey. The second toe of each hind foot was a sickle-shaped claw that could be flicked like a switchblade to rip into its victims. This claw is larger in other dromaeosaurids, while the skull is higher than the *Velociraptor* that *Adasaurus* closely resembles.

**HOW BIG IS IT?**

**CLAWS**
The killer toe claw was held off the ground while *Adasaurus* moved to prevent it getting damaged.

DINOSAUR

LATE CRETACEOUS

**TIMELINE (millions of years ago)**

| 540 | 505 | 438 | 408 | 360 | 280 | 248 | 208 | 146 | 65 | 1.8 to today |

## LATE CRETACEOUS

# Aeolosaurus

• **ORDER** • Saurischia • **FAMILY** • Titanosauria • **GENUS & SPECIES** • *Aeolosaurus rionegrinus, A. colhuehuapensis*

| VITAL STATISTICS | |
|---|---|
| Fossil Location | Argentina |
| Diet | Herbivorous |
| Pronunciation | EE-oh-loh-SAWR-us |
| Weight | 10 tonnes (11 tons) |
| Length | 15m (49ft) |
| Height | Unknown |
| Meaning of name | 'Aeolus lizard' after the Greek and Roman god of the winds, because it was found in a windy location |

*Aeolosaurus* was a herbivorous dinosaur that was common in the Southern Hemisphere in the Late Cretacean Period. It roamed in herds, consuming large quantities of plant matter to sustain a bulky body.

**WHERE IN THE WORLD?**

Remains come from three separate rock formations in Rio Negro Province, Argentina.

**SMALL TEETH**
*Aeolosaurus* cropped plant matter off with its small teeth, swallowing it whole as it was unable to chew.

### FOSSIL EVIDENCE
Various incomplete skeletons have been discovered showing this was a long-necked, long-tailed quadrupedal plant-eater possibly inhabiting the swampy lowlands and coastal planes of Argentina. Pieces of armour about 15cm (6in) in diameter suggest that its back was at least partially covered by protective plates. A major difference between it and other sauropods is the presence of forward-facing barbs on the tail vertebrae. This implies that Aeolosaurus was able to rise on its hind legs, propped up by its tail, so that it could reach higher conifer branches.

DINOSAUR

LATE CRETACEOUS

**HOW BIG IS IT?**

**ROBUST LEGS**
It stood on its four robust, columnar legs most of the time to support its hefty body.

**TIMELINE (millions of years ago)**

| 540 | 505 | 438 | 408 | 360 | 280 | 248 | 208 | 146 | 65 | 1.8 to today |

# Alamosaurus

• ORDER • Saurischia • FAMILY • Titanosauridae • GENUS & SPECIES • *Alamosaurus sanjuanensis*

## VITAL STATISTICS

| | |
|---|---|
| Fossil Location | USA |
| Diet | Herbivorous |
| Pronunciation | Al-uh-moe-SAWR-us |
| Weight | 30 tonnes (33 tons) |
| Length | 16m (53ft) |
| Height | 6m (20ft) |
| Meaning of name | 'Alamo lizard' after the Ojo Alamo trading post near the first discovery |

For many millions of years the record of sauropods in North America is blank. Then *Alamosaurus* appears as fossils in the Late Cretaceous possibly because its ancestors migrated across a landbridge of the isthmus of Panama from South America and flourished. One estimate suggests there were 350,000 living together in Texas alone.

### WHERE IN THE WORLD?

Finds have been uncovered since 1922 in New Mexico, Utah and Texas.

**Height Advantage**
*Alamosaurus* thrived in the southern USA because it was able to reach the leaves of trees that soared 27m (90ft) high in the warm climate.

## FOSSIL EVIDENCE

Numerous fragmentary skeletons and bones but no skulls have been found, showing how widespread and successful *Alamosaurus* became. The best specimens are juveniles, from which approximate adult sizes have been extrapolated. These huge herbivores probably roamed in herds, stripping leaves from tall trees to be digested with the help of gastroliths (swallowed stones) in their gizzards. They survived while sharing territory with predatory tyrannosaurs and other theropods and may well have been one of the last non-avian dinosaurs to face extinction.

**HOW BIG IS IT?**

DINOSAUR
LATE CRETACEOUS

**Tail**
The tail may have worked as a whip to deter predators. *Alamosaurus* may also have had some body armour.

**TIMELINE (millions of years ago)**

| 540 | 505 | 438 | 408 | 360 | 280 | 248 | 208 | 146 | 65 | 1.8 to today |

LATE CRETACEOUS

# Albertosaurus

• **ORDER** • Saurischia • **FAMILY** • Tyrannosauridae • **GENUS & SPECIES** • *Albertosaurus sarcophagus*

| VITAL STATISTICS | |
|---|---|
| Fossil Location | North America |
| Diet | Carnivorous |
| Pronunciation | Al-BER-toe-SAWR-us |
| Weight | 1.3–1.7 tonnes (1.4–1.9 tons) |
| Length | 9m (30ft) |
| Height | 3.4m (11ft) at the hips |
| Meaning of name | 'Alberta lizard' after the Canadian province where many were found |

This theropod was a close relative of *Tyrannosaurus rex*, which appeared a few million years later. It was half the size of its later distant cousin and a speedy runner, possibly capable of 40–48kmh (25–30mph).

**WHERE IN THE WORLD?**

Alberta in Canada, plus Montana and Wyoming in the western USA.

**JAW**
The lower jaw had 14–16 saw-edged teeth and the upper jaw 17–19. Each tooth had a replacement growing underneath ready to replace broken or worn-out ones.

**EYES**
While its body structure was similar to *T. rex*, its eyes were positioned more towards the sides, making it harder for it to judge distances.

**FOSSIL EVIDENCE**
The finding of various individuals aged from two to 28 years of age on the same site has led some palaeontologists to suggest that *Albertosaurus* hunted in packs, possibly family groups, with the speedier juveniles herding prey towards the larger, slower elders. Its two-fingered arms were not very useful being too short to reach its mouth, so it would have led attacks with its gaping jaw. *Albertosaurus* also possibly had a keen sense of smell, which would have allowed it to scavenge for dead meat, hunting by scent.

DINOSAUR

LATE CRETACEOUS

**HOW BIG IS IT?**

**TIMELINE (millions of years ago)**

| 540 | 505 | 438 | 408 | 360 | 280 | 248 | 208 | 146 | 65 | 1.8 to today |

LATE CRETACEOUS

# Alectrosaurus

• **ORDER** • Saurischia • **FAMILY** • Tyrannosauridae • **GENUS & SPECIES** • *Alectrosaurus olseni*

| VITAL STATISTICS | |
|---|---|
| FOSSIL LOCATION | Mongolia |
| DIET | Carnivorous |
| PRONUNCIATION | Ah-LECK-troh-SAWR-us |
| WEIGHT | 1.5 tonnes (1.7 tons) |
| LENGTH | 5m (17ft) |
| HEIGHT | Unknown |
| MEANING OF NAME | 'Unmarried lizard' because it was first thought to be unlike other dinosaurs |

This is one of the oldest known members of the tyrannosaur family, predating T. rex by about 20 million years.

### WHERE IN THE WORLD?

All remains have been found in the Gobi Desert region of Mongolia.

**TEETH**
*Alectrosaurus'* main weapon was its short and very sharp teeth, with which it would probably have led attacks on its prey.

### FOSSIL EVIDENCE

Fossils of this animal were first discovered in 1923 and it was reconstructed with long arms using bones found nearby. These were later found to be those of a segnosaur. *Alectrosaurus* finally took on its rightful upright shape of a tyrannosaur, with its large head, powerful legs, tiny arms and stiff, pointed tail. However, it is half the size of later tyrannosaurs with a smaller head and a more slender build.

**HIND LEGS**
The hind legs are slim compared to those of its close relatives, so it would have been lighter but less powerful.

**HOW BIG IS IT?**

DINOSAUR

LATE CRETACEOUS

**TIMELINE (millions of years ago)**

| 540 | 505 | 438 | 408 | 360 | 280 | 248 | 208 | 146 | 65 | 1.8 to today |

LATE CRETACEOUS

# Alioramus

• ORDER • Saurischia • FAMILY • Tyrannosauridae • GENUS & SPECIES • *Alioramus remotus*

From the little that is known about this off-shoot of the tyrannosaur family, it is clear it was very different from its relations, being smaller and having more teeth.

### VITAL STATISTICS

| | |
|---|---|
| FOSSIL LOCATION | Mongolia |
| DIET | Carnivorous |
| PRONUNCIATION | AL-ee-uh-RAY-mus |
| WEIGHT | 1800kg (4000lb) |
| LENGTH | 5–6m (16.5–20ft) |
| HEIGHT | 2m (7ft) |
| MEANING OF NAME | 'Different branch' because it diverged from other tyrannosaurs |

**SKULL**
The skull was 45cm (18in) long and held 76 or 78 teeth, far more than any other tyrannosaur, but the jaw was weaker so its cousins had a stronger bite

### WHERE IN THE WORLD?

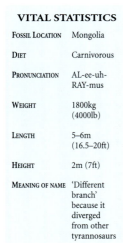

The only remains were found in Mongolia's Gobi Desert in the early 1970s.

### FOSSIL EVIDENCE

A few jawbones, some other skull parts and a trio of foot bones are all that have been found of *Alioramus*, making it difficult to describe with confidence, especially as the remains may be of a juvenile. The four bony knobs on the snout and a further two above the eyes were too small for defensive purposes and may have been for display – there is speculation that only males had them. *Alioramus* forms the Asian part of the tyrannosaurid family, along with *Tarbosaurus*.

**HOW BIG IS IT?**

**SNOUT**
The pronounced bony ridge on its snout may have been for display or for ritual head-butting contests

DINOSAUR

LATE CRETACEOUS

**TIMELINE** (millions of years ago)

| 540 | 505 | 438 | 408 | 360 | 280 | 248 | 208 | 146 | 65 | 1.8 to today |

236

LATE CRETACEOUS

# Alvarezsaurus

• ORDER • Saurischia • FAMILY • Alvarezsauridae • GENUS & SPECIES • *Alvarezsaurus calvoi*

### VITAL STATISTICS

| | |
|---|---|
| FOSSIL LOCATION | Argentina |
| DIET | Carnivorous |
| PRONUNCIATION | Al-vuh-rez-SAWR-us |
| WEIGHT | 20kg (44lb) |
| LENGTH | 2m (7ft) |
| HEIGHT | 1.4m (4.6ft) |
| MEANING OF NAME | 'Alvarez lizard' after the Argentine historian Gregorio Alvarez |

This is one of the creatures that give palaeontologists a headache. *Alvarezsaurus* has been classified at times as a non-avian theropod dinosaur and an early flightless bird. It may represent an important link between the two.

### WHERE IN THE WORLD?

Found in the Bajo de la Carpa Formation in Neuquén, Argentina.

**TAIL**
*Alvarezsaurus* had an amazingly long tail, much like those seen in some very agile modern lizards.

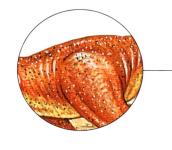

**COMPACT BODY**
There are practically no spines on the back vertebrae, so the body is compact with no ridges down its back, like a bird.

### FOSSIL EVIDENCE

The weirdness of this animal and the lack of a complete skull or forelimbs among the remains explain the classification problem. *Alvarezsaurus* had very long, slender legs ending in long feet, short arms, and an extended, thin, flat tail that made up more than half of its length. The long, flexible S-shaped neck ended in a small skull with small unserrated teeth in the snout. It couldn't fly, but it was very fast and agile.

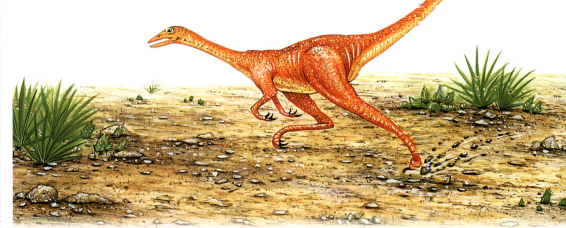

DINOSAUR

LATE CRETACEOUS

**HOW BIG IS IT?**

**TIMELINE (millions of years ago)**

| 540 | 505 | 438 | 408 | 360 | 280 | 248 | 208 | 145 | 65 | 1.8 to today |

237

LATE CRETACEOUS

# Anatotitan

• **ORDER** • Ornithischia • **FAMILY** • Hadrosauridae • **GENUS & SPECIES** • *Anatotitan copei, A. longiceps*

| VITAL STATISTICS | |
|---|---|
| Fossil Location | USA |
| Diet | Herbivorous |
| Pronunciation | uh-NAT-uh-tie-tin |
| Weight | 7.3 tonnes (8 tons) |
| Length | 10m (33ft) |
| Height | 2.5m (8ft) at the hips |
| Meaning of name | 'Giant duck' after its broad bill |

The hadrosaurs were duck-billed dinosaurs and *Anatotitan* had the most duck-like broad, flat snout of them all. A slow-moving dinosaur with few defences, it would have needed keen senses to avoid predators.

**WHERE IN THE WORLD?**

Found in the states of South Dakota and Montana in western USA.

**TOOTHLESS BEAK**
Behind the toothless beak were 720 closely packed cheek teeth arranged in several rows for grinding down plant material.

**FOSSIL EVIDENCE**
At least five specimens, some complete, have been found – in 1904 one was swapped for a revolver! *Anatotitan* was among the last of the hadrosaurs to evolve, and had a larger and longer (up to 1.1m [46in]) skull than its relatives. Some palaeontologists argue that the skulls were crushed during preservation and that this is actually a species of *Edmontosaurus*. It had knobs over the eyes, but no crest. The 'duck' description is inaccurate: the snout was more like that of a horse and the mouth was probably not sensitive like a duck's.

DINOSAUR

LATE CRETACEOUS

**HOW BIG IS IT?**

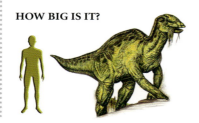

**LIMBS**
*Anatotitan* probably would have walked on its four limbs to feed, and then may have lifted itself up onto two three-hoofed hind legs for faster movement.

**TIMELINE** (millions of years ago)

| 540 | 505 | 438 | 408 | 360 | 280 | 248 | 208 | 146 | 65 | 1.8 to today |

LATE CRETACEOUS

# Anchiceratops

• **ORDER** • Ornithischia • **FAMILY** • Ceratopsidae • **GENUS & SPECIES** • *Anchiceratops ornatus*

## VITAL STATISTICS

| | |
|---|---|
| FOSSIL LOCATION | Canada |
| DIET | Herbivorous |
| PRONUNCIATION | AN-key-SAIR-ah-tops |
| WEIGHT | 2470kg (5400lb) |
| LENGTH | 4.5–6m (15–20ft) |
| HEIGHT | 2.6m (9ft) |
| MEANING OF NAME | 'Near horn face' because it was originally thought to be closely related to another horned dinosaur, *Centrosaurus* |

*Anchiceratops* is one of the rarest chasmosaurines, a group of advanced ceratopsians with three horns, parrot-like beaks and frilled shields. They roamed in herds, rooting up plant matter and watching out for *T. rex*, which were likely able to pierce their leathery hides.

**NECK FRILL**
Its most distinctive feature is the shield-like rectangular scalloped neck frill topped by six backwards-pointing points. Two curved horns project from beneath the frill.

### WHERE IN THE WORLD?

All remains are from the Horseshoe Canyon and Dinosaur Park Formations of Alberta, Canada.

## FOSSIL EVIDENCE

Six skulls have been found, and it is assumed that the rest of the body followed the shape of other ceratopsians. All specimens were found near marine sediments, suggesting *Anchiceratops* could have lived in estuaries away from its relatives. The frill is unusual because of its large size, edged by triangular bony projections. Such frills may have had defensive or display benefits, may have been used for thermoregulation, or support for large neck and chewing muscles. The skull supported two large and one small horn.

**HOW BIG IS IT?**

**STRONG BEAK**
The massive and strong beak was used to shear low-lying plants, probably mostly conifers, cycads and ferns, but possibly also the newly evolved flowering plants.

DINOSAUR

LATE CRETACEOUS

**TIMELINE (millions of years ago)**

| 540 | 505 | 438 | 408 | 360 | 280 | 248 | 208 | 143 | 65 | 1.8 to today |

LATE CRETACEOUS

# Anserimimus

• ORDER • Saurischia • FAMILY • Ornithomimidae • GENUS & SPECIES • *Anserimimus planinychus*

| VITAL STATISTICS | |
|---|---|
| Fossil Location | Mongolia |
| Diet | Carnivorous |
| Pronunciation | AN-ser-i-MIME-us |
| Weight | 62kg (137lb) |
| Length | 1m (3ft) |
| Height | Unknown |
| Meaning of name | 'Goose mimic', 'anser' being the generic name of several species of goose and ornithomimosaurs have traditionally been named after different types of birds |

The ornithimids were so named because they mimic ostriches by being long-legged and flightless. In fact their skulls are more like those of New Zealand's extinct ground-dwelling birds. *Anserimimus* was a strong-armed species of the group.

**WHERE IN THE WORLD?**

The specimens were recovered from the Nemegt Formation of Mongolia.

**BEAK**
*Anserimimus* used its beak to pick food up and may have had a comb-like structure to filter food particles in its mouth.

**FOSSIL EVIDENCE**
Only parts of an incomplete forelimb and hind limb have been found of this species. These show several differences from other ornithimids. The hand claws are not curved, but long and straight, while the forelimb is more powerful, featuring attachments for large arm muscles. These suggest it dug for food, perhaps insects and dinosaur eggs and possibly even roots (some palaeontologists believe it supplemented its meat diet). The long shins and feet gave it speed, probably its only defence against attack.

**FINGERS**
The fingers were positioned close together, almost like hoofs, and the claws were slightly flattened, features that are not fully explained.

**HOW BIG IS IT?**

DINOSAUR

LATE CRETACEOUS

**TIMELINE** (millions of years ago)

540 | 505 | 438 | 408 | 360 | 280 | 248 | 208 | 146 | 65 | 1.8 to today

LATE CRETACEOUS

# Antarctosaurus

• ORDER • Saurischia • FAMILY • Antarctosauridae • GENUS & SPECIES • *Antarctosaurus wichmannianus*

| VITAL STATISTICS | |
|---|---|
| FOSSIL LOCATION | South America and possibly India |
| DIET | Herbivorous |
| PRONUNCIATION | Ant-ARK-toe-SAWR-us |
| WEIGHT | 34 tonnes (37 tons) |
| LENGTH | 18m (60ft) |
| HEIGHT | 6m (20ft) |
| MEANING OF NAME | 'Opposite of north lizard' because it was in the Southern Hemisphere |

This was one of the largest South American sauropods and one of the most widespread dinosaurs in the Southern Hemisphere. It had a bulky body supported by tall legs.

**TEETH**
*Antarctosaurus* was a plant-eater with a few weak, peg-shaped teeth at the front of the jaws to harvest but not chew plant matter.

**WHERE IN THE WORLD?**

Remains have been found in Argentina and possibly also in India.

## FOSSIL EVIDENCE

The size of *Antarctosaurus* is not clear. The remains for the first find were scattered and may not be from the same animal. A later find is disputed, but includes a thigh bone measuring 2.35m (7ft 9in), which is double the size of the other and possibly the largest land animal of all time. The figures supplied here are the lower estimates. This is one of the few sauropods whose skull has been found. It reveals a 60cm (2ft) long head with large eyes and a few peg-like teeth at the front of the jaws.

**HOW BIG IS IT?**

**BODY**
The bulky body of this giant probably contained gastroliths (swallowed stones) to crush and grind down the enormous quantities of fibrous greenery it consumed.

DINOSAUR
LATE CRETACEOUS

**TIMELINE (millions of years ago)**

| 540 | 505 | 438 | 408 | 360 | 280 | 248 | 208 | 146 | 65 | 1.8 to today |

LATE CRETACEOUS

# Aralosaurus

• ORDER • Ornithischia • FAMILY • Hadrosauridae • GENUS & SPECIES • *Aralosaurus tuberiferus*

| VITAL STATISTICS | |
|---|---|
| Fossil Location | Kazakhstan |
| Diet | Herbivorous |
| Pronunciation | AR-a-lo-SAWR-US |
| Weight | 5 tonnes (5.5 tons) |
| Length | 6–9m (20–30ft) |
| Height | Unknown |
| Meaning of name | 'Aral lizard' because it was found near the Aral Sea |

One of the frustrations of partial finds is the difficulty in identifying and classifying the animal. Equipped with the back half of a skull, we know Aralosaurus was hadrosaurid, but after that there are more questions than answers.

**WHERE IN THE WORLD?**

Near the Aral Sea of Kazakhstan.

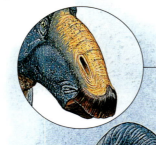

**Nasal Peak**
The nasal peak may have been used as a weapon in head-butting duels between rival males competing for food or mates.

## FOSSIL EVIDENCE

There is no skeleton for *Aralosaurus*, just a skull missing the front of the snout and the lower jaw. This reveals that the animal had a deeply hooked nasal arch in front of the eyes, which was far more developed than those of other hadrosaurids. The hook was both high and wide, and *Aralosaurus* had particularly large nostril openings beneath it. Some believe these would have been used to inflate loose skin like a balloon, enabling it to bellow loudly as a warning to predators or an invitation to mates.

**HOW BIG IS IT?**

**Teeth**
*Aralosaurus* had an exceptional number of teeth: as many as 30 rows, each.

DINOSAUR

LATE CRETACEOUS

**TIMELINE (millions of years ago)**

| 540 | 505 | 438 | 408 | 360 | 280 | 248 | 208 | 146 | 65 | 1.8 to today |

# Archaeornithomimus

**LATE CRETACEOUS**

• FAMILY • Ornithomimidae (disputed) • GENUS & SPECIES • *Archaeornithomimus asiaticus* • ORDER • Saurischia

## VITAL STATISTICS

| | |
|---|---|
| FOSSIL LOCATION | China and Kazakhstan |
| DIET | Omnivorous (possibly carnivorous) |
| PRONUNCIATION | AHR-kee-or-NITH-oh-MIME-us |
| WEIGHT | 50kg (110lb) |
| LENGTH | 3.4m (11ft) |
| HEIGHT | 1.8m (6ft) |
| MEANING OF NAME | 'Ancient *Ornithomimus*' because it was originally named *Ornithomimus* and then found to be a more primitive relative |

This was a predecessor to the better-known ornithomimid, *Ornithomimus*. Both are bird-like theropods with light-weight skeletons and slender limbs that allowed them to move quickly to find food and escape predators.

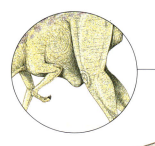

**BALANCED BODY**
*Archaeornithomimus* was well balanced, with a small head on a long neck, a slim body, and long, slender limbs. It would have been able to turn quickly.

### WHERE IN THE WORLD?

Fossils have been found in the Nei Mongol province of China and in Kazakhstan.

## FOSSIL EVIDENCE

Two pieces of evidence suggest this was a speedy beast. The legs were elongated because the tibia and fibula are long in relation to the femur, while analysis of the distance between footprints suggests that this dinosaur could have reached speeds of 70kmh (43mph). It was light and agile, its long tail enabling it to balance as it sprinted. While sometimes described as a carnivore, *Archaeornithomimus* may well have been an omnivore that ate small mammals and plants as well as fruit and eggs.

**TRIPLE CLAW**
It had three straight claws on its three fingers, with which it could possibly grab small mammals and dig for eggs.

### HOW BIG IS IT?

DINOSAUR
LATE CRETACEOUS

## TIMELINE (millions of years ago)

| 540 | 505 | 438 | 408 | 360 | 280 | 248 | 208 | 146 | 65 | 1.8 to today |

LATE CRETACEOUS

# Arrhinoceratops

• ORDER • Ornithischia • FAMILY • Ceratopsidae • GENUS & SPECIES • *Arrhinoceratops brachyops*

| VITAL STATISTICS | |
|---|---|
| FOSSIL LOCATION | Canada |
| DIET | Herbivorous |
| PRONUNCIATION | Ay-RYE-no-SER-uh-tops |
| WEIGHT | 3.5 tonnes (3.9 tons) |
| LENGTH | 6m (20ft) |
| HEIGHT | 2.1m (7ft) |
| MEANING OF NAME | 'No nose-horn face' based on an incorrect description |

This dinosaur has been saddled with a misleading name, since its 1925 discoverer decided it had no nose-horn, a view that was corrected in the 1970s. *Arrhinoceratops* is among the last of the long-frilled ceratopsians.

**WHERE IN THE WORLD?**

Remains were collected from the Horseshoe Canyon Formation in Alberta, Canada.

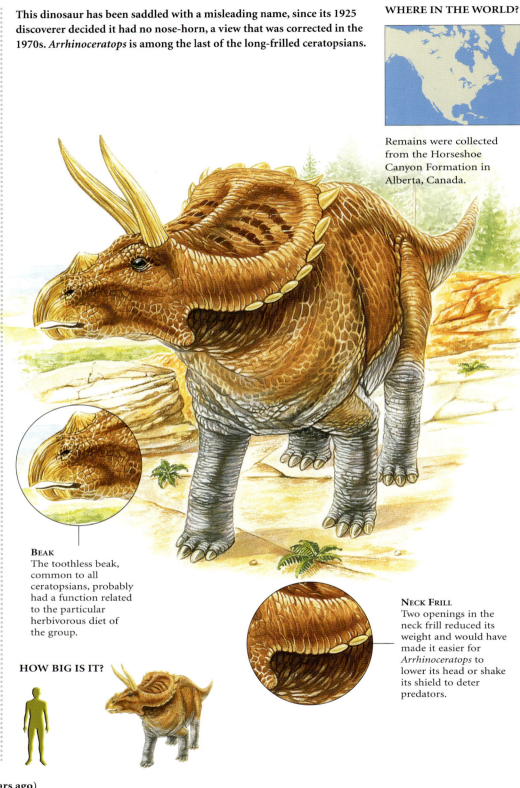

**FOSSIL EVIDENCE**
Only a skull has so far been ascribed to *Arrhinoceratops*, much of which had been crushed and distorted over time. This made the pattern of bones extremely difficult to understand, and of course the construction of the rest of the body is guesswork based on other advanced ceratopsians. Its familiar feature is the broad neck frill. It also has two fairly long brow horns and the short, blunt nose horn that was originally missed. The skull is 1.5m (4ft 6in) long and the face is short compared to other dinosaurs of its type.

**BEAK**
The toothless beak, common to all ceratopsians, probably had a function related to the particular herbivorous diet of the group.

**NECK FRILL**
Two openings in the neck frill reduced its weight and would have made it easier for *Arrhinoceratops* to lower its head or shake its shield to deter predators.

**HOW BIG IS IT?**

DINOSAUR

LATE CRETACEOUS

**TIMELINE (millions of years ago)**

| 540 | 505 | 438 | 408 | 360 | 280 | 248 | 208 | 146 | 65 | 1.8 to today |

LATE CRETACEOUS

# Aublysodon

• ORDER • Saurischia • FAMILY • Tyrannosauridae • GENUS & SPECIES • *Aublysodon mirandus*

## VITAL STATISTICS

| | |
|---|---|
| FOSSIL LOCATION | North America, China |
| DIET | Carnivorous |
| PRONUNCIATION | Aw-BLIS-oh-don |
| WEIGHT | 80kg (176lb) |
| LENGTH | 4.5m (15ft) |
| HEIGHT | 1.7m (6ft) |
| MEANING OF NAME | Probably 'Backwards flow tooth' for the way the cutting edges of the front teeth are swept back |

The tale of *Aublysodon* is the story of dinosaur hunting: in the wild early days, a new genus could be named from one tooth, and as palaeontology became more sophisticated these early descriptions were challenged.

**TEETH**
The good condition of many tyrannosaur teeth found has led some to suggest it was a scavenger rather than a hunter.

### WHERE IN THE WORLD?

Remains have been found in many American states, western Canada and China.

## FOSSIL EVIDENCE

*Aublysodon* was named in 1868 after a carnivore-like tooth with an unusual D-shaped cross-section. The Greek-derived, confusing name was based on inaccurate textbook information from the time. The tooth was lost, but since then, similar teeth and a partial skull have been found. Depending on who you believe, these either show that *Aublysodon* was quite a common small, primitive Late Cretaceous predator, or was a typical juvenile specimen of other tyrannosaurs and suggest that *Aublysodon* was never a species in its own right.

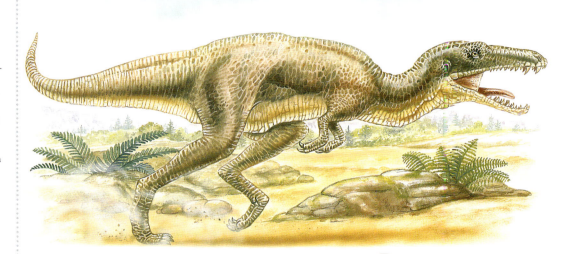

**BODY SHAPE**
Aublysodon would have been about one-third of the length of its cousin *Tyrannosaurus rex*, and very similar in shape.

### HOW BIG IS IT?

DINOSAUR

LATE CRETACEOUS

## TIMELINE (millions of years ago)

540 | 505 | 438 | 408 | 360 | 280 | 248 | 208 | 146 | 65 | 1.8 to today

LATE CRETACEOUS

# Avaceratops

• ORDER • Ornithischia • FAMILY • Ceratopsidae • GENUS & SPECIES • *Avaceratops lammersi*

*Avaceratops* was one of the smallest of the ceratopsids. It was a horned dinosaur with a bony frill around its face and a single horn on the nose. It possibly roamed in herds, feeding on low-lying vegetation.

## VITAL STATISTICS

| | |
|---|---|
| FOSSIL LOCATION | USA |
| DIET | Herbivorous |
| PRONUNCIATION | AY-vah-SAIR-uk-tops |
| WEIGHT | 1200kg (2650lb) |
| LENGTH | 2–4m (7–13ft) |
| HEIGHT | 1.3m (4ft) |
| MEANING OF NAME | 'Ava's horned face' after the wife of its finder, Eddie Cole |

### WHERE IN THE WORLD?

The remains come from the Judith River Formation of Montana, northwestern USA.

**FRILL**
The frill is short and thick and would have been heavy because there were no hollows to reduce its weight.

## FOSSIL EVIDENCE

An almost complete skeleton of a juvenile has been found, although it lacks the roof of the skull where the horn cores were. It looks rather like a *Triceratops* without its pair of horns and, like that animal, it lacked the weight-reducing hollows (fenestrae) in the frill, which were common in other ceratopsians. So it may have been a close relative of *Triceratops*.

**HOW BIG IS IT?**

**BEAK**
Food was collected with the strong, sharp, toothless beak and broken down by batteries of shearing cheek teeth in the jaw.

DINOSAUR

LATE CRETACEOUS

**TIMELINE (millions of years ago)**

| 540 | 505 | 438 | 408 | 360 | 280 | 248 | 208 | 146 | 65 | 1.8 to today |

LATE CRETACEOUS

# Bagaceratops

• ORDER • Ornithischia • FAMILY • Protoceratopsidae • GENUS & SPECIES • *Bagaceratops rozhdestvenskyi*

## VITAL STATISTICS

| | |
|---|---|
| FOSSIL LOCATION | Mongolia |
| DIET | Herbivorous |
| PRONUNCIATION | BAG-uh-CER-uh-TOPS |
| WEIGHT | 22kg (50lb) |
| LENGTH | 1m (3ft) |
| HEIGHT | 50cm (20in) |
| MEANING OF NAME | 'Small horned face' as it is smaller than its relatives |

*Bagaceratops* appears to be a smaller, more primitive version of *Protoceratops*, but cannot be one of its ancestors because it walked the planet after, rather than before, its lookalike.

**TEETH**
There were ten grinding teeth on each side of the jaw, each with a replacement growing underneath, so breakages were not a problem.

**WHERE IN THE WORLD?**

All the remains have been found preserved in desert sandstone in Mongolia.

## FOSSIL EVIDENCE

There are five complete and 17 fragmentary skulls of *Bagaceratops*, young and old, enabling palaeontologists to build a clear picture of its appearance and how it developed. It was about the size of a pig and had a small triangular frill forming a shield around its neck. This was a desert-dweller, possibly roaming in herds and uprooting low-level plants and roots with its toothless sharp beak – it probably could not stand on its hind legs. Its close relatives had no horns, but *Bagaceratops* had a small bony bump on its snout.

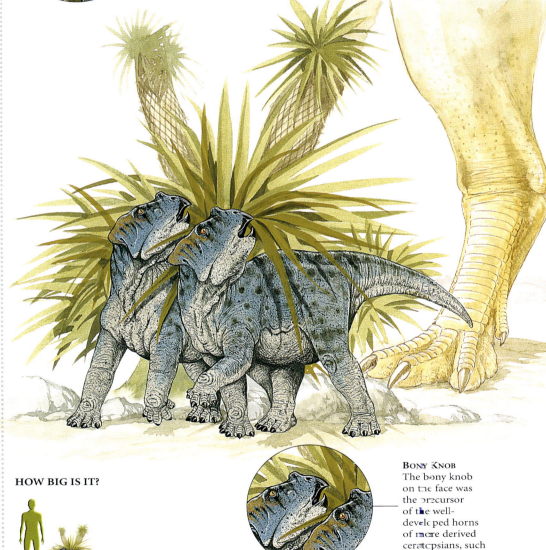

**BONY KNOB**
The bony knob on the face was the precursor of the well-developed horns of more derived ceratopsians, such as *Centrosaurus* and *Triceratops*.

**HOW BIG IS IT?**

DINOSAUR
LATE CRETACEOUS

**TIMELINE (millions of years ago)**

| 540 | 505 | 438 | 408 | 360 | 280 | 248 | 208 | 146 | 65 | 1.8 to today |

LATE CRETACEOUS

# Borogovia

• ORDER • Saurischia • FAMILY • Troodontidae • GENUS & SPECIES • *Borogovia gracilicrus*

| VITAL STATISTICS | |
|---|---|
| FOSSIL LOCATION | Mongolia |
| DIET | Carnivorous |
| PRONUNCIATION | Bor-oh-GOH-vee-a |
| WEIGHT | 13kg (29lb) |
| LENGTH | 2m (7ft) |
| HEIGHT | 70cm (28in) |
| MEANING OF NAME | Named after the imaginary 'borogoves' in Lewis Carroll's poem *Jabberwocky* |

'All mimsy were the borogoves' says Lewis Carroll's nonsense poem *Jabberwocky*, after which this animal was named, and it almost adds to our knowledge of it because the remains are so scant.

**WHERE IN THE WORLD?**

The remains were all found in Mongolia.

**FEATHERED TAIL**
This was a bird-like dinosaur and parts of its body and tail may have been feathered.

**THIN TOES**
The thin toes suggest that *Borogovia* had slender arms and legs so it was fast-moving and, like other troodonts, quick-thinking with its relatively large brain.

**FOSSIL EVIDENCE**
Only parts of a leg and a foot have been found, so most of what is known about *Borogovia* derives from studies of *Saurornithoides* and *Tochisaurus*, animals to which it is similar and whose remains are found at the same site. Some palaeontologists suggest they are the same animal. *Borogovia* is a lightly built theropod with long, wiry legs, far slimmer than *Saurornithoides*. Each foot has a sickle-like killing claw. This claw is smaller and straighter than those found on other troodontids – they seem to have reduced in size as the group evolved.

DINOSAUR

LATE CRETACEOUS

**HOW BIG IS IT?**

**TIMELINE (millions of years ago)**

540 | 505 | 438 | 408 | 360 | 280 | 248 | 208 | 146 | 65 | 1.8 to today

LATE CRETACEOUS

# Brachyceratops

• ORDER • Ornithischia • FAMILY • Ceratopsidae • GENUS & SPECIES • *Brachyceratops montanensis*

| VITAL STATISTICS | |
|---|---|
| FOSSIL LOCATION | North America |
| DIET | Herbivorous |
| PRONUNCIATION | BRACK-ee-SAIR-a-tops |
| WEIGHT | Unknown |
| LENGTH | Up to 4m (13ft) |
| HEIGHT | Unknown |
| MEANING OF NAME | 'Short horned head', to describe its face |

The discovery of five young dinosaurs in the same spot left palaeontologists excited but baffled. What were the adults like? Might these be babies of a known species?

**SKULL**
The skull had small bumps over the eyes, rather than the horns that were a feature of better-known ceratopsians such as *Triceratops*.

**WHERE IN THE WORLD?**

The first remains were discovered in the Two Medicine Formation in Montana, USA. Others have been discovered in Alberta, Canada.

**FOSSIL EVIDENCE**
The first find was made in 1913 and revealed the mixed-up and incomplete remains of five animals, each about 1.5m (5ft) long. The horns were not fully fused to the skull, indicating that these were juveniles. It is very difficult to assess adult size from such young specimens. The face is short, as is the frill on the back of the skull, which also lacks the weight-saving hollows seen in most adults of this group. Some experts believe these are juveniles of some other horned dinosaur.

DINOSAUR
LATE CRETACEOUS

**HOW BIG IS IT?**

**FRILLS**
One of the values of the frills that all ceratopsians have is that it makes the animal appear much larger than it is – probably useful in deterring attacks.

**TIMELINE (millions of years ago)**

| 540 | 505 | 438 | 408 | 360 | 280 | 248 | 208 | 116 | 65 | 1.8 to today |

LATE CRETACEOUS

# Brachylophosaurus

• ORDER • Ornithischia
• FAMILY • Hadrosauridae • GENUS & SPECIES • *Brachylophosaurus canadensis*

## VITAL STATISTICS

| | |
|---|---|
| FOSSIL LOCATION | North America |
| DIET | Herbivorous |
| PRONUNCIATION | BRACK-ee-LOAF-ah-SAW-rus |
| WEIGHT | 1300-2700kg (2900-5950lb) |
| LENGTH | 7m (23ft) |
| HEIGHT | 2.8m (9ft) |
| MEANING OF NAME | 'Short crested lizard' after its notable crest feature |

In 2000 one of the most extraordinary dinosaur finds was made when a mummified young *Brachylophosaurus* appeared out of the Montana sandstone. The find revealed the skin, muscles and even the last meal the animal ate.

**CREST**
This animal had a broad, flat crest with a spike above its eyes. This spike, being solid, was possibly used for ritual head-pushing contests.

### WHERE IN THE WORLD?

Fossils and other remains are from Montana's Judith River Formation and the Oldman Formation of Alberta, Canada.

**BEAK**
The beak of the upper jaw was wider than other hadrosaurs'. *Brachylophosaurus* also had longer forelimbs.

## FOSSIL EVIDENCE

*Brachylophosaurus* was already known as a crested hadrosaur when the mummified remains were discovered in 2000. About 80 per cent of the skin and muscles were intact, as well as the half-digested remains of 40 different types of plants and some evergreen wood in its gut. The juvenile dinosaur probably got trapped on a sandbar and was naturally mummified by the hot air and covered in sand. Studies of other remains of this crested dinosaur have shown it was unusually prone to tumours.

### HOW BIG IS IT?

DINOSAUR

LATE CRETACEOUS

**TIMELINE** (millions of years ago)

| 540 | 505 | 438 | 408 | 360 | 280 | 248 | 208 | 146 | 65 | 1.8 to today |

LATE CRETACEOUS

# Centrosaurus

• **ORDER** • Ornithischia • **FAMILY** • Ceratopsidae • **GENUS & SPECIES** • *Centrosaurus apertus*, *C. brinkmani*

## VITAL STATISTICS

| | |
|---|---|
| Fossil Location | Canada |
| Diet | Herbivorous |
| Pronunciation | SEN-tro-SAWR-us |
| Weight | 3 tonnes (3 tons) |
| Length | 6m (20ft) |
| Height | 3.5m (11ft) |
| Meaning of name | 'Pointed lizard' after the hooked horns on its neck shield |

Extensive piles of *Centrosaurus* bones found together suggest that this horned dinosaur travelled in very large herds, probably as a defence against predators.

**Frill**
Holes in the frill reduced its weight and may have provided anchors for the strong jaw muscles needed to deal with its diet of low-lying plants.

### WHERE IN THE WORLD?

Finds were made in Canada's Red Deer River and Dinosaur Provincial Parks, both in Alberta.

## FOSSIL EVIDENCE

*Centrosaurus* was named after the two inwardly directed, hook-shaped prongs that point down from its frill shield. It was given this identity before the big, single horn on its snout was discovered. The horn was sometimes straight, sometimes curved and could measure up to 46cm (18in). Combined with the sharply hooked spikes on its frill, these weapons were capable of inflicting serious wounds on attackers. The gigantic bonebeds found in Dinosaur Provincial Park in Alberta may have formed after a large herd drowned while trying to cross a flooded river.

**HOW BIG IS IT?**

DINOSAUR
LATE CRETACEOUS

**Hoofed Limbs**
*Centrosaurus* is thought to have had thick, strong, hoofed hind and forelimbs to carry it in its quest for food.

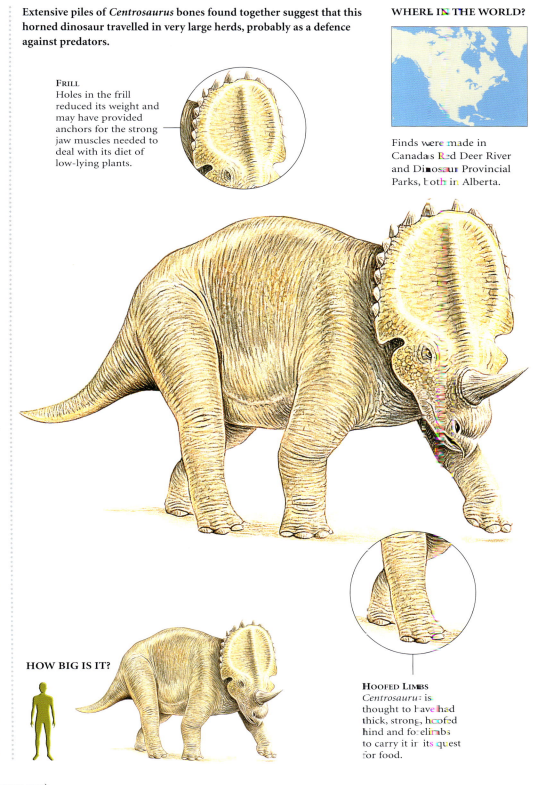

**TIMELINE (millions of years ago)**

| 540 | 505 | 438 | 408 | 360 | 280 | 248 | 208 | 146 | 65 | 1.8 to today |

251

LATE CRETACEOUS

# Chasmosaurus

• FAMILY • Ceratopsidae • GENUS & SPECIES • Several species within the genus *Chasmosaurus*

• ORDER • Ornithischia

| VITAL STATISTICS | |
|---|---|
| Fossil Location | Canada, USA |
| Diet | Herbivorous |
| Pronunciation | KAS-mo-SAWR-us |
| Weight | 3.6 tonnes (4 tons) |
| Length | 5–6m (16–20ft) |
| Height | 3m (10ft) |
| Meaning of name | 'Chasm lizard' because of the holes in its frill |

What was *Chasmosaurus*' vast triangular frill for? It would have looked fearsome, especially if the animals lined up facing attackers, but it was too fragile to offer any real protection.

**WHERE IN THE WORLD?**

Over a long swathe of land from Alberta, Canada, to Texas, USA.

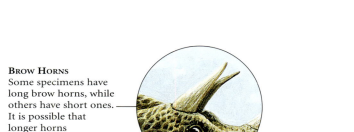

**BROW HORNS**
Some specimens have long brow horns, while others have short ones. It is possible that longer horns identified males.

**FOSSIL EVIDENCE**
*Chasmosaurus* was a common ceratopsian that roamed in herds. Its distinguishing feature was a huge neck frill about 1.4m (5ft) long and 1m (3ft) wide. Although it looked fearsome, the frill probably had little practical defensive use. It may have been more important for display and possibly thermo-regulation. *Chasmosaurus* had three facial horns, two large brow horns and one shorter, wider horn on the nose.

**FOSSILIZED SKIN**
Fossilized skin remains show that *Chasmosaurus* had evenly spaced, five- or six-sided bony knobs, each with a diameter of 5cm (2in).

**HOW BIG IS IT?**

DINOSAUR

LATE CRETACEOUS

**TIMELINE (millions of years ago)**

| 540 | 505 | 438 | 408 | 360 | 280 | 248 | 208 | 146 | 65 | 1.8 to today |

LATE CRETACEOUS

# Chirostenotes

• ORDER • Saurischia • FAMILY • Caenagnathidae • GENUS & SPECIES • *Chirostenotes pergracilis, C. elegans*

## VITAL STATISTICS

| | |
|---|---|
| FOSSIL LOCATION | Canada |
| DIET | Possibly omnivorous |
| PRONUNCIATION | KIE-roh-STEN-oh-teez |
| WEIGHT | 30kg (66lb) |
| LENGTH | 2m (7ft) |
| HEIGHT | 70cm (28in) |
| MEANING OF NAME | 'Narrow hand' after its unusual hands, the first part found |

*Chirostenotes* is the skeleton in the dinosaur family cupboard – literally! Only partial remains were known until a specimen that had been gathering dust for 60 years was rediscovered in 1988 and identified.

### WHERE IN THE WORLD?

Remains have been found in the Dinosaur Park Formation in Alberta, Canada.

### FOSSIL EVIDENCE

Although the hands, feet and jaws had been discovered (in 1924, 1932 and 1936 respectively), it took the reconstruction of the 'lost' skeleton in 1988 to prompt the realization that the partial remains were from the same beast. *Chirostenotes* was a lightly built bird-like theropod with three slender, clawed fingers on each hand, the middle one being bigger than the others. *Chirostenotes* would have been a quick, agile predator. Primarily a carnivore, it may have been omnivorous. Various species have been found; some were larger and heavier than the measurements given here.

DINOSAUR

LATE CRETACEOUS

**CREST**
The head was topped with a crest similar to that of the *Oviraptor*, which lived two million years later, suggesting that the two dinosaurs are related.

### HOW BIG IS IT?

**HANDS**
The three-fingered hands seem particularly well suited to grabbing fish, snatching eggs and prying insects out of bark.

**TIMELINE (millions of years ago)**

| 540 | 505 | 438 | 408 | 360 | 280 | 248 | 208 | 146 | 65 | 1.8 to today |

LATE CRETACEOUS

# Conchoraptor

• ORDER • Saurischia • FAMILY • Oviraptoridae • GENUS & SPECIES • *Conchoraptor gracilis*

**This was a primitive oviraptorid, a group of dinosaurs considered bird-like because of their general build, with shoulders featuring a wishbone, the presence of a beak, and other skeletal features.**

| VITAL STATISTICS | |
|---|---|
| FOSSIL LOCATION | Mongolia |
| DIET | Possibly omnivorous |
| PRONUNCIATION | KONK-oh-RAP-tor |
| WEIGHT | 6kg (13lb) |
| LENGTH | 1.5m (5ft) |
| HEIGHT | 50cm (20in) |
| MEANING OF NAME | 'Shellfish thief' because of its assumed mollusc diet |

### WHERE IN THE WORLD?

A few fragmentary finds have been made in the Nemegt Formation of Mongolia.

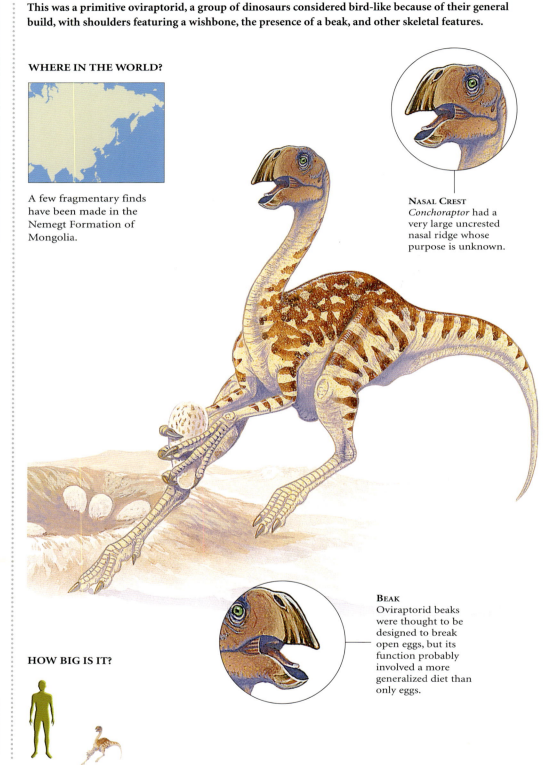

**NASAL CREST** *Conchoraptor* had a very large uncrested nasal ridge whose purpose is unknown.

**BEAK** Oviraptorid beaks were thought to be designed to break open eggs, but its function probably involved a more generalized diet than only eggs.

### FOSSIL EVIDENCE

Very few remains of this dinosaur have been found, but the anatomy of the hands has led to it being classified as its own genus. Some see it as a transitional 'missing link' between *Oviraptor* and the small theropod *Ingenia*. *Conchoraptor* lacks their large crests of other oviraptorids – this feature leading some palaeontologists to believe it is simply a juvenile *Oviraptor*. It had a powerful toothless jaw and bony bumps in its mouth, suggesting it could crush both eggs and molluscs, although it was probably omnivorous.

### HOW BIG IS IT?

DINOSAUR

LATE CRETACEOUS

**TIMELINE (millions of years ago)**

| 540 | 505 | 438 | 408 | 360 | 280 | 248 | 208 | 146 | 65 | 1.8 to today |

LATE CRETACEOUS

# Corythosaurus

• ORDER • Ornithischia • FAMILY • Hadrosauridae • GENUS & SPECIES • *Corythosaurus casuarius*

This was one of the most abundant duck-billed dinosaurs of western North America, best known for its splendid crest that not only looked impressive but may have helped it to make trumpeting calls.

### VITAL STATISTICS

| | |
|---|---|
| FOSSIL LOCATION | Canada, Western USA |
| DIET | Herbivorous |
| PRONUNCIATION | Co-RITH-oh-SAWR-us |
| WEIGHT | 5 tonnes (5.5 tons) |
| LENGTH | 9m (30ft) |
| HEIGHT | 2m (6.5ft) at the hips |
| MEANING OF NAME | 'Helmet lizard' because its crest is like those worn by Corinthian soldiers |

### WHERE IN THE WORLD?

Numerous remains have been found in Alberta, Canada, and Montana, USA.

### FOSSIL EVIDENCE

Around 20 skulls have been found, often in the same location, which suggests that *Corythosaurus* travelled in groups. The elaborate crest was not only decorative: it was hollow and contains nasal passages that may have allowed it to make loud, low-pitched resonating calls. The crest may also have helped with cooling, enhanced *Corythosaurus*' sense of smell, and played a role in its courtship rituals. Some skin remains have survived through natural mummification. These reveal a rough surface of tiny bumps with larger scales on the belly.

**EYES**
Large eyes on either side of the head allowed it to see in two different directions at once, helping it avoid predators.

**HIND LEGS**
*Corythosaurus* could stand on its hind legs and stretch its neck to pull food from high branches with its toothless beak. The food was then ground up by hundreds of cheek teeth.

### HOW BIG IS IT?

DINOSAUR

LATE CRETACEOUS

**TIMELINE** (millions of years ago)

| 540 | 505 | 438 | 408 | 360 | 280 | 248 | 208 | 146 | 65 | 1.8 to today |

## LATE CRETACEOUS

# Diceratus

• **ORDER** • Ornithischia • **FAMILY** • Ceratopsidae • **GENUS & SPECIES** • *Diceratus hatcheri*

| VITAL STATISTICS | |
|---|---|
| Fossil Location | USA |
| Diet | Herbivorous |
| Pronunciation | Die-SAIR-ah-tops |
| Weight | 11 tonnes (12.1 tons) |
| Length | Up to 9m (30ft) |
| Height | 2.7m (9ft) |
| Meaning of name | 'Two-horned' after the horns on its face |

You could forgive *Diceratus* for having an identity crisis. Considered to be a type of Triceratops for years, it was eventually recognized as a genus in its own right, but a bug had taken its original name.

### WHERE IN THE WORLD?

The only specimen was found in Wyoming, eastern USA.

**Frill**
Skin-covered holes in the frill would have reduced its overall weight and provided anchors for the large muscles needed to power the jaw.

### FOSSIL EVIDENCE

All descriptions of this creature are based on one poorly preserved skull with its lower jaw, discovered in 1905. This horned dinosaur was thought to be a very large type of Triceratops for decades, but there are significant differences. It has a rounded stump where the nose horn should be, and the brow horns are almost vertical, while the neck frills are structured differently. By the time this was understood, the original name of *Diceratops* had been given to a group of insects. The dinosaur was renamed *Diceratus* in 2008.

DINOSAUR

LATE CRETACEOUS

### HOW BIG IS IT?

**Legs**
*Diceratus* would have moved slowly on four strong legs, probably reaching only 2–4kmh (1–2.5mph) as it browsed low-lying plants.

**TIMELINE (millions of years ago)**

| 540 | 505 | 438 | 408 | 360 | 280 | 248 | 208 | 146 | 65 | 1.8 to today |

LATE CRETACEOUS

# Dravidosaurus

• **ORDER** • Plesiosauria  **FAMILY** • Unranked  **GENUS & SPECIES** • *Dravidosaurus blandfordi*

| VITAL STATISTICS | |
|---|---|
| FOSSIL LOCATION | India |
| DIET | Carnivorous |
| PRONUNCIATION | Druh-VID-oh-SAWR-us |
| WEIGHT | 900kg (2000lb) |
| LENGTH | 3m (10ft) |
| HEIGHT | 1.2m (4ft) |
| MEANING OF NAME | 'Dravidanadu lizard' after the region where it was found |

The trouble with identifying animals from isolated bones is that it is easy to get misled and go in the wrong direction. A jumble of fossils found in India illustrates this dinosaur dilemma.

**MISTAKEN IDENTITY**
When *Dravidosaurus* was thought to be a stegosaur, it was illustrated like the one seen here. Now the scrappy remains seem to be from a plesiosaur, much like *Libonectes* on p330.

## FOSSIL EVIDENCE
The finding of a partial skull, a tooth, some thin, triangular armour plates, a spike and a few other bones led some palaeontologists to the conclusion that they had found the last of the stegosaurs. Even more remarkably, it was living in India, where no other stegosaurs are known, tens of millions of years after its fellow plated dinosaurs died out. This view was revised and *Dravidosaurus* is now classified as a long-necked swimming reptile – and the armoured plates are thought to be poorly preserved parts of a limb. The issue remains unresolved.

**ARMOURED PLATES**
What were once interpreted as the plates of a terrestrial stegosaur are now considered more likely the heavily weathered bones of a marine reptile.

**HOW BIG IS IT?**

PREHISTORIC ANIMAL
LATE CRETACEOUS

**WHERE IN THE WORLD?**

The remains were found in the province of Tamil Nadu in southern India.

**TIMELINE (millions of years ago)**

| 540 | 505 | 438 | 408 | 360 | 280 | 248 | 208 | 146 | 65 | 1.8 to today |

LATE CRETACEOUS

# Dromaeosaurus

• ORDER • Saurischia • FAMILY • Dromaeosauridae • GENUS & SPECIES • Several species within the genus *Dromaeosaurus*

**This was probably an intelligent, quick and agile wolf-sized predator with keen senses and vicious teeth and claws. It may well have hunted in packs to take on much larger prey.**

| VITAL STATISTICS | |
|---|---|
| Fossil Location | Canada, USA |
| Diet | Carnivorous |
| Pronunciation | DROH-mee-oh-SAWR-us |
| Weight | 15kg (33lb) |
| Length | 1.8m (6ft) |
| Height | 70cm (2ft) |
| Meaning of name | 'Running lizard' because of its swiftness |

**FOSSIL EVIDENCE**

*Dromaeosaurus* gave its name to a family of dinosaurs, but the remains are scant (just a partial skull and a few foot bones) and much of the data on it is taken from its relatives. It walked, ran and leapt on powerful hind legs, dispatching prey with a sickle-like toe claw and a strong jaw equipped with backwards-curving teeth. Its skull was larger than later dromaeosaurids and its teeth were more robust – better for gripping than ripping. It may well have been partly feathered.

**TEETH**
Remains of the serrated teeth show heavy wear, suggesting they wore down on the bones of its prey.

**STIFFENED TAIL**
The tail was stiffened by a network of bony rods but was flexible enough to bend somewhat side to side.

**HOW BIG IS IT?**

**WHERE IN THE WORLD?**

First found in the Judith River Formation of Canada, with scant further remains in Montana, USA.

DINOSAUR
LATE CRETACEOUS

**TIMELINE (millions of years ago)**

540 | 505 | 438 | 408 | 360 | 280 | 248 | 208 | 146 | 65 | 1.8 to today

LATE CRETACEOUS

# Dryptosaurus

• ORDER • Saurischia • FAMILY • Dryptosauridae • GENUS & SPECIES • *Dryptosaurus aquilunguis*

### VITAL STATISTICS

| | |
|---|---|
| FOSSIL LOCATION | USA |
| DIET | Carnivorous |
| PRONUNCIATION | DRIP-to-SAW-rus |
| WEIGHT | Over 1 tonne (just under 1 ton) |
| LENGTH | 4.5–6m (15–20ft) |
| HEIGHT | 1.8m (6ft) at the hips |
| MEANING OF NAME | 'Tearing lizard' because of its large claws |

### FOSSIL EVIDENCE

*Dryptosaurus* was a primitive tyrannosauroid. The best specimen, found almost 150 years ago, consists of a partial lower jaw with teeth, two humeri, a left leg minus foot, vertebrae and fragments. It was first classified as megalosaurid and later assigned to its own family, the Dryptosauridae. Later yet it was found to be a coelurosaur, a very broad group in which its exact placement remained uncertain. A recent discovery in Alabama of the more complete *Appalachiosaurus* made it clear that closely related *Dryptosaurus* was a primitive tyrannosauroid like its southern cousin.

DINOSAUR

LATE CRETACEOUS

The New Jersey Devil may be a myth, but New Jersey was once the home to medium-sized tyrannosauroid with tearing claws and ripping teeth.

Very little is known of this animal's skeleton, but being a tyrannosauroid means it had a large, toothy skull and that it was a bipedal hunter. It had powerful hand and foot claws and a long tail held aloft to balance the front part of the body. It probably only had two fingers on each hand, like other late tyrannosaurs.

**TEETH**
Teeth of carnivorous dinosaurs, like those of most reptiles, are not good for separating species, but their serrated design clearly shows their predatory purpose.

**CLAWS**
*Dryptosaurus* had enormous, curved hand claws, for which its species name *aquilunguis* ('eagle claw') is aptly descriptive.

**HOW BIG IS IT?**

### WHERE IN THE WORLD?

The only good specimen of *Dryptosaurus* was found in New Jersey in 1866, but individual teeth found subsequently in North Carolina suggest that this genus was more widespread in Eastern North America.

### DID YOU KNOW?
This drawing shows *Dryptosaurus* killing a pachycephalosaur, a group as yet unknown from rocks with *Dryptosaurus* fossils. But pachycephalosaurs are commonly found with other tyrannosauroids around the world, so fossils may turn up some day.

### TIMELINE (millions of years ago)

| 540 | 505 | 438 | 408 | 360 | 280 | 248 | 208 | 146 | 65 | 1.8 to today |

LATE CRETACEOUS

# Edmontosaurus

• ORDER • Ornithischia • FAMILY • Hadrosauridae • GENUS & SPECIES • *Edmontosaurus regalis*

| VITAL STATISTICS | |
|---|---|
| FOSSIL LOCATION | Western North America |
| DIET | Herbivorous |
| PRONUNCIATION | Ed-MON-toh-SAWR-us |
| WEIGHT | 3.9 tonnes (3.5 tons) |
| LENGTH | 13m (43ft) |
| HEIGHT | Unknown |
| MEANING OF NAME | 'Edmonton lizard' after the Edmonton Rock Formation in Canada, where its fossils were found |

*Edmontosaurus* had a skull like a duck's: flat and broadening out into a beak. Some have suggested that it had loose flaps of skin on its face, which it may have inflated to generate bellowing calls.

**BEAK**
Its beak was toothless, but it had 60 rows of cheek teeth holding about 1000 teeth. It chewed by grinding food against its tooth batteries.

### WHERE IN THE WORLD?

*Edmontosaurus* was found in what is now Western North America.

### FOSSIL EVIDENCE

Some specimens of *Edmontosaurus* include skin impressions, which show that its skin was scaly and leathery, with tubercles, or tiny bumps, along its neck, down its back and along its tail. Its stomach contents have also been fossilized, indicating that it ate the needles, seeds and twigs of conifers. One specimen, displayed at the Denver Museum of Nature and Science (USA), has bite marks at the top of its tail. The only dinosaur large enough to have launched such an attack on *Edmontosaurus* is *Tyrannosaurus rex*.

DINOSAUR

LATE CRETACEOUS

**HOW BIG IS IT?**

**FRONT LEGS**
*Edmontosaurus* could probably walk both bipedally and quadrupedally: its front limbs had hooves on two digits and weight-bearing pads.

**TIMELINE (millions of years ago)**

| 540 | 505 | 438 | 408 | 360 | 280 | 248 | 208 | 146 | 65 | 1.8 to today |

# Einiosaurus

• ORDER • Ornithischia • FAMILY • Ceratopsidae • GENUS & SPECIES • *Einiosaurus procurvicornis*

## VITAL STATISTICS

| | |
|---|---|
| FOSSIL LOCATION | USA |
| DIET | Herbivorous |
| PRONUNCIATION | Eye-nee-o-SAWR-us |
| WEIGHT | 2–2.2 tonnes (1.8–2 tons) |
| LENGTH | 5–6m (16–20ft) |
| HEIGHT | Unknown |
| MEANING OF NAME | 'Bison lizard' named to honour the Black Feet tribe, on whose land the fossils were found and refer to the idea that ceratopsians were the "buffaloes" of the Cretaceous |

Setting *Einiosaurus* apart from other horned dinosaurs is its forward-curving nose horn, often likened to a can opener. It looked intimidating, but *Einiosaurus's* best defence was that it probably lived in herds.

**SPIKES**
Two spikes at the top of its bony frill intimidated predators, though they probably offered little practical defence.

### WHERE IN THE WORLD?

To date, *Einiosaurus* has been located only in Montana, USA.

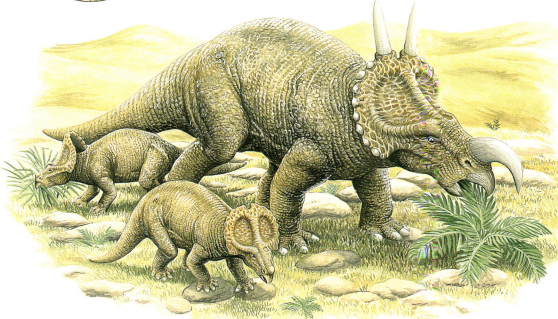

**LEGS**
It is not clear whether *Einiosaurus* stood upright, like an elephant, or adopted a more sprawling posture, like a lizard.

## FOSSIL EVIDENCE

*Einiosaurus* was discovered comparatively recently, in 1985, prompting an excavation that was not completed until 1989. Its formal description came in 1995. The discovery was made in the Two Medicine Formation, Montana, a deposit rich in dinosaur specimens and one of the most important in the world. The evidence suggests that *Einiosaurus* lived in a semi-arid climate, experiencing a long dry season. Three skulls have been found, suggesting that the distinctive nose horn was a feature only of mature adults.

DINOSAUR
LATE CRETACEOUS

**HOW BIG IS IT?**

**TIMELINE (millions of years ago)**

| 540 | 505 | 438 | 408 | 360 | 280 | 248 | 208 | 146 | 65 | 1.8 to today |

LATE CRETACEOUS

# Elasmosaurus

• **ORDER** • Plesiosauria • **FAMILY** • Elasmosauridae
• **GENUS & SPECIES** • Several species within the genus *Elasmosaurus*

| VITAL STATISTICS | |
|---|---|
| Fossil Location | Asia, North America |
| Diet | Carnivorous |
| Pronunciation | Eh-LAZZ-mo-SAWR-us |
| Weight | 4.4 tonnes (4 tons) |
| Length | 14m (46ft) |
| Height | Unknown |
| Meaning of name | 'Plate lizard' in reference to its plate-like shoulder bones |

### WHERE IN THE WORLD?

*Elasmosaurus* was found in Japan and in the inland sea covering the western part of North America.

Distinguished by its long neck, *Elasmosaurus* had up to 75 vertebrae. (By contrast, most modern mammals have between seven and eight.) Living in the open water, it breathed air, the way modern dolphins do.

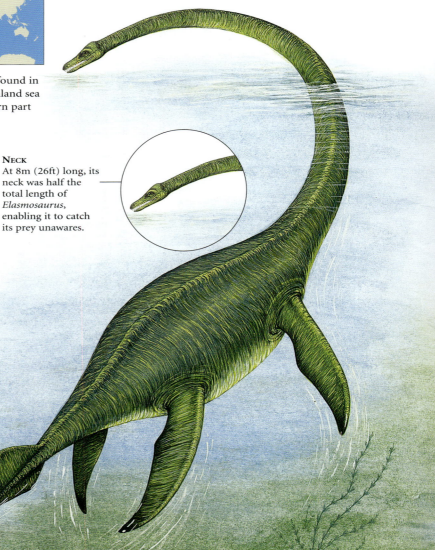

**Neck**
At 8m (26ft) long, its neck was half the total length of *Elasmosaurus*, enabling it to catch its prey unawares.

### FOSSIL EVIDENCE

*Elasmosaurus* was first reconstructed in 1868 by the palaeontologist Edward Cope, who mistakenly identified its long neck as a tail. The weight of the neck is such that *Elasmosaurus* could probably only lift its head above the water. Combined with the evidence of its four rigid flippers, this suggests that *Elasmosaurus* lived on the open seas, never coming to land. In water, it could move only slowly.

PREHISTORIC ANIMAL

LATE CRETACEOUS

### HOW BIG IS IT?

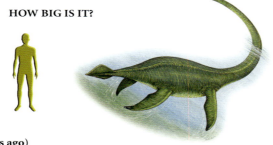

**Head**
Its small head limited what it could swallow, and *Elasmosaurus* probably survived on small fish, squid and ammonites.

**TIMELINE (millions of years ago)**

| 540 | 505 | 438 | 408 | 360 | 280 | 248 | 208 | 146 | 65 | 1.8 to today |

LATE CRETACEOUS

# Elmisaurus

• ORDER • Saurischia  **FAMILY** • Caenagnathidae  **GENUS & SPECIES** • *Elmisaurus rarus*

## VITAL STATISTICS

| | |
|---|---|
| FOSSIL LOCATION | Canada, Mongolia |
| DIET | Carnivorous |
| PRONUNCIATION | ELM-ee-SAWR-us |
| WEIGHT | 32kg (71lb) |
| LENGTH | 2m (7ft) |
| HEIGHT | Unknown |
| MEANING OF NAME | 'Hindfoot lizard' because of the peculiar fusing of the bones of its hindfoot |

*Elmisaurus* walked on two legs, and had hands with three digits on long arms. A member of the oviraptor group, it probably had a short, parrot-like skull and may have been feathered.

**ARMS**
*Elmisaurus* had arms that were longer and hands that were thinner than its relatives, the coelurosaurs.

### WHERE IN THE WORLD?

To date, *Elmisaurus* has been located only in Mongolia and Canada.

**HEAD**
No skull is yet known, but recent analysis suggests that it was probably more like the one in the *Ingenia* illustration on p273.

## FOSSIL EVIDENCE

*Elmisaurus* was discovered in Mongolia, and subsequent finds in North America provide evidence that non-avian dinosaurs migrated. Its bones are so thin that they have not been well preserved in the fossil record, and only its hands, feet and leg bones have been discovered. Elongated feet and shins suggest that it ran swiftly, while it probably used the three fingers on its clawed hands to grab insects and small animals. Similarities to the fossils of other dinosaurs enable palaeontologists to speculate on its general appearance.

DINOSAUR
LATE CRETACEOUS

**HOW BIG IS IT?**

**TIMELINE (millions of years ago)**

| 540 | 505 | 438 | 408 | 360 | 280 | 248 | 208 | 146 | 65 | 1.8 to today |

263

LATE CRETACEOUS

# Erlikosaurus

• ORDER • Saurischia • FAMILY • Therizinosauridae • GENUS & SPECIES • *Erlikosaurus andrewsi*

*Erlikosaurus* was a therizinosaur, an unusual family of dinosaurs, its members characterized by large claws on their hands. What these were used for is unclear.

| VITAL STATISTICS | |
|---|---|
| Fossil Location | Mongolia |
| Diet | Possibly omnivorous |
| Pronunciation | ER-lik-oh-SAWR-us |
| Weight | 160kg (353lb) |
| Length | 5–6m (16–20ft) |
| Height | Unknown |
| Meaning of name | 'Erlik's lizard' after Erlik, the king of the dead, from a Buddhist sect in Mongolia |

**Mouth**
A toothless beak and cheek teeth, which were small and leaf-shaped, suggest that *Erlikosaurus* was a herbivore.

**Hands**
Three fingers with long claws suggest that *Erlikosaurus* was a predator. Certainly they are suitable for grasping and slashing.

## FOSSIL EVIDENCE

*Erlikosaurus* was the first therizinosaur for which a skull was found, but it is otherwise known only from fragmentary specimens. Its tail and legs were short, which do not seem suitable for a predator. It may have been herbivorous, reaching up to grab vegetation in its claws. Comparison with its relatives, including the discovery of a pelvis possibly shaped for sitting and vertebrae designed to hold its body in an upright position, suggest that therizinosaurs as a group evolved from predators into herbivores.

DINOSAUR
LATE CRETACEOUS

### HOW BIG IS IT?

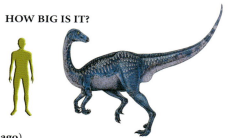

### WHERE IN THE WORLD?

To date, *Erlikosaurus* has been located only in Mongolia.

**TIMELINE (millions of years ago)**

| 540 | 505 | 438 | 408 | 360 | 280 | 248 | 208 | 146 | 65 | 1.8 to today |

264

LATE CRETACEOUS

# Euoplocephalus

• ORDER • Ornithischia • FAMILY • Ankylosauridae • GENUS & SPECIES • *Euoplocephalus tutus*

| VITAL STATISTICS | |
|---|---|
| FOSSIL LOCATION | Canada, USA |
| DIET | Herbivorous |
| PRONUNCIATION | YOU-oh-plo-SEF-ah-lus |
| WEIGHT | 2–3.3 tonnes (1.8–3 tons) |
| LENGTH | 6m (20ft) |
| HEIGHT | Unknown |
| MEANING OF NAME | 'Well armoured head' after the armoured plates on its skull |

Its entire head and body were covered with armour that was studded with spikes, and *Euoplocephalus* even had bony shutters that slid over its eyes for protection. Horns projected from the back of its skull.

### WHERE IN THE WORLD?

*Euoplocephalus* was found in North America, specifically Alberta, Canada and Montana, USA.

### FOSSIL EVIDENCE

*Euoplocephalus* has been found in several locations, making it one of the most common dinosaurs found in North America. Only isolated specimens have been found, suggesting that it lived a solitary life, although there is fossil evidence that some of its relatives lived in herds. The armour plating running across its back offered *Euoplocephalus* good protection. Indeed, it was probably vulnerable only if flipped over on its back: an analysis of dinosaur bones in Alberta, Canada, reveal no bite marks on *Euoplocephalus* or any of its armoured relatives.

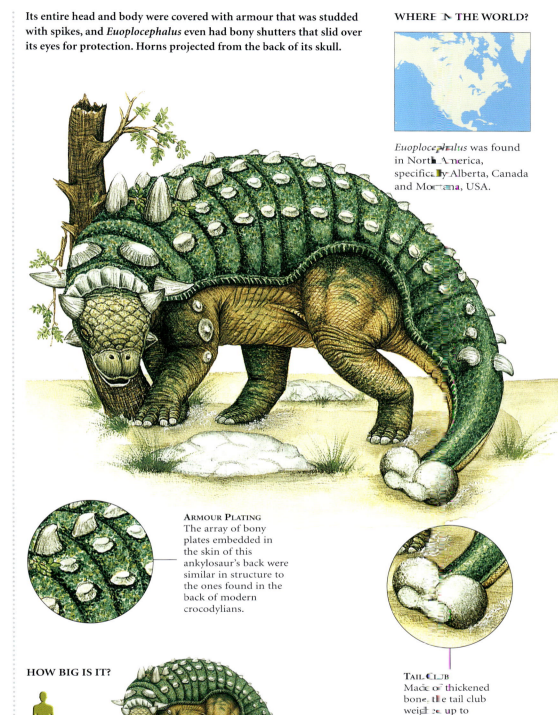

**ARMOUR PLATING**
The array of bony plates embedded in the skin of this ankylosaur's back were similar in structure to the ones found in the back of modern crocodylians.

**TAIL CLUB**
Made of thickened bone, the tail club weighed up to 20kg (44lb) and could deliver a powerful blow.

### HOW BIG IS IT?

DINOSAUR
LATE CRETACEOUS

**TIMELINE (millions of years ago)**

| 540 | 505 | 438 | 408 | 360 | 280 | 248 | 208 | 146 | 65 | 1.8 to today |

LATE CRETACEOUS

# Garudimimus

• ORDER • Saurischia • FAMILY • Garudimidae • GENUS & SPECIES • *Garudimimus brevis*

*Garudimimus* was a primitive ornithomimid, or ostrich mimic dinosaur. It had large powerful eyes, and may have relied on these to detect predators; it lacks obvious defences.

| VITAL STATISTICS | |
|---|---|
| FOSSIL LOCATION | Mongolia |
| DIET | Possibly omnivorous |
| PRONUNCIATION | Ga-ROOD-uh-MIME-us |
| WEIGHT | 85kg (187lb) |
| LENGTH | 3.5–4m (11–13ft) |
| HEIGHT | Unknown |
| MEANING OF NAME | 'Garuda mimic' after Garuda, a monstrous bird from Asian mythology |

**WHERE IN THE WORLD?**

To date, *Garudimimus* has been located only in Mongolia.

**MOUTH**
A toothless beak suggests a diet of plants, but sharp cheek teeth suggest *Garudimimus* may also have eaten insects and small mammals.

## FOSSIL EVIDENCE

*Garudimimus* is known only from incomplete fragments. It was originally thought that *Garudimimus* had a horn, but further analysis suggests that this was simply a misidentified bone fragment. Like a modern ostrich, it had an upright stance and long legs. Fossilized tracks of its relatives suggest that they could run at speeds of 35kmh (22mph). *Garudimimus* itself may not have been so swift: its legs seem less well-adapted for running, as were its feet, which had a small first toe that is not found in later relatives.

**HOW BIG IS IT?**

DINOSAUR

LATE CRETACEOUS

**HANDS**
Its hands were not adapted for grasping, but may have been useful for digging, presumably in search of prey.

**TIMELINE (millions of years ago)**

| 540 | 505 | 438 | 408 | 360 | 280 | 248 | 208 | 146 | 65 | 1.8 to today |

# Goyocephale

• ORDER • Ornithischia • FAMILY • Pachycephalosauridae • GENUS & SPECIES • *Goyocephale lattimorei*

LATE CRETACEOUS

## VITAL STATISTICS

| | |
|---|---|
| FOSSIL LOCATION | Mongolia |
| DIET | Herbivorous |
| PRONUNCIATION | GOH-yoh-SEF-ah-lee |
| WEIGHT | 47kg (15lb) |
| LENGTH | 2–3m (7–10ft) |
| HEIGHT | Unknown |
| MEANING OF NAME | 'Decorated head' in reference to the spikes and knobs on its head |

*Goyocephale* was a swift runner, speed being its best defence. Prominent canine-like teeth may have intimidated potential predators, and some protection might also have come from gathering in herds.

### WHERE IN THE WORLD?

*Goyocephale* was located in what is now the Gobi Desert, Mongolia.

### HEAD
The surface of its head was rough and bumpy, and a bony shelf jutted from the back of its flat skull.

## FOSSIL EVIDENCE

In the 1960s, the Joint Polish–Mongolian Expeditions to the Gobi Desert found many new dinosaurs. Among them was *Goyocephale*, which was not formally described until 1982. It is one of the best known pachycephalosaurs: not just its thick skull has been found but also its limbs and tail. It was lightly built, but its vertebrae are reinforced by bony tendons. These may have been of particular advantage to adult males, reducing the stress of the head-butting contests they likely fought in order to establish dominance.

### TAIL
Extending its stiffened tail for balance, the light-weight *Goyocephale* ran swiftly on its slim hind legs.

### HOW BIG IS IT?

DINOSAUR

LATE CRETACEOUS

**TIMELINE (millions of years ago)**

| 540 | 505 | 438 | 408 | 360 | 280 | 248 | 208 | 146 | 65 | 1.8 to today |

LATE CRETACEOUS

# Hadrosaurus

• **ORDER** • Ornithischia • **FAMILY** • Hadrosauridae • **GENUS & SPECIES** • *Hadrosaurus foulkii*

*Hadrosaurus* possibly lived in herds, moving through the conifer forests and swamps near the coast. It may have been a good swimmer, venturing quite far from shore, or remains of it found in marine sediments may have washed out from rivers.

## VITAL STATISTICS

| | |
|---|---|
| Fossil Location | USA |
| Diet | Herbivorous |
| Pronunciation | HAD-ro-SAWR-us |
| Weight | 2–3 tonnes (1.9–2.7 tons) |
| Length | 7–10m (23–33ft) |
| Height | Unknown |
| Meaning of name | 'Bulky lizard' in reference to its powerful build |

### WHERE IN THE WORLD?

*Hadrosaurus* lived along the northeastern coast of modern North America.

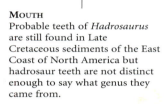

**Mouth**
Probable teeth of *Hadrosaurus* are still found in Late Cretaceous sediments of the East Coast of North America but hadrosaur teeth are not distinct enough to say what genus they came from.

## FOSSIL EVIDENCE

*Hadrosaurus* was the first duck-billed dinosaur to be found. It was also the first almost complete non-avian dinosaur skeleton to be discovered in the world; large bones were first excavated in 1838, but it was not until 1858 that the full skeleton was dug up. (Only the head is missing, and indeed the skull of *Hadrosaurus* has not been discovered to this day.) In 1868, *Hadrosaurus* became the first dinosaur skeleton to be put on display. It caused a sensation, offering to the public proof of the existence of non-avian dinosaurs.

HOW BIG IS IT?

**Hind Legs**
Hind legs significantly longer than its front legs suggest that *Hadrosaurus* walked upright, using its front legs for support only while feeding.

DINOSAUR

LATE CRETACEOUS

**TIMELINE (millions of years ago)**

| 540 | 505 | 438 | 408 | 360 | 280 | 248 | 208 | 146 | 65 | 1.8 to today |

LATE CRETACEOUS

# Homalocephale

• ORDER • Ornithischia • FAMILY • Homalocephalidae • GENUS & SPECIES • *Homalocephale calathocercos*

*Homalocephale* was a boneheaded dinosaur, but unlike its relatives had a flat-topped head. Fossil specimens found with very worn teeth suggest that this herbivore fed on tough vegetation.

## VITAL STATISTICS

| | |
|---|---|
| FOSSIL LOCATION | Mongolia |
| DIET | Herbivorous |
| PRONUNCIATION | HOME-ah-low-SEF-ah-lee |
| WEIGHT | Unknown |
| LENGTH | 3m (10ft) |
| HEIGHT | Unknown |
| MEANING OF NAME | 'Level head' because of its flat head |

**HEAD**
*Homalocephale's* head was flatter than many of its relatives in the bonehead group, and was covered with bumps and knobs, perhaps to attract a mate.

### WHERE IN THE WORLD?

To date, *Homalocephale* has been located only in Mongolia.

## FOSSIL EVIDENCE

A bone-headed dinosaur, *Homalocephale* probably did not head-butt its opponents as its thick-headed cousins likely did. Analysis of the skull shows that it was not solid but porous and fragile. The most it probably did, according to palaeontologist Mark Goodwin, was to engage in head-shoving contests. *Homalocephale* also had an unusually broad pelvis, which may have allowed it to brace itself to withstand such pushing. Alternatively, some palaeontologists believe that the wide hips allowed females to give birth to live young.

DINOSAUR

LATE CRETACEOUS

### HOW BIG IS IT?

**NOSE**
Analysis of its skull shows a large area for its olfactory nerve, suggesting that *Homalocephale* had a keen sense of smell.

**TIMELINE (millions of years ago)**

| 540 | 505 | 438 | 408 | 360 | 280 | 248 | 208 | 146 | 65 | 1.8 to today |

LATE CRETACEOUS

# Hypacrosaurus

• **ORDER** • Ornithischia • **FAMILY** • Hadrosauridae • **GENUS & SPECIES** • *Hypacrosaurus altispinus*

| VITAL STATISTICS | |
|---|---|
| FOSSIL LOCATION | Canada, USA |
| DIET | Herbivorous |
| PRONUNCIATION | Hi-PACK-roe-SAWR-us |
| WEIGHT | 1.4 tonnes (1.3 tons) |
| LENGTH | 9m (30ft) |
| HEIGHT | Unknown |
| MEANING OF NAME | 'Almost the top lizard' because it was almost as large as *Tyrannosaurus rex* |

*Hypacrosaurus* foraged on all four legs, but walked and ran on two legs.

**CREST**
A hollow crest may have acted as a resonating chamber, which would have enabled *Hypacrosaurus* to make sounds. Alternatively, it may have enhanced its ability to smell.

**WHERE IN THE WORLD?**

*Hypacrosaurus* was found in the west of North America, in what is now Alberta, Canada, and Montana, USA.

**FOSSIL EVIDENCE**
The first specimen of *Hypacrosaurus* was found near Alberta, Canada, in 1912. Several decades later, in the 1990s, a deposit in Devil's Coulee, in southwestern Alberta, revealed much information about young *Hypacrosaurus*. A nest was found, and inside, lined up in rows, were eight eggs, each the size of a cantaloupe. It is possible that the nest was originally covered over with sand or plant matter. Hatchlings were found nearby. This suggests *Hypacrosaurus* created nesting sites, in which multiple nests could be guarded.

**HOW BIG IS IT?**

**MOUTH**
Forty rows of tightly packed cheek teeth allowed it to grind up its food, which may have included pine needles, fruit, twigs and possibly even flowering plants.

DINOSAUR

LATE CRETACEOUS

**TIMELINE** (millions of years ago)

| 540 | 505 | 438 | 408 | 360 | 280 | 248 | 208 | 146 | 65 | 1.8 to today) |

# LATE CRETACEOUS

# Hypselosaurus

• **ORDER** • Saurischia • **FAMILY** • Titanosauridae • **GENUS & SPECIES** • *Hypselosaurus priscus*

| VITAL STATISTICS | |
|---|---|
| FOSSIL LOCATION | France, Spain |
| DIET | Herbivorous |
| PRONUNCIATION | HIP-sel-oh-SAWR-us |
| WEIGHT | 9.9 tonnes (9 tons) |
| LENGTH | 12m (39ft) |
| HEIGHT | Unknown |
| MEANING OF NAME | 'High lizard' in reference to its height and long limbs |

Palaeontologists are still not sure what *Hypselosaurus* looked like. It was certainly small by comparison with other sauropods, and its legs were unusually thick. It may have had some form of armour.

**NECK**
The reason for its remarkably long neck is unclear, but it may have enabled *Hypselosaurus* to reach plant life unavailable to others.

### WHERE IN THE WORLD?

*Hypselosaurus* was widespread across Western Europe, in what is now Spain and France.

### FOSSIL EVIDENCE

The first non-avian dinosaur eggs ever found were probably those of *Hypselosaurus*. However, there are some palaeontologists who argue that the eggs belong to the flightless bird *Gargantuavis*. The bumpy-surfaced eggs were 30cm (12in) long and had a volume of 2 litres (half a gallon). At about twice the size of a modern ostrich egg, this was unusually large. The eggs were found in a crater-like nest, set up in a line. Did *Hypselosaurus* carefully nudge the eggs into line after laying, or did it lay its eggs while walking?

DINOSAUR
LATE CRETACEOUS

**HOW BIG IS IT?**

**TEETH**
*Hypselosaurus* had small, peg-like teeth used for cropping the plant material, but not for chewing since the jaws were not designed for it.

**TIMELINE** (millions of years ago)

| 540 | 505 | 438 | 408 | 360 | 280 | 248 | 208 | 146 | 65 | 1.8 to today |

271

LATE CRETACEOUS

# Indosuchus

• **ORDER** • Saurischia • **FAMILY** • Abelisauridae • **GENUS & SPECIES** • *Indosuchus raptorius*

| VITAL STATISTICS | |
|---|---|
| Fossil Location | India |
| Diet | Carnivorous |
| Pronunciation | In-doh-SOOK-us |
| Weight | 1.1 tonnes (1 ton) |
| Length | 6m (20ft) |
| Height | Unknown |
| Meaning of name | 'Indian crocodile' |

Little is known about it, but Indosuchus walked on two legs and may have been up to 6m (20ft) long. It had a narrow crest on its skull, flattened on top.

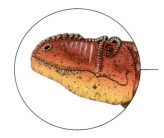

**Teeth**
Its serrated teeth were continually replaced; this was a necessity because teeth could be lost while tearing into flesh or crunching bones.

**WHERE IN THE WORLD?**

To date, *Indosuchus* has been located in Madhya Pradesh, India.

**Arms**
Its arms were noticeably short, but probably helped hold its victims steady as it sunk in its claws.

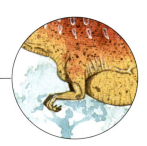

**FOSSIL EVIDENCE**
Only fragments of its skull and skeleton have been found, making Indosuchus a difficult dinosaur to clarify. Careful analysis of the fossil specimens show a likeness to *Abelisaurus*, a dinosaur for which some skull fragments were found in 1985 in Argentina. Both were predators, with large jaws filled with serrated teeth. The head resembles that of *Tyrannosaurus rex*, although Indosuchus had more teeth, and shorter ones. Presumably one of the more dominant predators, Indosuchus may also have been a scavenger.

DINOSAUR
LATE CRETACEOUS

**HOW BIG IS IT?**

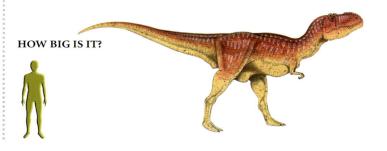

**TIMELINE (millions of years ago)**

| 540 | 505 | 438 | 408 | 360 | 280 | 248 | 208 | 146 | 65 | 1.8 to today |

# Ingenia

• ORDER • Saurischia • FAMILY • Oviraptoridae • GENUS & SPECIES • *Ingenia yanshini*

The light-weight *Ingenia* had powerful legs, making it an agile and swift runner. Its strong hands had three short fingers with small claws, useful perhaps for clutching branches while it ate leaves.

## VITAL STATISTICS

| | |
|---|---|
| FOSSIL LOCATION | Mongolia |
| DIET | Possibly omnivorous |
| PRONUNCIATION | In-JEE-ne-ah |
| WEIGHT | 40kg (88lb) |
| LENGTH | 1.5–2m (5–7ft) |
| HEIGHT | Unknown |
| MEANING OF NAME | Named after the Ingeni-Khobur depression in Mongolia where it was found |

### WHERE IN THE WORLD?

To date, *Ingenia* has been found only in southwestern Mongolia.

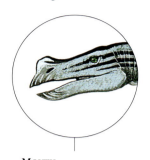

**MOUTH**
Its large toothless beak may have enabled it to crush eggs or chew small bones, but some believe that *Ingenia's* diet included leaves.

**SKULL**
Its small skull (10.5cm/4in long) was smaller than that of its relative *Oviraptor*.

## FOSSIL EVIDENCE

*Ingenia* is a member of the oviraptorid family – a name now shown to be inappropriate. The first oviraptorid was found alongside eggs that were misidentified as belonging to *Protoceratops*, and was assumed to have died while stealing the eggs. The eggs were later shown to be its own, suggesting that oviraptorids had parental instincts. The fossil record for other oviraptorids includes nests and eggs, suggesting that *Ingenia* may have brooded like modern birds. Twelve pairs of eggs were found arranged in three layers.

DINOSAUR
LATE CRETACEOUS

### HOW BIG IS IT?

**TIMELINE (millions of years ago)**

LATE CRETACEOUS

# Jaxartosaurus

• ORDER • Ornithischia • FAMILY • Hadrosauridae • GENUS & SPECIES • *Jaxartosaurus aralensis*

*Jaxartosaurus* was a herbivore that cropped plants with its toothless beak, chewing them with large flat-topped cheek teeth. Adults had crests, though the female's was probably smaller.

## VITAL STATISTICS

| | |
|---|---|
| FOSSIL LOCATION | Kazakhstan, China |
| DIET | Herbivorous |
| PRONUNCIATION | Jax-SAHR-toh-SAWR-us |
| WEIGHT | Unknown |
| LENGTH | 8–9m (26–30ft) |
| HEIGHT | Unknown |
| MEANING OF NAME | 'Jaxartes lizard' after Jaxartes, the ancient name for the Syr Darya River near where the fossil was found in Kazakhstan |

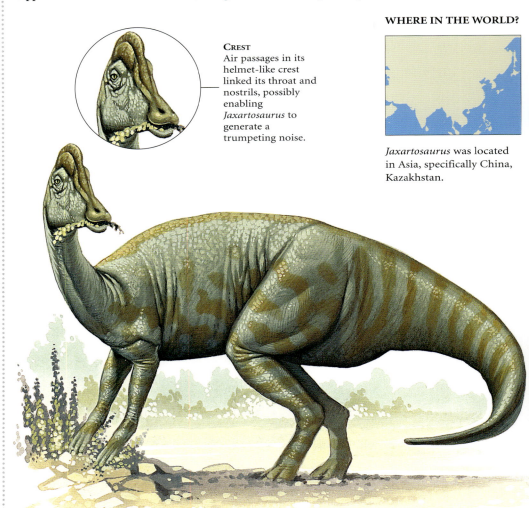

**CREST**
Air passages in its helmet-like crest linked its throat and nostrils, possibly enabling *Jaxartosaurus* to generate a trumpeting noise.

**WHERE IN THE WORLD?**

*Jaxartosaurus* was located in Asia, specifically China, Kazakhstan.

**LEGS**
*Jaxartosaurus* possibly fed on all fours, but perhaps when it sensed danger it rose onto its hind legs to run, stiffening its tail for balance.

## FOSSIL EVIDENCE

Little is known about *Jaxartosaurus*. It is known only from partial fragments, including the skull roof and braincase. In common with other duck-billed dinosaurs, it had a series of hinges and joints in its skull, which may have helped to protect its brain – and also its teeth – stopping them from breaking under impact. One joint is similar to that found on a modern crocodile, which acts as a shock absorber when the jaws snap shut. Fossilized teeth show extensive wear, evidence of constant chewing.

DINOSAUR
LATE CRETACEOUS

**HOW BIG IS IT?**

**TIMELINE (millions of years ago)**

| 540 | 505 | 438 | 408 | 360 | 280 | 248 | 208 | 146 | 65 | 1.8 to today |

LATE CRETACEOUS

# Leptoceratops

• ORDER • Ornithischia • FAMILY • Leptoceratopsidae • GENUS & SPECIES • *Leptoceratops gracilis*

## VITAL STATISTICS

| | |
|---|---|
| FOSSIL LOCATION | Canada, western USA |
| DIET | Herbivorous |
| PRONUNCIATION | LEP-to-SAIR-ah-tops |
| WEIGHT | 68kg (150lb) |
| LENGTH | 1.8m (6ft) |
| HEIGHT | Unknown |
| MEANING OF NAME | 'Slender-horned face' |

*Leptoceratops* was a primitive horned dinosaur, but it lived alongside its larger relatives. Its small size probably allowed it to feed on vegetation that the larger herbivores could not reach.

## WHERE IN THE WORLD?

To date, *Leptoceratops* has been discovered in Alberta, Canada, and Wyoming, USA.

**BEAK**
To defend itself against predators, *Leptoceratops* used its parrot-like beak, which could snap the arm of *Troodon* or a young Tyrannosaur.

## FOSSIL EVIDENCE

The first fossils of *Leptoceratops* were found in 1910, in the badlands near Alberta. Part of the skeleton had been eroded, but even so, the skeletons of two individuals were eventually excavated. *Leptoceratops* had a large head, and several skulls have been found. Its front legs are shorter than its hind legs, suggesting that it could stand – even walk – upright. Lacking few obvious natural defences, *Leptoceratops* was vulnerable, so it is no surprise that bonebeds have been found, suggesting that it travelled in herds.

**FRILL**
*Leptoceratops* had a short frill around its neck, which was probably too short to offer any protection.

**HOW BIG IS IT?**

DINOSAUR

LATE CRETACEOUS

**TIMELINE (millions of years ago)**

LATE CRETACEOUS

# Magyarosaurus

• ORDER • Saurischia • FAMILY • Unranked • GENUS & SPECIES • *Magyarosaurus dacus*

| VITAL STATISTICS | |
|---|---|
| Fossil Location | Romania (in a region formerly part of Hungary) |
| Diet | Herbivorous |
| Pronunciation | MAG-yar-o-SAWR-us |
| Weight | 900kg (1984lb) |
| Length | 6–8m (20–26ft) |
| Height | Unknown |
| Meaning of name | 'Magyar lizard' after the Magyars, the predominant ethnic group of Hungary |

**WHERE IN THE WORLD?**

*Magyarosaurus* was located in the Hunedoara region of Romania, at a time when much of Europe was under water.

One-quarter the size of its larger relatives, *Magyarosaurus* was a dwarf sauropod – and one of the last. It was light-weight by comparison, with slender limbs. Little else is known about it.

**Mouth**
Its small teeth were once thought to have been suitable for eating aquatic plants, but it is now known that *Magyarosaurus* was a land dweller.

**Neck**
Its long neck prompted speculation that *Magyarosaurus* lived in water, needing the support, but now all sauropods are known to be land dwellers.

**FOSSIL EVIDENCE**
It is not yet clear what conclusions can be drawn from the fossil evidence for *Magyarosaurus*. No more than 8m (26ft) long, it is the smallest known member of the titanosaur family. This may be because it lived on small islands, at a time when most of Europe was underwater: limited vegetation and fewer predators therefore restricted its size. Some scientists have identified three species, all coming from the same general region. However, some palaeontologists argue that these fossils may be the remains of separate dinosaurs.

**HOW BIG IS IT?**

| DINOSAUR |
| LATE CRETACEOUS |

**TIMELINE (millions of years ago)**

| 540 | 505 | 438 | 408 | 360 | 280 | 248 | 208 | 146 | 65 | 1.8 to today |

LATE CRETACEOUS

# Majungasaurus

• ORDER • Saurischia • FAMILY • Abelisauridae
• GENUS & SPECIES • *Majungasaurus crenatissimus*

*Majungasaurus* was a predator that probably caught its victims in its broad snout, holding on until they were subdued – like modern cats. It was the top predator in its environment.

## VITAL STATISTICS

| | |
|---|---|
| FOSSIL LOCATION | Madagascar |
| DIET | Carnivorous |
| PRONUNCIATION | Mah-JOONG-ah-THOL-us |
| WEIGHT | Unknown |
| LENGTH | 8–9m (26–30ft) |
| HEIGHT | Unknown |
| MEANING OF NAME | 'Majunga lizard' after the Majunga district of northern Madagascar, where the fossil was discovered |

### WHERE IN THE WORLD?

*Majungasaurus* was located in Madagascar, which was then joined to South America and India by a land bridge.

**SKULL**
*Majungasaurus* had a roughly textured skull, thickened bone on top of its snout, and a single rounded horn above its eyes.

**LEGS**
Its front limbs are not complete in the fossil record but are noticeably short, while its hind legs were longer and stocky.

## FOSSIL EVIDENCE

When discovered, some early *Majungasaurus* material was misidentified from a skull fragment as a boneheaded dinosaur and named *Majungatholus*. The mistake was corrected in 1998 with the discovery of one of the most complete dinosaur skulls ever found. Now known to be a theropod, *Majungasaurus* is also one of the few dinosaurs for which there is evidence of cannibalism. It may simply have scavenged carcasses, rather than actively pursuing its own kind, or it may have cannibalized defeated rivals, but numerous bones have been found with tooth marks.

DINOSAUR

LATE CRETACEOUS

**HOW BIG IS IT?**

**TIMELINE (millions of years ago)**

| 540 | 505 | 438 | 408 | 360 | 280 | 248 | 208 | 146 | 65 | 1.8 to today |

LATE CRETACEOUS

# Mandschurosaurus

• **ORDER** • Ornithischia • **FAMILY** • Hadrosauridae • **GENUS & SPECIES** • *Mandschurosaurus amurensis*

| VITAL STATISTICS | |
|---|---|
| Fossil Location | China |
| Diet | Herbivorous |
| Pronunciation | Mand-CHOOR-o-SAWR-us |
| Weight | Unknown |
| Length | 8m (26ft) |
| Height | 4.9m (16ft) |
| Meaning of name | 'Manchurian lizard' after Manchuria, China, where the fossil was discovered |

The first Chinese dinosaur to be named, *Mandschurosaurus* was a flat-headed dinosaur with no crest.

**Beak**
*Mandschurosaurus* used its toothless beak to snip plant material, grinding it up with cheek teeth.

**WHERE IN THE WORLD?**

To date, Mandschurosaurus has been located only in Manchuria, China.

## FOSSIL EVIDENCE

*Mandschurosaurus* is known from an incomplete skeleton discovered in 1914. It is a hadrosaur, one of the duck-billed dinosaurs, which were once poorly understood. Early misconceptions demonstrate how difficult it is to analyze fossils. Hadrosaurs were believed to live in water, their tail providing propulsion, while webbing on their feet was thought to make them good swimmers. Further study, however, indicated that this 'webbing' was a foot pad, and bony tendons in the tail made it too stiff to provide propulsion.

DINOSAUR

LATE CRETACEOUS

**HOW BIG IS IT?**

**Tail**
Its narrow tail was stiffened by bony tendons, which provided balance when it rose onto its hind legs to run.

**TIMELINE (millions of years ago)**

| 540 | 505 | 438 | 408 | 360 | 280 | 248 | 208 | 146 | 65 | 1.8 to today |

LATE CRETACEOUS

# Microceratus

• **ORDER** • Ornithischia • **FAMILY** • Protoceratopsidae • **GENUS & SPECIES** • *Microceratus gobiensis*

*Microceratus* is the smallest known horned dinosaur. It had a frill and a horny beak and was a very fast runner so it could easily elude an attacker.

## VITAL STATISTICS

| | |
|---|---|
| FOSSIL LOCATION | Southern Mongolia, northern China |
| DIET | Herbivorous |
| PRONUNCIATION | MY-cro-SAIR-ah-tops |
| WEIGHT | 4–7kg (9–15lb) |
| LENGTH | 80cm (32in) |
| HEIGHT | Unknown |
| MEANING OF NAME | 'Tiny horned face' |

**BEAK**
With its parrot-like beak, *Microceratus* probably fed on the predominant plants of its time, including conifers, ferns and cycads, cropping the leaves and needles.

## FOSSIL EVIDENCE

The fossil evidence for small dinosaurs is generally very rare but often better preserved and more complete than that for much larger creatures. They are rarer because they are more fragile but they are more apt to be complete because larger carcasses tend to be scattered and weather more before burial. *Microceratus* was no longer than 80cm (32in); its relative *Protoceratops* was 1.8m (6ft). It was first described in 1953 as *Microceratops*, but its name had already been assigned to an ichneumon wasp. A new name – *Microceratus* – was therefore created for it in 2008.

DINOSAUR
LATE CRETACEOUS

**HOW BIG IS IT?**

**HIND LEGS**
*Microceratus* was bipedal, and shins twice as long as its thighs – as well as long narrow feet – made it a swift runner.

**WHERE IN THE WORLD?**

*Microceratus* was found in what is now China and Mongolia.

**TIMELINE (millions of years ago)**

| 540 | 505 | 438 | 408 | 360 | 280 | 248 | 208 | 146 | 65 | 1.8 to today |

LATE CRETACEOUS

# Montanoceratops

• ORDER • Ornithischia  **FAMILY** • Leptoceratopsidae  **GENUS & SPECIES** • *Montanoceratops cerorhynchos*

| VITAL STATISTICS | |
|---|---|
| Fossil Location | USA, Canada |
| Diet | Herbivorous |
| Pronunciation | Mon-TAN-oh-SAIR-ah-tops |
| Weight | 450kg (992lb) |
| Length | 1.8–3m (6–10ft) |
| Height | Unknown |
| Meaning of name | 'Montana horned face' after the state of Montana, USA, where the fossil was found |

*Montanoceratops* was a medium-sized horned dinosaur. It had a small frill and nose horn, a powerful jaw and a parrot-like beak.

**HIND LEGS**
Its hind legs were longer than its front legs, making *Montanoceratops* back slope slightly forward.

**WHERE IN THE WORLD?**

*Montanoceratops* was found in Montana, USA, and Alberta, Canada.

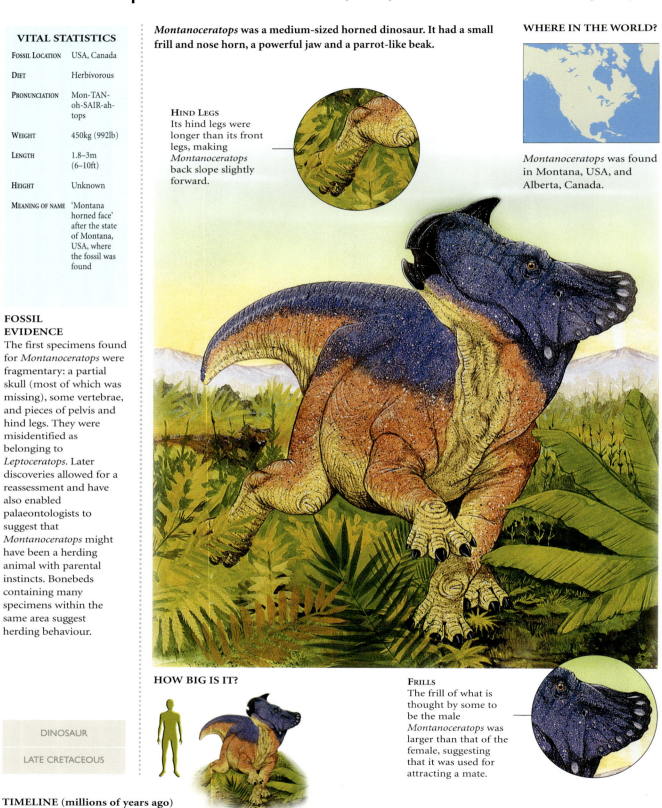

**FOSSIL EVIDENCE**
The first specimens found for *Montanoceratops* were fragmentary: a partial skull (most of which was missing), some vertebrae, and pieces of pelvis and hind legs. They were misidentified as belonging to *Leptoceratops*. Later discoveries allowed for a reassessment and have also enabled palaeontologists to suggest that *Montanoceratops* might have been a herding animal with parental instincts. Bonebeds containing many specimens within the same area suggest herding behaviour.

DINOSAUR

LATE CRETACEOUS

**HOW BIG IS IT?**

**FRILLS**
The frill of what is thought by some to be the male *Montanoceratops* was larger than that of the female, suggesting that it was used for attracting a mate.

**TIMELINE (millions of years ago)**

| 540 | 505 | 438 | 408 | 360 | 280 | 248 | 208 | 146 | 65 | 1.8 to today |

# LATE CRETACEOUS

# Nanotyrannus

• ORDER • Saurischia • FAMILY • Tyrannosauridae • GENUS & SPECIES • *Nanotyrannus lancensis*

## VITAL STATISTICS

| | |
|---|---|
| FOSSIL LOCATION | Western USA |
| DIET | Carnivorous |
| PRONUNCIATION | NAN-oh-tie-RAN-us |
| WEIGHT | 900kg (2000lb) |
| LENGTH | 5–6m (16–20ft) |
| HEIGHT | Unknown |
| MEANING OF NAME | 'Tiny tyrant', a reference to how small it is compared to other tyrannosaurs |

Like *Tyrannosaurus rex*, *Nanotyrannus* had eyes facing forwards, which meant that it could judge size, distance and angles of attack. It may have lain in wait, then ambushed its prey.

**TEETH**
Its sharp teeth enabled it to rip off chunks of flesh but were no good for chewing, so *Nanotyrannus* swallowed its food whole.

## WHERE IN THE WORLD?

*Nanotyrannus* has been located in Montana and South Dakota, USA.

## FOSSIL EVIDENCE

The fossil evidence for Nanotyrannus has divided palaeontologists. It was originally known from only one skull, and some features are puzzling: the fact that some bones have not yet fused prompts some palaeontologists to suggest that it is simply a juvenile *Tyrannosaurus rex*. In 2001, a more complete specimen was found. A study of growth rings in its bone does seem to confirm that it is not an adult *Nanotyrannus* but a juvenile *Tyrannosaurus*.

**HOW BIG IS IT?**

**ARMS**
*Nanotyrannus* had arms so short that they could not reach its own mouth. They may simply have been used to hold down prey.

TIMELINE (millions of years ago)

| 540 | 505 | 438 | 408 | 360 | 280 | 248 | 208 | 146 | 65 | 1.8 to today |

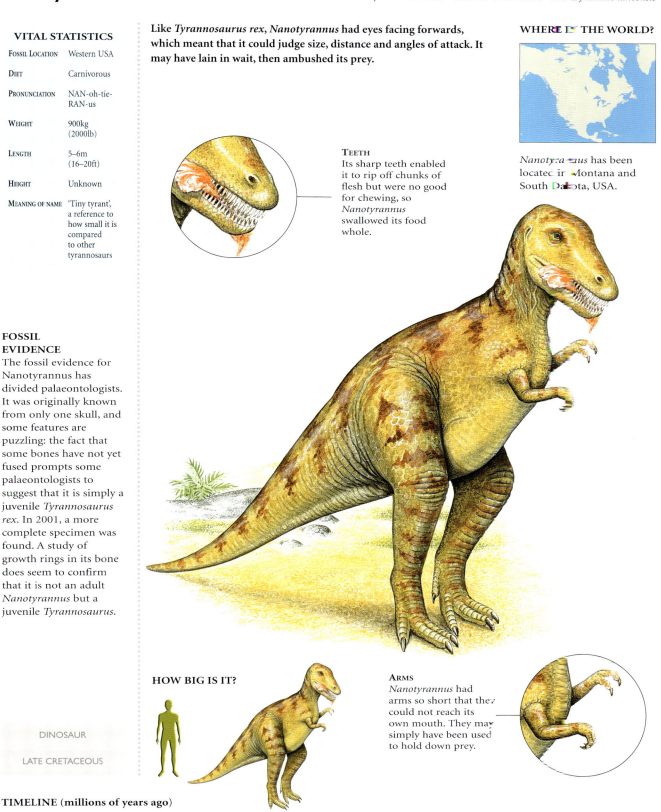

LATE CRETACEOUS

# Nanshiungosaurus

• **ORDER** • Saurischia • **FAMILY** • Therizinosauridea • **GENUS & SPECIES** • *Nanshiungosaurus brevispinus. N. bohlini*

| VITAL STATISTICS | |
|---|---|
| Fossil Location | China |
| Diet | Possibly omnivorous |
| Pronunciation | NAN-shee-ung-ah-SAWR-us |
| Weight | 600kg (1323lb) |
| Length | 4m (13ft) |
| Height | Unknown |
| Meaning of name | 'Nanshiung lizard' after the area in China where it was discovered |

Two species of this therizinosaur have been found – *Nanshiungosaurus brevispinus* and *N. bohlini* – both in China, and both dating from the Late Cretaceous Period. The fossil specimens are fragmentary.

### WHERE IN THE WORLD?

To date, *Nanshiungosaurus* has been located in what is now China.

**HANDS**
As a therizinosaur, *Nanshiungosaurus* probably had claws on its hands, either for catching lizards and small mammals or for taking hold of branches.

## FOSSIL EVIDENCE

The fossil evidence for *Nanshiungosaurus* is sparse. To date, only an incomplete pelvis and vertical column have been found. In other words, there is no skull, no limbs and no tail, so assumptions about its appearance and lifestyle are based on comparisons with its relatives. A therizinosaur, *Nanshiungosaurus* walked on its hind legs, on feet that had four toes. Its forearms had a remarkable range of motion, enabling it to reach forward. What separates it from its relatives are cervical ribs that have shrunk and fused to the cervical vertebrae.

**HOW BIG IS IT?**

**BODY**
Assuming a likeness to its cousin *Beipiaosaurus*, palaeontologists believe that *Nanshiungosaurus* was probably covered with feathers.

DINOSAUR
LATE CRETACEOUS

**TIMELINE (millions of years ago)**

| 540 | 505 | 438 | 408 | 360 | 280 | 248 | 208 | 146 | 65 | 1.8 to today |

LATE CRETACEOUS

# Nemegtosaurus

• ORDER • Saurischia • FAMILY • Nemegtosauridae • GENUS & SPECIES • *Nemegtosaurus mongoliensis*

*Nemegtosaurus* was a titanosaur, a member of the family of enormous sauropods. Little is known about it, but it may have had armour like its relatives. Titanosaurs, like all non-avian dinosaurs, did not survive beyond the Late Cretaceous.

## VITAL STATISTICS

| | |
|---|---|
| FOSSIL LOCATION | Mongolia |
| DIET | Herbivorous |
| PRONUNCIATION | NAY-meg-toe-SAWR-us |
| WEIGHT | Unknown |
| LENGTH | 12m (39ft) |
| HEIGHT | Unknown |
| MEANING OF NAME | 'Nemegt lizard' after the Nemegt Valley in Mongolia, where it was found |

### WHERE IN THE WORLD?

*Nemegtosaurus* was located in what is now the Gobi Desert of southern Mongolia.

**MOUTH**
With its blunt teeth, *Nemegtosaurus* probably stripped foliage from plants and swallowed them without chewing.

**BODY**
If *Nemegtosaurus* did have armour, this would have been located to protect its bulky body. Its gut was vast, to process all the plant material it ate.

## FOSSIL EVIDENCE

The fossil remains of other sauropods of the Late Cretaceous tend to be missing their skulls. The opposite is true for *Nemegtosaurus*, which is known from a partial skull and part of its neck. Its head has been likened to that of *Diplodocus*, but any relationship is doubtful because *Diplodocus* lived during the late Jurassic Period, many millions of years earlier. Like other sauropods, it had peg-like teeth at the front of its jaws for stripping plant material. Its long neck allowed it to reach through the forests for foliage.

### HOW BIG IS IT?

DINOSAUR
LATE CRETACEOUS

**TIMELINE (millions of years ago)**

| 540 | 505 | 438 | 408 | 360 | 280 | 248 | 208 | 146 | 65 | 1.8 to today |

LATE CRETACEOUS

# Neuquensaurus

• ORDER • Saursichia • FAMILY • Titanosauridae • GENUS & SPECIES • *Neuquensaurus australis*

*Neuquensaurus* was a sauropod that lived in small herds. To meet its daily requirements, it spent much of its time eating, and possibly migrated seasonally to keep up with changes in the food supply.

## VITAL STATISTICS

| | |
|---|---|
| FOSSIL LOCATION | Argentina |
| DIET | Herbivorous |
| PRONUNCIATION | NEH-oo-ken-SAW-rus |
| WEIGHT | Unknown |
| LENGTH | 10–15m (33–49ft) |
| HEIGHT | Unknown |
| MEANING OF NAME | 'Neuquén lizard' after Neuquén Province, Argentina, where the first fossil was found |

**OSTEODERMS**
Embedded in its back were bony oval osteoderms, which offered protection against (and may even have intimidated) large predators.

### WHERE IN THE WORLD?

*Neuquensaurus* was located in Argentina and Uruguay, South America.

## FOSSIL EVIDENCE

A titanosaurid sauropod, *Neuquensaurus* was protected by oval scutes on its back. Titanosaurs have been found on every continent except Australia, and some palaeontologists believe that their armour may have been an adaptation that was key to their survival. Many sauropods had died out by the end of the Jurassic Period, perhaps because a lack of body armour made them vulnerable to carnivores. The titanosaurs, by contrast, not only survived but were distributed worldwide, rather than being restricted (as many dinosaurs were) to certain geographic areas.

DINOSAUR
LATE CRETACEOUS

**HOW BIG IS IT?**

**BODY**
*Neuquensaurus* needed a vast gut for processing the plant material it ate, which were likely broken down by gastroliths (small stones) it swallowed.

**TIMELINE (millions of years ago)**

| 540 | 505 | 438 | 408 | 360 | 280 | 248 | 208 | 146 | 65 | 1.8 to today |

LATE CRETACEOUS

# Nipponosaurus

• ORDER • Ornithischia • FAMILY • Hadrosauridae • GENUS & SPECIES • *Nipponosaurus sachalinensis*

The first dinosaur to be found on Japanese soil, *Nipponosaurus* was a small duck-billed dinosaur. Walking on two legs, it had a broad head carried on a short neck, and a long beak.

### VITAL STATISTICS

| | |
|---|---|
| FOSSIL LOCATION | Japan |
| DIET | Herbivorous |
| PRONUNCIATION | Ni-PON-oh-SAWR-us |
| WEIGHT | Unknown |
| LENGTH | 7.6m (25ft) |
| HEIGHT | Unknown |
| MEANING OF NAME | 'Japanese lizard' because its fossil was found on an island that was in Japanese possession at the time |

**CREST**
A low, dome-shaped crest, which was connected to its nose and throat via air passages, possibly allowed it to generate sound.

### WHERE IN THE WORLD?

*Nipponosaurus* was found in the south of Sakhalin Island, a region that is now part of Russia.

## FOSSIL EVIDENCE

*Nipponosaurus* was discovered in 1934, during the construction of a hospital, and further material from the same specimen was recovered in 1937. It has been identified from a poorly preserved, incomplete skeleton, including fragments of the skull, teeth, parts of the forelimbs and most of the hind limbs. The specimen is thought to be about 60 per cent complete. Even so, *Nipponosaurus* is a dinosaur about which relatively little is known. Some palaeontologists argue that the specimen is, in fact, a juvenile of *Jaxartosaurus*, another hadrosaur.

**HOW BIG IS IT?**

DINOSAUR
LATE CRETACEOUS

**MOUTH**
Its beak was toothless but it had hundreds of small teeth, for grinding the plant material it ate.

**TIMELINE (millions of years ago)**

| 540 | 505 | 438 | 408 | 360 | 280 | 248 | 208 | 146 | 65 | 1.8 to today |

LATE CRETACEOUS

# Noasaurus

• ORDER • Saurischia • FAMILY • Abelisauridae • GENUS & SPECIES • Noasaurus leali

| VITAL STATISTICS | |
|---|---|
| Fossil Location | Argentina |
| Diet | Carnivorous |
| Pronunciation | NOH-ah-SAWR-us |
| Weight | 15kg (33lb) |
| Length | 1.8–2.4m (6–8ft) |
| Height | Unknown |
| Meaning of name | 'Northwestern Argentine lizard', in reference to where it was discovered ('NOA' abbreviates the Spanish 'noroeste Argentina') |

Lightly built, *Noasaurus* was an efficient predator, and may have hunted in packs to attack juvenile sauropods.

**LEGS**
Long legs made *Noasaurus* a swift runner. One of the fastest of its time, it was estimated to have reached speeds of 56kmh (35mph).

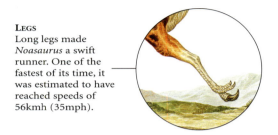

**WHERE IN THE WORLD?**

*Noasaurus* was located in what is now northwestern Argentina.

## FOSSIL EVIDENCE

*Noasaurus* demonstrates how difficult the analysis of fossil evidence can be. It was first thought to have a retractable claw on the second toe of each foot, a feature found on *Dromaeosaurus*, a theropod not closely related. However, further analysis has led palaeontologists to conclude that this claw actually belonged on its hand. When originally described, *Noasaurus* was assigned to the family of Noasauridae, but with a lower jaw similar to that of *Abelisaurus* it has now been reassigned to the Abelisauridae.

**HAND CLAWS**
*Noasaurus* had relatively long arms, meaning that it could make good use of its large claws to attack.

**HOW BIG IS IT?**

DINOSAUR

LATE CRETACEOUS

**TIMELINE (millions of years ago)**

| 540 | 505 | 438 | 408 | 360 | 280 | 248 | 208 | 146 | 65 | 1.8 to today |

LATE CRETACEOUS

# Opisthocoelicaudia

• ORDER • Saurischia • FAMILY • Saltasauridae • GENUS & SPECIES • *Opisthocoelicaudia skarzynskii*

## VITAL STATISTICS

| | |
|---|---|
| FOSSIL LOCATION | Mongolia |
| DIET | Herbivorous |
| PRONUNCIATION | Oh-PIS-tho-SEEL-ih-CAWD-ee-ah |
| WEIGHT | 16.5 tonnes (15 tons) |
| LENGTH | 12m (39ft) |
| HEIGHT | Unknown |
| MEANING OF NAME | 'Hollow-backed tail' because of the opisthocoelous (hollow-behind) structure of the tail vertebrae |

### WHERE IN THE WORLD?

*Opisthocoelicaudia* was located in Mongolia, in what is now the Gobi Desert.

Most sauropods had an arched profile, but *Opisthocoelicaudia* held its body almost completely straight from its neck to its tail. It might have been able to rear up to reach leaves, resting on its legs and tail.

**TAIL**
Thanks to unique joints and considerable muscle attachments, the tail of *Opisthocoelicaudia* slanted upwards, not downwards, like other sauropods.

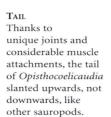

## FOSSIL EVIDENCE

Neither the head nor neck of this creature has been found, but there are tooth marks on the pelvis and femur. This suggests scavengers fed on its carcass. Not far from the fossil site, the skull of *Nemegtosaurus* was found, and some palaeontologists believe that the specimens are the same dinosaur. The vertebrae of *Opisthocoelicaudia* are, however, unique: the side facing the end of the tail cups inwards; the side facing the front of the animal cups outwards, perhaps enabling *Opisthocoelicaudia* to use its tail as a prop.

### HOW BIG IS IT?

**HIND LEGS**
An extra vertebra in its pelvic area and a strong hip socket might have allowed *Opisthocoelicaudia* to stand on its hind legs, unlike other sauropods.

DINOSAUR
LATE CRETACEOUS

## TIMELINE (millions of years ago)

| 540 | 505 | 438 | 408 | 360 | 280 | 248 | 208 | 146 | 65 | 1.8 to today |

LATE CRETACEOUS

# Pachyrhinosaurus

• ORDER • Ornithischia • FAMILY • Ceratopsidae • GENUS & SPECIES • *Pachyrhinosaurus canadensis*

| VITAL STATISTICS | |
|---|---|
| Fossil Location | Canada |
| Diet | Herbivorous |
| Pronunciation | Pack-ee-RINE-oh-SAWR-us |
| Weight | 2 tonnes (1.8 tons) |
| Length | 2m (7ft) |
| Height | Unknown |
| Meaning of name | 'Thick-nosed lizard', a reference to the lump of bone on its nose |

When threatened, *Pachyrhinosaurus* may have charged its attacker, just as the modern rhinoceros does. With large bosses on its skull, it was equipped to butt and shove – putting up a formidable defence.

**Nose**
*Pachyrhinosaurus* had a large bony pad on its nose, believed by some palaeontologists to support a nose horn missing from the fossil record.

**WHERE IN THE WORLD?**

*Pachyrhinosaurus* was found in Alberta, Canada.

## FOSSIL EVIDENCE

It is from the fossil evidence that we know *Pachyrhinosaurus* probably travelled in herds. In 1972, a huge group was found at Pipestone Creek, Alberta, in a bonebed from which 14 skulls and 3500 bones were removed. Ranging from juvenile to adult specimens, they suggest that *Pachyrhinosaurus* cared for its young. The animals were probably trying to cross a river, or were overcome by a flash flood. Certainly, their carcasses were washed downriver to provide a meal for scavengers, as indicated by several findings of teeth.

**Frill**
Horns grew from the edge of its neck frill, and one small spike protruded from the centre.

**HOW BIG IS IT?**

DINOSAUR

LATE CRETACEOUS

**TIMELINE (millions of years ago)**

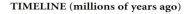

# Panoplosaurus

• ORDER • Ornithischia • FAMILY • Nodosauridae • GENUS & SPECIES • *Panoplosaurus mirus*

LATE CRETACEOUS

### VITAL STATISTICS

| | |
|---|---|
| FOSSIL LOCATION | Canada, western USA |
| DIET | Herbivorous |
| PRONUNCIATION | PAN-oh-ploh-SAWR-us |
| WEIGHT | 3.9 tonnes (3.5 tons) |
| LENGTH | 5.5–7m (18–23ft) |
| HEIGHT | 1.2m (4ft) |
| MEANING OF NAME | 'Fully armoured lizard' in reference to its spiky armour |

*Panoplosaurus* was seemingly not well equipped to fight, lacking the tail club found on other ankylosaurs. However, armour embedded in its skin and spikes on its neck, side, tail and shoulders probably put off most predators.

### WHERE IN THE WORLD?

*Panoplosaurus* was located in Alberta, Canada, and Montana, USA.

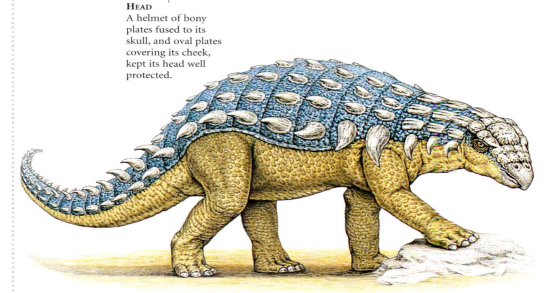

**HEAD**
A helmet of bony plates fused to its skull, and oval plates covering its cheek, kept its head well protected.

### FOSSIL EVIDENCE

*Panoplosaurus* is known from two partial specimens, one of which features an almost intact skull. The skull suggests that *Panoplosaurus* had cheeks, for preventing food from falling out of its mouth as it ate. In fact, so detailed is the skull that it has been used to model the skulls of other armoured dinosaurs whose skulls have not been preserved. Also in the fossil record are parts of its armour, and similarities to the armour of *Edmontonia* (its contemporary, also found in North America) suggest the two were closely related.

**BEAK**
Weighed down by its armour *Panoplosaurus* browsed on low-lying plants, shearing them with its toothless beak.

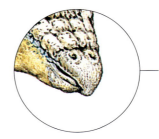

### HOW BIG IS IT?

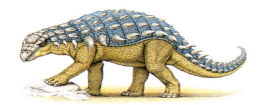

DINOSAUR

LATE CRETACEOUS

**TIMELINE (millions of years ago)**

| 540 | 505 | 438 | 408 | 360 | 280 | 248 | 208 | 146 | 65 | 1.8 to today |

LATE CRETACEOUS

# Parksosaurus

• **ORDER** • Ornithischia • **FAMILY** • Hypsilophodontidae • **GENUS & SPECIES** • *Parksosaurus warreni*

| VITAL STATISTICS | |
|---|---|
| Fossil Location | Canada |
| Diet | Herbivorous |
| Pronunciation | PARK-soh-SAWR-us |
| Weight | 70kg (230lb) |
| Length | 2m (7ft) |
| Height | Unknown |
| Meaning of name | 'Parks' lizard' in honour of William Arthur Parks, a Canadian palaeontologist |

*Parksosaurus* was a small dinosaur that walked upright, its long tail balanced by its long neck. It probably had good vision, its large eyes protected by a ring of bone.

**HIND LEGS**
With no obvious defences, *Parksosaurus* probably relied on flight to survive.

### WHERE IN THE WORLD?

To date, *Parksosaurus* has been located only in Alberta, Canada.

### FOSSIL EVIDENCE

*Parksosaurus* is known from an incomplete skeleton. At death, the dinosaur fell on its left side, and most of the right side has not been preserved. The head was separated from its body and the neck was also lost. A curious feature is the presence in its pectoral girdle of a suprascapula, a bone more commonly found in lizards but that is found in other ornithopods. The size of *Parksosaurus* has been estimated by comparison with the bones of *Thescelosaurus*, its contemporary in North America.

| DINOSAUR |
|---|
| LATE CRETACEOUS |

**HOW BIG IS IT?**

**JAWS**
Wide jaws held unusual teeth, with low, rounded ridges, presumably for grinding plants. Such teeth have not been found on other hypsilophodonts.

**TIMELINE** (millions of years ago)

| 540 | 505 | 438 | 408 | 360 | 280 | 248 | 208 | 146 | 65 | 1.8 to today |

LATE CRETACEOUS

# Pentaceratops

• **ORDER** • Ornithischia • **FAMILY** • Ceratopsidae • **GENUS & SPECIES** • *Pentaceratops sternbergii*

*Pentaceratops* had one of the largest skulls of any land-based vertebrate, and its frill and horns served to intimidate both predators and rivals. A herbivore, it could chew well with its cheek teeth – unlike many dinosaurs.

## VITAL STATISTICS

| | |
|---|---|
| Fossil Location | Southwestern USA |
| Diet | Herbivorous |
| Pronunciation | PEN-ta-SAIR-ah-tops |
| Weight | 6.6 tonnes (6 tons) |
| Length | 7.5m (25ft) |
| Height | Unknown |
| Meaning of name | 'Five-horned face' in reference to its three face horns and two cheek spikes |

## FOSSIL EVIDENCE

*Pentaceratops* is known from nine skulls and several skeletons. A herbivore, it probably fed on the dominant plants of its time: conifers, cycads and ferns. Towards the end of the Cretaceous, the fossil record demonstrates a progressive decline in biodiversity, and it is some time after the last *Pentaceratops,* that an ecological crisis, perhaps caused by the impact of a meteor strike, seems to have blocked sunlight, affecting those plants that depended on photosynthesis. As their food sources became scarce, the herbivores died out.

DINOSAUR

LATE CRETACEOUS

**Spikes**
*Pentaceratops* took its name from the large horns on its brow, a short nose horn and two spikes protruding from under its eyes.

**How big is it?**

**Frill**
Reaching halfway down its back, *Pentaceratops*' frill was edged with bony knobs. It was no burden since large fenestrae, or holes, reduced its weight.

## WHERE IN THE WORLD?

*Pentaceratops* was located in what is now New Mexico, USA.

**TIMELINE (millions of years ago)**

| 540 | 505 | 438 | 408 | 360 | 280 | 248 | 208 | 146 | 65 | 1.8 to today |

LATE CRETACEOUS

# Pinacosaurus

• **ORDER** • Ornithischia • **FAMILY** • Ankylosauridae • **GENUS & SPECIES** • *Pinacosaurus grangeri*

| VITAL STATISTICS | |
|---|---|
| Fossil Location | Mongolia, China |
| Diet | Herbivorous |
| Pronunciation | PIN-ah-co-SAWR-us |
| Weight | Unknown |
| Length | 5.5m (18ft) |
| Height | Unknown |
| Meaning of name | 'Plank lizard', after the small, flat armoured plates on its head |

Though armoured, *Pinacosaurus* was lightly built. With a long skull, it had between two and five additional holes near each nostril; their purpose remains unclear. *Pinacosaurus* was probably a herbivore, though living in a desert region.

### WHERE IN THE WORLD?

*Pinacosaurus* was located in Asia, specifically in Mongolia and China.

**SKULL**
Small bony plates protected the top of its skull. Separated in juveniles, they grew together into one sheet as *Pinacosaurus* aged.

### FOSSIL EVIDENCE

*Pinacosaurus* is the best known of the Asian ankylosaurs: among the specimens are five skulls and one skeleton, which is almost complete. The first specimen was discovered in the Gobi Desert, specifically the Djadochta Formation, a deposit that also provides evidence of the habitat at the time. With its sand dunes and few sources of fresh water, the region was little different from the desert of today. Two groups of juveniles have been found huddled together, probably the victims of a sandstorm. These discoveries suggest that *Pinacosaurus* gathered in herds based on age group.

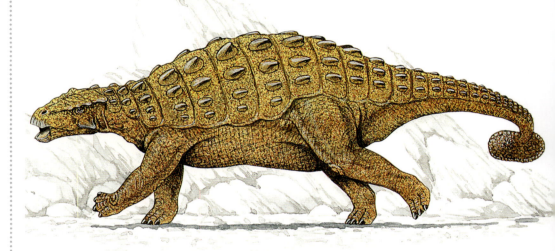

**TAIL**
At the end of its tail, a bony club somewhat like a double-edged axe dealt any predator a painful blow.

**HOW BIG IS IT?**

DINOSAUR

LATE CRETACEOUS

**TIMELINE (millions of years ago)**

| 540 | 505 | 438 | 408 | 360 | 280 | 248 | 208 | 146 | 65 | 1.8 to today |

LATE CRETACEOUS

# Prenocephale

• ORDER • Ornithischia • FAMILY • Pachycephalosauridae • GENUS & SPECIES • Prenocephale prenes

| VITAL STATISTICS | |
|---|---|
| FOSSIL LOCATION | Mongolia |
| DIET | Herbivorous |
| PRONUNCIATION | PREN-oh-SEF-al-ee |
| WEIGHT | 135kg (298lb) |
| LENGTH | 2.4m (8ft) |
| HEIGHT | Unknown |
| MEANING OF NAME | 'Sloping head' |

The head-butting behaviour of pachycephalosaurs has been debated for a long time and continues to this day. Recent research seems to show that they were incapable of absorbing the impacts, but even more recent research suggests that they could.

**WHERE IN THE WORLD?**

*Prenocephale* certainly lived in what is now Mongolia, but may also have lived in North America.

**EYES**
Large eye sockets suggest that *Prenocephale* had good eyesight, which was a necessity if, as some palaeontologists believe, it also caught insects to eat.

**FOSSIL EVIDENCE**
*Prenocephale* is known only from skulls and a few bones. Assumptions about its appearance are therefore based on features common to all boneheads. As is the case with most dinosaurs, the specific details of its diet are unknown. However, teeth that are set less wide than its relative, *Stegoceras*, suggest different feeding preferences, and its front teeth were suitable for shredding fruit and seeds as well as plants. *Prenocephale* may have had an extensive range: some palaeontologists believe that it should be identified with *Sphaerotholus*, a bonehead found in North America.

DINOSAUR

LATE CRETACEOUS

**HOW BIG IS IT?**

**HEAD**
On top of its skull, *Prenocephale* had a high, rounded dome. Protecting the braincase, it extended to the tiny frill at the back.

**TIMELINE** (millions of years ago)

| 540 | 505 | 438 | 408 | 360 | 280 | 248 | 208 | 146 | 65 | 1.8 to today |

293

LATE CRETACEOUS

# Prosaurolophus

• ORDER • Ornithischia • FAMILY • Hadrosauridae • GENUS & SPECIES • *Prosaurolophus maximus*

*Prosaurolophus* was a large-headed duck-bill. Its beak was ideal for clipping the flowering plants, shrubs and fruits it ate. The back of its mouth contained thousands of teeth for grinding up food.

| VITAL STATISTICS | |
|---|---|
| Fossil Location | Canada, USA |
| Diet | Herbivorous |
| Pronunciation | PROH-saw-ROL-off-us |
| Weight | 2 tonnes (1.8 tons) |
| Length | 8m (26ft) |
| Height | Unknown |
| Meaning of name | 'Before Saurolophus' because it was originally believed to be the ancestor of Saurolophus |

## WHERE IN THE WORLD?

*Prosaurolophus* was located in Alberta, Canada, and Montana, USA.

## FOSSIL EVIDENCE

*Prosaurolophus* was described in 1916 by Barnum Brown, the palaeontologist who had described *Saurolophus* four years earlier. The two dinosaurs may not have been as closely related as their names imply; a review in 2004 links *Prosaurolophus* more closely to *Gryposaurus*. A bonebed discovered in Montana suggests that *Prosaurolophus* sometimes lived together. These individuals are assumed to have congregated by a water source during a time of drought. In general, though, the fossil record in Alberta indicated that *Prosaurolophus* lived in a warm climate, with wet and dry seasons.

DINOSAUR

LATE CRETACEOUS

**Crest**
Above its eyes was a small triangular crest, which may have helped to distinguish individuals, as well as males from females.

**Hind Legs**
Though it walked on all fours, *Prosaurolophus* could probably rise up onto its hind legs, resting against its tail for balance.

**HOW BIG IS IT?**

TIMELINE (millions of years ago)

| 540 | 505 | 438 | 408 | 360 | 280 | 248 | 208 | 146 | 65 | 1.8 to today |

# Quaesitosaurus

• ORDER • Saurischia • FAMILY • Nemegtosauridae • GENUS & SPECIES • *Quaesitosaurus orientalis*

## VITAL STATISTICS

| | |
|---|---|
| FOSSIL LOCATION | Mongolia |
| DIET | Herbivorous |
| PRONUNCIATION | Kwee-SIT-oh-SAWR-us |
| WEIGHT | Unknown |
| LENGTH | Up to 23m (75ft) |
| HEIGHT | Unknown |
| MEANING OF NAME | 'Unusual lizard' after its unusual skull |

Relatively little is known about *Quaesitosaurus*. But it is interesting to note that it and its close relative and neighbour *Nemegtosaurus* are both mainly known from skulls — the rarest parts of sauropods.

**MOUTH**
Soft peg-like teeth were adapted for stripping soft plant material, which may have included aquatic plants. It swallowed its food whole, without chewing.

## WHERE IN THE WORLD?

To date, *Quaesitosaurus* has been located only in Mongolia.

**NECK**
Although no neck is yet known for *Quaesitosaurus*, we know by comparison with other sauropods that it was a long one.

## FOSSIL EVIDENCE

*Quaesitosaurus* is known only from a partial skull, found in the Gobi Desert, a region that was semi-arid during the Late Cretaceous. Large ear openings suggest that it may have had good hearing. In most cases, however, assumptions about its appearance and behaviour must be based on what is known about other sauropods. It probably travelled in herds, probably hatched from eggs, and was likely among the least intelligent of the dinosaurs. All that said, it is possible that *Quaesitosaurus* should be identified as *Nemegtosaurus*.

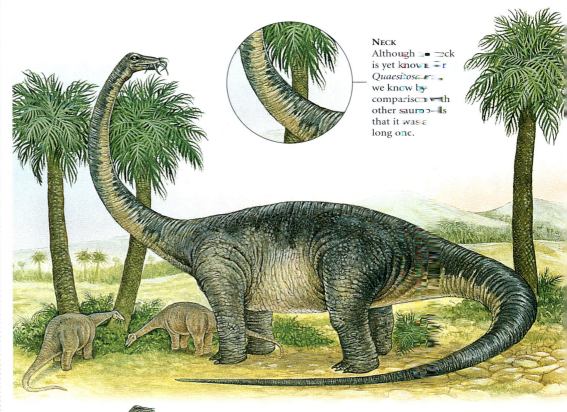

DINOSAUR
LATE CRETACEOUS

**HOW BIG IS IT?**

**TIMELINE (millions of years ago)**

| 540 | 505 | 438 | 408 | 360 | 280 | 248 | 208 | 146 | 65 | 1.8 to today |

LATE CRETACEOUS

# Rhabdodon

• ORDER • Ornithischia • FAMILY • Rhabdodontidae • GENUS & SPECIES • *Rhabdodon priscus*

| VITAL STATISTICS | |
|---|---|
| FOSSIL LOCATION | Europe |
| DIET | Herbivorous |
| PRONUNCIATION | RAB-doe-don |
| WEIGHT | 450kg (993lb) |
| LENGTH | 4m (13ft) |
| HEIGHT | Unknown |
| MEANING OF NAME | 'Fluted teeth' in reference to the shape of its teeth |

With its stocky body and short front limbs, *Rhabdodon* walked upright. However, it probably dropped onto all fours to feed on the shrubs, cycads and ferns that probably made up its diet.

### WHERE IN THE WORLD?

*Rhabdodon* was located in Europe, specifically France, Spain and Romania.

**HANDS**
With its short arms and five-fingered hands, *Rhabdodon* may have grasped low-lying plants, pulling them towards its mouth.

### FOSSIL EVIDENCE

Described in 1869, *Rhabdodon* was one of the first dinosaurs to be discovered. Since then, palaeontologists have debated its classification: currently it is recognized as either an iguanodont or a hypsilophodont, and some palaeontologists argue that it is a missing link between the two. It seems, however, to be most closely related to *Tenontosaurus*, a primitive iguanodont. A specimen of *Rhabdodon* found on Hateg Island in what is now Romania is smaller than those found in France and Spain, which may simply be an adaptation to the environment.

**TAIL**
When running, its long tail kept *Rhabdodon* balanced. Rather than pursuing a straight line, it may have weaved left and right to avoid predators.

### HOW BIG IS IT?

DINOSAUR
LATE CRETACEOUS

**TIMELINE** (millions of years ago)

| 540 | 505 | 438 | 408 | 360 | 280 | 248 | 208 | 146 | 65 | 1.8 to today |

# LATE CRETACEOUS

# Saurolophus

• ORDER • Ornithischia • FAMILY • Hadrosauridae • GENUS & SPECIES • *Saurolophus osborni*

| VITAL STATISTICS | |
|---|---|
| FOSSIL LOCATION | Southwestern Canada, Mongolia |
| DIET | Herbivorous |
| PRONUNCIATION | SAWR-oh-LOHF-us |
| WEIGHT | 2.9 tonnes (2.6 tons) |
| LENGTH | 9–12m (30–39ft) |
| HEIGHT | Unknown |
| MEANING OF NAME | 'Crested lizard' in reference to the small crest on top of its head |

*Saurolophus* walked primarily on two legs, balanced by its tail. The top bill of its beak curved upwards, and this herbivore relied on its cheek teeth to grind the twigs, seeds and conifer needles it likely ate.

## WHERE IN THE WORLD?

*Saurolophus* was located on two continents – Asia and North America.

**CREST**
*Saurolophus* had a spike-like crest projecting up and back from its skull. Found also on juveniles, it was about 13cm (5in) long in adults.

## FOSSIL EVIDENCE

Some paleontologists have suggested that *Saurolophus* had loose skin on top of its snout. It may have inflated this, either as a means of display or to generate sound. To create a loud bellow, they suggest *Saurolophus* breathed out to deflate the balloon: the air travelling through the crest and out of its nostrils generated the sound, which may have been used to attract a mate or to intimidate. Specimens have been found in Asia and North America, helping confirm a one-time land connection between the two continents.

DINOSAUR
LATE CRETACEOUS

## HOW BIG IS IT?

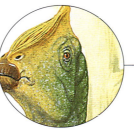

**Eyes**
*Saurolophus* was one of the first hadrosaurs discovered to have a ring of bone supporting its eyes, known as a sclerotic ring.

## TIMELINE (millions of years ago)

| 540 | 505 | 438 | 408 | 360 | 280 | 248 | 208 | 146 | 65 | 1.8 to today |

LATE CRETACEOUS

# Saurornithoides

• ORDER • Saurischia • FAMILY • Troodontidae • GENUS & SPECIES • *Saurornithoides mongoliensis*

| VITAL STATISTICS | |
|---|---|
| FOSSIL LOCATION | Mongolia |
| DIET | Carnivorous |
| PRONUNCIATION | SAWR-or-nith-OY-dees |
| WEIGHT | 13–27kg (29–60lb) |
| LENGTH | 2–3.5m (7–11ft) |
| HEIGHT | Unknown |
| MEANING OF NAME | 'Lizard bird form' because its skull resembled that of a bird with teeth |

*Saurornithoides* had one of the largest brains, relative to its body weight, of any dinosaur. It probably had a highly developed sense of hearing, and it is possible that its eyesight, already acute, was also good in low light.

**EYES**
Its large eyes were in sockets that were placed forward, rather than on the sides of its head, giving this predator binocular vision.

**FEET**
On the second toe of each foot, *Saurornithoides* had a large, scythe-like claw for slashing its prey.

### WHERE IN THE WORLD?

*Saurornithoides* was located in Mongolia, in the region that is now the Gobi Desert.

### FOSSIL EVIDENCE

Before it was described in 1924, *Saurornithoides* was believed to be a primitive bird, hence its name. Further study suggests that it was, in fact, a non-avian dinosaur. It is very similar to *Troodon*, suggesting that a landbridge connected Asia to North America when their common ancestor lived. Both dinosaurs had large eye sockets, and palaeontologists speculate that both had good night vision. This offered a considerable advantage: their prey probably included cold-blooded reptiles, whose activity slows as the temperature drops in the evening.

DINOSAUR

LATE CRETACEOUS

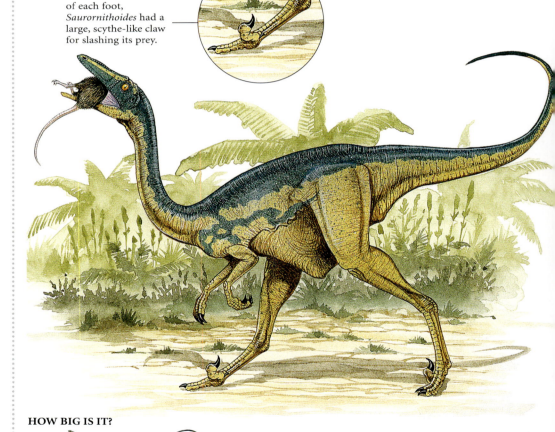

**HOW BIG IS IT?**

**TIMELINE** (millions of years ago)

| 540 | 505 | 438 | 408 | 360 | 280 | 248 | 208 | 146 | 65 | 1.8 to today |

# Secernosaurus

• ORDER • Ornithischia • FAMILY • Hadrosauridae? • GENUS & SPECIES • *Secernosaurus koerneri*

## VITAL STATISTICS

| | |
|---|---|
| FOSSIL LOCATION | Argentina |
| DIET | Herbivorous |
| PRONUNCIATION | See-SIR-no-SAWR-us |
| WEIGHT | Unknown |
| LENGTH | 3m (10ft) |
| HEIGHT | Unknown |
| MEANING OF NAME | 'Separated lizard' because while other duck-billed specimens were found in North America, this was found in South America |

A small duck-billed dinosaur, *Secernosaurus* probably gathered in herds. This offered a degree of protection while it foraged for the ferns, conifers and flowering plants that probably made up its diet.

**MOUTH**
*Secernosaurus* had rows of cheek teeth that were self-sharpening, grinding against each other as it chewed on tough plant material.

## FOSSIL EVIDENCE

*Secernosaurus* is known only from a partial skull and a few pelvic bones, including the ilium. These were discovered in 1923, but it was not until 1979 that they were studied and identified. *Secernosaurus* was the first duck-billed dinosaur to be found in South America. The fragmentary nature of the fossil remains means that palaeontologists do not know whether it had a crest or not. It was identified as a hadrosaur from its ilium, but some palaeontologists now believe that the bone was deformed in the fossilization process.

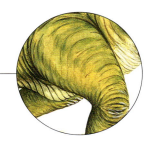

**HIND LEGS**
From the size of its hip bones, we know that Secernosaurus was a small dinosaur. It could probably walk as well on two legs as on four.

## HOW BIG IS IT?

DINOSAUR

LATE CRETACEOUS

### WHERE IN THE WORLD?

*Secernosaurus* was located in South America, specifically Argentina.

## TIMELINE (millions of years ago)

| 540 | 505 | 438 | 408 | 360 | 280 | 248 | 208 | 146 | 65 | 1.8 to today |

LATE CRETACEOUS

# Shantungosaurus

• **ORDER** • Ornithischia • **FAMILY** • Hadrosauridae • **GENUS & SPECIES** • *Shantungosaurus giganteus*

| VITAL STATISTICS | |
|---|---|
| Fossil Location | China |
| Diet | Herbivorous |
| Pronunciation | SHAHN-DUNG-oh-SAWR-us |
| Weight | Up to 17.6 tonnes (16 tons) |
| Length | 12–15m (39–49ft) |
| Height | Unknown |
| Meaning of name | 'Shantung lizard' after Shantung province, China, where it was found |

A herbivore, *Shantungosaurus* had a toothless beak, sheathed with horn, and jaws packed with hundreds of small teeth. Foraging on coastal plains and floodplains, it was remarkably similar to the North American *Edmontosaurus*, only larger.

**NOSTRILS**
Around its nostrils, *Shantungosaurus* had a large hole that may have been covered by a loose skin flap, which could be inflated to generate sound.

### WHERE IN THE WORLD?

*Shantungosaurus* was located in Asia, specifically in what is now Shantung, China.

**FOSSIL EVIDENCE**
*Shantungosaurus* was discovered in 1964 and described in 1973. It is known from the incomplete remains of five individuals. All were found in the same fossil bed, their bones mixed together, suggesting that *Shantungosaurus* lived in herds. Presumably this was a strategy that offered a degree of protection from tyrannosaurs, the only predators large enough to attack it. Its best defence was probably to flee, though, and the fossil evidence suggests it would have risen onto its powerful hind legs to do so.

DINOSAUR
LATE CRETACEOUS

**HOW BIG IS IT?**

**HIND LEGS**
Perhaps the largest hadrosaur, *Shantungosaurus* depended on hind legs that were stout and well muscled to support its weight.

**TIMELINE** (millions of years ago)

| 540 | 505 | 438 | 408 | 360 | 280 | 248 | 208 | 146 | 65 | 1.8 to today |

LATE CRETACEOUS

# Stegoceras

• ORDER • Ornithischia • FAMILY • Pachycephalosauridae • GENUS & SPECIES • *Stegoceras validum*

*Stegoceras* was a bone-headed dinosaur, with a fringe of bony knobs around the back of its skull. Foraging on low-lying plants, it may have lived in herds, scattering when attacked.

## VITAL STATISTICS

| | |
|---|---|
| Fossil Location | Canada, western USA |
| Diet | Herbivorous |
| Pronunciation | steh-GAH-ser-us |
| Weight | 78kg (170lb) |
| Length | 2m (7ft) |
| Height | Unknown |
| Meaning of name | 'Roofed horn' because it was mistakenly assumed to be a horned dinosaur when it was named |

**SKULL**
The domed skull of *Stegoceras* was 8cm (3in) thick. It grew thicker with age, and was possibly thicker on males.

### WHERE IN THE WORLD?

*Stegoceras* was located in North America, specifically Montana, USA, and Alberta, Canada.

## FOSSIL EVIDENCE

There are relatively few fossil remains of *Stegoceras*. This may indicate that it lived in upland areas where few fossils of any kind have been preserved. (In fact, when *Stegoceras* was first discovered, it was identified with *Troodon*, a mistake corrected after better specimens were found.) One of the few boneheads for which more than one skull has been discovered, it is the species on which palaeontologists have based their assumptions about other boneheads. The skull of one probable male has been found in which the dome had overgrown the shelf around the skull.

DINOSAUR

LATE CRETACEOUS

**HOW BIG IS IT?**

**TEETH**
Its curved teeth with serrated edges were not suited for tough, fibrous plants, suggesting that *Stegoceras* ate leaves and fruit instead.

## TIMELINE (millions of years ago)

| 540 | 505 | 438 | 408 | 360 | 280 | 248 | 208 | 146 | 65 | 1.8 to today |

LATE CRETACEOUS

# Struthiomimus

• ORDER • Saurischia • FAMILY • Ornithomidae • GENUS & SPECIES • Struthiomimus altus

| VITAL STATISTICS | |
|---|---|
| Fossil Location | Western Canada, northeastern USA |
| Diet | Possibly omnivorous |
| Pronunciation | STROOTH-ee-o-MIME-us |
| Weight | 150kg (33lb) |
| Length | 3–4m (10–13ft) |
| Height | Unknown |
| Meaning of name | 'Ostrich mimic' in reference to its ostrich-like appearance |

Possibly a sprinter, *Struthiomimus* may have reached very high speeds for a dinosaur. Running was probably its best defence, though it may have lashed out at attackers with the claws on its feet.

**WHERE IN THE WORLD?**

*Struthiomimus* was located in North America, specifically New Jersey, USA, and Alberta, Canada.

**Body**
With a long neck, long legs, small head and large eyes, *Struthiomimus* looked like an ostrich. It may also have had feathers.

**FOSSIL EVIDENCE**
What did this creature eat? The fossil evidence is inconclusive. Its second and third fingers were of equal length, and were unable to function independently, so it seems *Struthiomimus* can have used its hands only as a hook, for grasping branches. However, so many specimens have been found at the Red Deer River site in Alberta that it seems it cannot have relied on a specialized diet and may have been an omnivore.

**Legs and Feet**
*Struthiomimus* was built for running. Its shins were longer than its thighs, and the bones in its feet were long.

**HOW BIG IS IT?**

DINOSAUR

LATE CRETACEOUS

**TIMELINE** (millions of years ago)

| 540 | 505 | 438 | 408 | 360 | 280 | 248 | 208 | 146 | 65 | 1.8 to today |

# Stygimoloch

• ORDER • Ornithischia  **FAMILY** • Pachycephalosauridae • **GENUS & SPECIES** • *Stygimoloch spinifer*

| VITAL STATISTICS | |
|---|---|
| FOSSIL LOCATION | Western USA |
| DIET | Herbivorous |
| PRONUNCIATION | STIH-jee-MO-lok |
| WEIGHT | 78kg (172lb) |
| LENGTH | 2–3m (7–10ft) |
| HEIGHT | Unknown |
| MEANING OF NAME | 'Demon of the Styx' because the thorny horns on its head gave it a demonic appearance and because it was found in the Hell Creek Formation alluded to by the River Styx |

*Stygimoloch* had a small domed skull that was slightly flattened. From either side projected bumpy nodules and clusters of three or four horns. These were surprisingly fragile, so it is unlikely they were used for defence.

**HORNS**
*Stygimoloch*'s horns were for display only. Two backwards-facing spikes at the back of the head grew up to 15cm (6in) long.

**MOUTH**
The teeth at the back of its mouth suggest *Stygimoloch* was a herbivore, but at the front it had sharp incisors, like those on a carnivore.

## FOSSIL EVIDENCE

The first specimens of *Stygimoloch* were discovered in the late 1800s, but there were too few to make a description possible. Only in 1982, after more discoveries, was this bone-headed dinosaur formally described. Then, in 1995, came a remarkable find: a complete skeleton. In fact, this was the first pachycephalosaur to be found with both its body and head intact. It was this specimen that queried the assumption that bone-headed dinosaurs engaged in head-butting contests; its neck was seemingly too weak to withstand the impact.

DINOSAUR
LATE CRETACEOUS

## HOW BIG IS IT?

## WHERE IN THE WORLD?

*Stygimoloch* was located in the USA, specifically Wyoming and Montana.

**TIMELINE** (millions of years ago)

| 540 | 505 | 438 | 408 | 360 | 280 | 248 | 208 | 146 | 65 | 1.8 to today |

LATE CRETACEOUS

# Talarurus

• ORDER • Ornithischia • FAMILY • Ankylosauridae • GENUS & SPECIES • *Talarurus plicatospineus*

| VITAL STATISTICS | |
|---|---|
| FOSSIL LOCATION | Mongolia |
| DIET | Herbivorous |
| PRONUNCIATION | TAL-a-RU-rus |
| WEIGHT | 0.8–1.1 tonnes (0.7–1 ton) |
| LENGTH | 5.7m (19ft) |
| HEIGHT | Unknown |
| MEANING OF NAME | 'Basket tail' in reference to the wicker-like appearance of the bones in its tail |

*Talarurus* was well protected by thick plates and hollow spines across its back and hips. However, any predator not put off by this risked a crippling blow from its tail club.

**WHERE IN THE WORLD?**

*Talarurus* was located in Mongolia, specifically the southeastern parts of what is now the Gobi Desert.

**SNOUT**
At the front of its snout, *Talarurus'* nostrils joined together, creating a single opening. The advantage of this remains unclear.

**FOSSIL EVIDENCE**
*Talarurus* was first discovered in the 1950s and is known from at least five individual specimens, including an almost complete skeleton. It has yet to receive a complete description, but it probably roamed the fertile lands of a floodplain, like a hippopotamus, between 95 and 88 million years ago. To date it more accurately requires comparison with rocks of a similar age from another site. However, there are relatively few sites that show evidence of land-based life from the Late Cretaceous.

**LEGS**
This weighty ankylosaur carried itself on four short but stout limbs. Its broad feet were protected by small, hoof-like nails.

**HOW BIG IS IT?**

DINOSAUR

LATE CRETACEOUS

**TIMELINE** (millions of years ago)

| 540 | 505 | 438 | 408 | 360 | 280 | 248 | 208 | 146 | 65 | 1.8 to today |

# Therizinosaurus

· ORDER · Saurischia · FAMILY · Therizinosauridae · GENUS & SPECIES · Therizinosaurus cheloniformis

LATE CRETACEOUS

## VITAL STATISTICS

| | |
|---|---|
| Fossil Location | Mongolia |
| Diet | Probably herbivorous |
| Pronunciation | THEH-rih-ZEEN-o-SAW-rus |
| Weight | 3–7 tonnes (3.3–7.7 tons) |
| Length | 10m (32ft) |
| Height | Possibly above 5m (16ft) |
| Meaning of name | 'Scythe lizard' because of its huge claws |

### FOSSIL EVIDENCE

*Therizinosaurus* was the largest-known member of the therizinosaurs. The first fossils, just arms and claws, were discovered in Mongolia in the late 1940s by a joint Soviet–Mongolian fossil expedition. In the 1990s, other more complete therizinosaurs were unearthed in Asia, including ones with preserved feathers. These discoveries were followed by descriptions in 2001 and 2005 of the first species from the USA, one of which is represented by a bonebed containing possibly thousands of individuals. With the new specimens, it became clear that the therizinosaurs were a herbivorous branch of the mainly carnivorous non-avian theropods.

DINOSAUR
LATE CRETACEOUS

**With the largest-known claws in the history of life, *Therizinosaurus* cuts a scary and perplexing figure.**

Aside from some birds, therizinosaurs were the only herbivorous theropods. The most distinctive feature of *Therizinosaurus* was the three gigantic claws on each digit of its arms. The fossil claws were more than 1m (3ft) long. In life, with the nail part of the claw covering the bone, they were likely one-third larger still.

### WHERE IN THE WORLD?

So far, *Therizinosaurus* is only known from the Nemegt Formation of southwestern Mongolia. But other genera of therizinosaurs occur elsewhere in Mongolia, as well as in China and the US.

**Claws**
Some have speculated that the long claws of this group were used to rake in leafy branches or tear open ant nests. Their shape indicates that they were not likely used for killing.

**Toes**
The hind limbs of therizinosaurids had four weight-bearing toes, unlike other theropod groups, in which the first toe was reduced to a dewclaw.

HOW BIG IS IT?

**TIMELINE (millions of years ago)**

| 540 | 505 | 438 | 408 | 360 | 280 | 248 | 208 | 146 | 65 | 1.8 to today |

LATE CRETACEOUS

# Thescelosaurus

• ORDER • Ornithischia • FAMILY • Hypsilophodontidae • GENUS & SPECIES • *Thescelosaurus neglectus*

*Thescelosaurus* was one of the last dinosaurs, falling victim to the Cretaceous-Tertiary extinction event some 66 million years ago. Heavily built, with short arms and a long tail, it walked on its hind legs.

## VITAL STATISTICS

| | |
|---|---|
| FOSSIL LOCATION | Canada, western USA |
| DIET | Herbivorous |
| PRONUNCIATION | THES-ke-low-SAWR-us |
| WEIGHT | 300kg (661lb) |
| LENGTH | 3.4m (11ft) |
| HEIGHT | Unknown |
| MEANING OF NAME | 'Surprising lizard' because the first palaeontologists to examine it were surprised to have discovered a new dinosaur |

### WHERE IN THE WORLD?

*Thescelosaurus* was located in Wyoming, Montana and South Dakota, USA, and Alberta and Saskatchewan, in Canada.

**MOUTH**
Possibly browsing on cycads and flowering plants, *Thescelosaurus* had a beak made of horn, small pointed teeth in its jaw, and leaf-shaped cheek teeth.

### FOSSIL EVIDENCE

A specimen of *Thescelosaurus* found in 1993 has caused considerable controversy. A paper published in 2000 described how a CAT scan revealed the fossilized remains of a four-chambered heart, suggesting that *Thescelosaurus* was a warm-blooded animal. This would confirm the theory that non-avian dinosaurs were closer in physiology to birds and mammals than cold-blooded reptiles. However, some palaeontologists remain unconvinced that the fossil is, indeed, a heart, and reject the implications. The specimen, called 'Willo', is displayed at the North Carolina Museum of Natural Sciences.

DINOSAUR

LATE CRETACEOUS

### HOW BIG IS IT?

**LEGS**
Lacking apparent physical defences, *Thescelosaurus*' best strategy was probably to flee. Its powerful hind legs probably made it a good sprinter.

**TIMELINE** (millions of years ago)

| 540 | 505 | 438 | 408 | 360 | 280 | 248 | 208 | 146 | 65 | 1.8 to today |

# LATE CRETACEOUS

# Titanosaurus

• ORDER • Saurischia • FAMILY • Titanosauridae • GENUS & SPECIES • Titanosaurus indicus

One of the last of the giant sauropods to roam the Earth, *Titanosaurus* was protected by bony plates across its back.

## VITAL STATISTICS

| | |
|---|---|
| FOSSIL LOCATION | India |
| DIET | Herbivorous |
| PRONUNCIATION | Tie-TAN-oh-SAWR-us |
| WEIGHT | 9.9–15.4 tonnes (9–14 tons) |
| LENGTH | 12–18m (39–59ft) |
| HEIGHT | Unknown |
| MEANING OF NAME | 'Titanic lizard' in reference to the size of the vertebrae |

### WHERE IN THE WORLD?

Specimens attributed to *Titanosaurus* have been found in Argentina and Hungary, but the only true specimens are from India.

**TAIL**
It is unlikely that *Titanosaurus* used its long tail in defence; it was simply a counterbalance to its long neck.

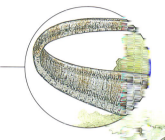

## FOSSIL EVIDENCE

*Titanosaurus* was found in India in 1870, when limb bones and a few vertebrae were discovered. For years afterwards, *Titanosaurus* was a 'wastebin taxon', that is, a species to which specimens are assigned because they have characteristics that fit nowhere else. However, subsequent discoveries have demonstrated that these characteristics are not unique and, indeed, belong to related dinosaurs whose identity has been satisfactorily established. What this means is that *Titanosaurus* is now considered by most scientists a *nomen dubium* (Latin for 'dubious name') as its characteristics cannot be distinguished from other dinosaurs.

DINOSAUR
LATE CRETACEOUS

### HOW BIG IS IT?

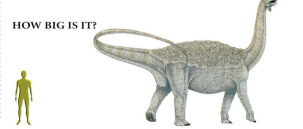

**BODY**
*Titanosaurus*' vast gut probably contained gastroliths (small stones) to help digest the vast amounts of vegetation it ate.

## TIMELINE (millions of years ago)

| 540 | 505 | 438 | 408 | 360 | 280 | 248 | 208 | 146 | 65 | 1.8 to today |

LATE CRETACEOUS

# Tsintaosaurus

• **ORDER** • Ornithischia • **FAMILY** • Hadrosauridae • **GENUS & SPECIES** • *Tsintaosaurus spinorhinus*

*Tsintaosaurus* was a large plant-eating dinosaur, it had a toothless beak and a powerful jaw filled with self-sharpening teeth.

| VITAL STATISTICS | |
|---|---|
| Fossil Location | China |
| Diet | Herbivorous |
| Pronunciation | Sin-taus-SAWR-us |
| Weight | 2700kg (6000lb) |
| Length | 10m (33ft) |
| Height | Unknown |
| Meaning of name | 'Qingdao lizard' after the Chinese city near where the fossil was found |

### WHERE IN THE WORLD?

To date, *Tsintaosaurus* has been located only in China.

**FOSSIL EVIDENCE**

*Tsintaosaurus* is another dinosaur that demonstrates how difficult the interpretation of fossil evidence can be. It was nicknamed 'Chinese unicorn' because it was reconstructed with a thin, hollow crest on its head jutting forwards like a unicorn's horn. Some palaeontologists, however, argued that this was either a snout bone that had been incorrectly placed, or even a bone from another creature. Later, a second specimen was found with the same crest. Now it is argued that the crest did not protrude but lay flat along the nose.

DINOSAUR

LATE CRETACEOUS

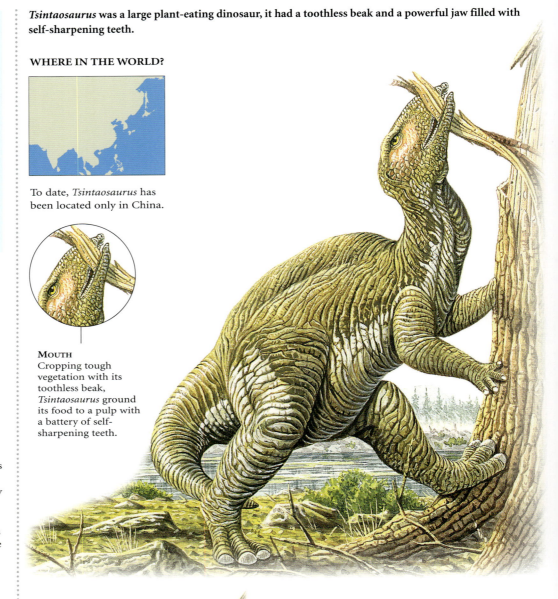

**Mouth**
Cropping tough vegetation with its toothless beak, *Tsintaosaurus* ground its food to a pulp with a battery of self-sharpening teeth.

### HOW BIG IS IT?

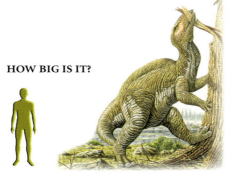

**Crest**
Even though the crest has been shown to be unlike a unicorn, it seems clear that the *Tsintaosaurus* did have a crest of presently unknown form.

**TIMELINE (millions of years ago)**

| 540 | 505 | 438 | 408 | 360 | 280 | 248 | 208 | 146 | 65 | 1.8 to today |

LATE CRETACEOUS

# Tylocephale

• ORDER • Ornithischia • FAMILY • Pachycephalosauridae • GENUS & SPECIES • *Tylocephale gilmorei*

### VITAL STATISTICS

| | |
|---|---|
| FOSSIL LOCATION | Mongolia |
| DIET | Herbivorous |
| PRONUNCIATION | TIE-low-SEF-ah-lee |
| WEIGHT | Unknown |
| LENGTH | 2.5m (8ft) |
| HEIGHT | Unknown |
| MEANING OF NAME | 'Swollen head' in reference to its thick skull |

The purpose of this bonehead's thickened dome is unclear. It probably did not provide protection, as *Tylocephale* had a small brain.

**WHERE IN THE WORLD?**

To date, *Tylocephale* has been located only in Mongolia.

**HEAD**
*Tylocephale* had one of the tallest domes of any pachycephalosaur. It flared out into a frill lined with bony bumps.

**MOUTH**
With sharp serrated teeth, boneheads probably enjoyed a mixed diet of leaves, seeds and fruit— and possibly even insects.

### FOSSIL EVIDENCE

*Tylocephale* was described in 1974 from an incomplete skull that was missing part of its snout and dome. It was one of several pachycephalosaurs that lived during the Late Cretaceous, including *Prenocephale* from central Asia, and *Stegoceras* and *Stygimoloch* from North America. The fossil evidence suggests that their common ancestor evolved in Asia, and emigrated via a landbridge to North America. Curiously, the evidence also suggests that these North American boneheads then migrated back to Asia, evolving into further species, including *Tylocephale*.

DINOSAUR

LATE CRETACEOUS

**HOW BIG IS IT?**

**TIMELINE** (millions of years ago)

| 540 | 505 | 438 | 408 | 360 | 280 | 248 | 208 | 146 | 65 | 1.8 to today |

LATE CRETACEOUS

# Tylosaurus

• ORDER • Squamata • FAMILY • Mosasauridae • GENUS & SPECIES • Various species within the genus *Tylosaurus*

*Tylosaurus* was one of the top marine predators of its time and the largest mosasaur in its environment.

| VITAL STATISTICS | |
|---|---|
| Fossil Location | USA, New Zealand |
| Diet | Carnivorous |
| Pronunciation | TY-lo-SORE-us |
| Weight | 10 tonnes (11 ton) |
| Length | 15m (49ft) |
| Height | Unknown |
| Meaning of name | 'Knob lizard' because its snout extended well beyond the foremost teeth |

### WHERE IN THE WORLD?

Fossils of *Tylosaurus* are very common in US states. Tylosaurs are also known from New Zealand.

**Snout**
The projecting snout of *Tylosaurus* may have been used as a ram to stun prey or in combat with others of its kind, but it was probably actually too fragile and sensitive for this purpose.

### FOSSIL EVIDENCE

Complete *Tylosaurus* fossils are common and show that the animal was often buried before it could be scavenged and disturbed. The first specimen was found near Monument Rocks, Kansas, in 1868. The rocks in this area are called the Niobrara Chalk and are famous for their abundant mosasaurs and other marine reptiles, fish, bivalves and other Late Cretaceous marine organisms. Even giant squid have been found here and may have been prey for *Tylosaurus*. The fine-grained nature of the chalk sediments, themselves composed of countless billions of microfossils, ensured the detailed preservation of the fossils found there.

### HOW BIG IS IT?

**Teeth**
*Tylosaurus*' teeth were replaced, as needed, throughout the life of the animal.

PREHISTORIC ANIMAL

LATE CRETACEOUS

**TIMELINE (millions of years ago)**

| 540 | 505 | 438 | 408 | 360 | 280 | 248 | 208 | 146 | 65 | 1.8 to today |

LATE CRETACEOUS

# Wannanosaurus

• ORDER • Ornithischia • FAMILY • Pachycephalosauridae • GENUS & SPECIES • *Wannanosaurus yansiensis*

## VITAL STATISTICS

| | |
|---|---|
| FOSSIL LOCATION | China |
| DIET | Herbivorous |
| PRONUNCIATION | Wah-NAN-oh-SAWR-us |
| WEIGHT | 1.5kg (3lb) |
| LENGTH | 60cm (24in) |
| HEIGHT | Unknown |
| MEANING OF NAME | 'Wannan lizard' after the Chinese province where the fossil was discovered |

Its diminutive size made *Wannanosaurus* vulnerable. It may have travelled alongside sauropods, relying on their intimidating size to deter predators. Its ultimate defence, though, was probably to flee or to hide in the undergrowth.

### WHERE IN THE WORLD?

To date, *Wannanosaurus* has been located only in China.

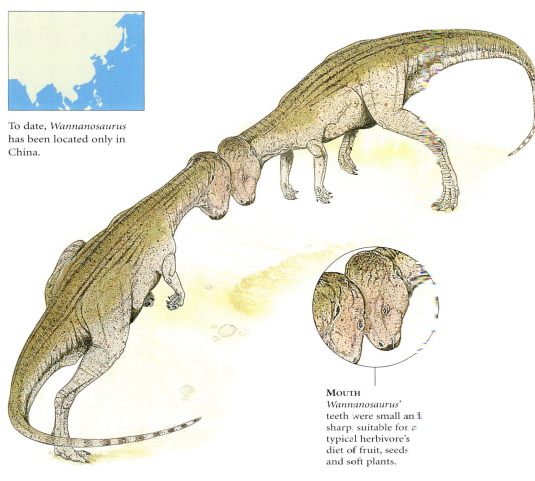

### FOSSIL EVIDENCE

*Wannanosaurus* is known from a single, incomplete skeleton. Fragments include a partial skull roof and lower jaw, an upper and a lower leg, and part of a rib. *Wannanosaurus* was the smallest known bonehead. The thickened skulls common to this family continue to divide palaeontologists. If these dinosaurs did not engage in head-butting contests – some analyses of both the skull and neck bones suggests they did not – what was the purpose of their thick skulls?

### MOUTH
*Wannanosaurus'* teeth were small and sharp, suitable for a typical herbivore's diet of fruit, seeds and soft plants.

### HIND LEGS
This small, relatively slow dinosaur walked on its hind legs, but may have dropped onto four legs when feeding.

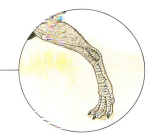

### HOW BIG IS IT?

DINOSAUR

LATE CRETACEOUS

## TIMELINE (millions of years ago)

540 | 505 | 438 | 408 | 360 | 280 | 248 | 208 | 146 | 65 | 1.8 to today

LATE CRETACEOUS

# Ankylosaurus

| VITAL STATISTICS | |
|---|---|
| Fossil Location | USA, Canada, Bolivia |
| Diet | Herbivorous |
| Pronunciation | an-KY-lo-SAW-rus |
| Weight | 4000kg (3.9 tons) |
| Length | 10.7m (35ft) |
| Height | 1.2m (3.9ft) |
| Meaning of name | 'Fused lizard' because of the many areas of fused bone in the skeleton |

*Ankylosaurus*, a leathery-skinned herbivore, was huge and covered in thick, oval-shaped armour plating. Even its eyes were protected from attack by bony plates. But *Ankylosaurus* was vulnerable on its unarmoured underside, so it could be injured or killed if flipped over onto its back. Despite its broad skull, *Ankylosaurus* was not one of the clever dinosaurs; its brain was tiny.

### WHERE IN THE WORLD?

Remains have been found in Montana, western USA, Alberta, Canada and Sucre, Bolivia, in South America.

**TAIL**
The armoured club on the end of *Ankylosaurus*' tail was used against any enemy that came within its range.

**FOSSIL EVIDENCE**
Despite its huge size and great weight, *Ankylosaurus* was likely fairly fast on its feet. It could run at a reasonable trot, to judge by the tracks of a close relative discovered in 1996 near Sucre in the Andes mountains of Bolivia. In the other locations where this dinosaur lived up to 65 million years ago, two *Ankylosaurus* skulls have been found, together with three part-skeletons. These skeletons included samples of *Ankylosaurus* armour and its club tail, which could be a damaging weapon when the dinosaur had to defend itself.

**FEROCIOUS APPEARANCE**
Although plant-eaters like *Ankylosaurus* were not usually as ferocious or bloodthirsty as carnivores like the dreaded *Tyrannosaurus rex*, they made a pretty terrifying sight for any other creature minded to attack them. Head-on, *Ankylosaurus* bristled with long, sharp spikes and its head was surrounded by them. One shake of that spiked head could do a lot of damage.

DINOSAUR

LATE CRETACEOUS

# LATE CRETACEOUS

• **ORDER** • Ornithischia • **FAMILY** • Ankylosauridae • **GENUS & SPECIES** • *Ankylosaurus magniventris*

## TAIL CLUB
The club at the end of *Ankylosaurus*' tail was made of large osteoderms (bony scales) that formed within the skin to make it very hard. These osteoderms were attached to the last few vertebrae in *Ankylosaurus*' tail and these were supported by the last seven tail vertebrae. Scientists have discovered thick tendons that were once attached to the vertebrae; combined with the rest of the tail structure, possibly allowed *Ankylosaurus* to exert enough force to break an assailant's bones.

**SPIKES**
Thick armour plating covered the topside of *Ankylosaurus*.

## CLUB VICTIM
The dinosaur in the background had scant chance against *Ankylosaurus* unless it flipped the ankylosaur over and exposed its unprotected underside. In fact, it looks as if the dinosaur is about to receive a blow from *Ankylosaurus*' swiftly swinging tail club.

## HOW BIG IS IT?

**TIMELINE (millions of years ago)**

| 540 | 505 | 438 | 408 | 360 | 280 | 248 | 208 | 146 | 65 | 1.8 to today |

LATE CRETACEOUS
# Ankylosaurus

## LATE CRETACEOUS

• **ORDER** • Ornithischia • **FAMILY** • Ankylosauridae • **GENUS & SPECIES** • *Ankylosaurus magniventris*

### THE DISCOVERY OF *ANKYLOSAURUS*

In 1906, a team led by Barnum Brown, the famous American fossil hunter, working at the Hell Creek Formation in the state of Montana, in northwestern USA, discovered the top of a skull, some vertebrae, part of a shoulder girdle, some ribs and some samples of body armour. These finds looked familiar to Brown: in 1900, he had found the skeleton of a theropod (two-footed) dinosaur while excavating the Lance Formation in Wyoming. More than 75 osteoderms were also unearthed and were later matched up with the remains of the 1906 discovery. Then, in 1910, Brown was digging at the Scollard Formation in Alberta, Canada, where he found the first known ankylosaur tail club, leg bones, ribs and more armour. It seemed evident that the specimens of 1900, 1906 and 1910 came from the same dinosaur. Barnum Brown, however, did not wait for the final discovery; in 1908 he had already named the dinosaur *Ankylosaurus*. *Ankylosaurus* was not Barnum Brown's only great find. In 1902, also at Hell Creek Formation, he discovered the remains of *Tyrannosaurus rex*, one of the most famous dinosaurs of them all.

LATE CRETACEOUS

# Nodosaurus

| VITAL STATISTICS | |
|---|---|
| Fossil Location | North America |
| Diet | Herbivorous |
| Pronunciation | No-doe-SORE-us |
| Weight | Unknown |
| Length | 4–6m (13–20ft) |
| Height | Unknown |
| Meaning of name | 'Knob lizard', for the knobby armour on its back |

*Nodosaurus* was one of the first armoured dinosaurs to be discovered in North America. It is a herbivorous dinosaur with knobby plates that covered its skin, these plates gave the dinosaur its name.

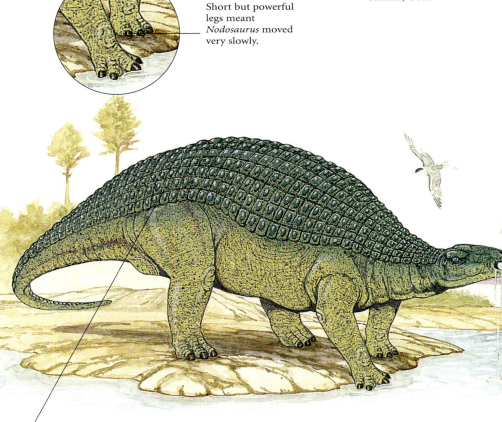

**WHERE IN THE WORLD?**

*Nodosaurus* has been found in Wyoming and Kansas, USA.

**LEGS**
Short but powerful legs meant *Nodosaurus* moved very slowly.

**FOSSIL EVIDENCE**
*Nodosaurus* was among one of the first Ankylosaurs that scientists studied. No complete fossils of *Nodosaurus* have ever been discovered which is why there is speculation as to whether this dinosaur had side spikes like other Nodosaurs. *Nodosaurus* was classified in 1989 on the basis of remains discovered in Wyoming and Kansas, USA. It is believed that *Nodosaurus* existed in the Late Cretaceous period.

**ARMOUR**
*Nodosaurus* had thick bony plates on its back which protected it against predators.

**IN THE FACE OF DANGER**
Some prehistoric instincts persist to this day. When present-day lizards are in danger of attack, they flatten themselves on the ground. This is what *Nodosaurus* may have done, not only to protect its vulnerable underside, but also to prevent an attacker from turning its body over to get a better chance of doing the dinosaur most damage. The lizard flattens itself, in part, so that its body camouflage will blend in more easily with the ground, but the principle is the same.

DINOSAUR

LATE CRETACEOUS

LATE CRETACEOUS

• **ORDER** • Nodosauria • **FAMILY** • Nodosauridae • **GENUS & SPECIES** • *Nodosaurus. textilis*

**SPIKES**
No side spikes have yet been found of *Nodosaurus*, but it may have had them like other nodosaurs had.

## DETECTIVE WORK

Because important parts of prehistoric creatures may be missing or badly damaged when their fossils are found, scientists often have to do detective work on those remains they have been able to discover. For example, the few, incomplete *Nodosaurus* specimens were not found with side spikes common in its close relatives. But since this type of armour was so common in its group, it is likely that *Nodosaurus* also had similar protection.

**HOW BIG IS IT?**

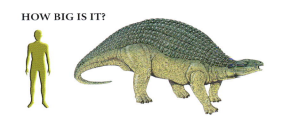

**TIMELINE (millions of years ago)**

| 540 | 505 | 438 | 408 | 360 | 280 | 248 | 208 | 146 | 65 | 1.8 to today |

317

LATE CRETACEOUS

# Saichania

| VITAL STATISTICS | |
|---|---|
| Fossil Location | Mongolia |
| Diet | Herbivorous |
| Pronunciation | Sigh-COON-ee-ah |
| Weight | 1800kg (3968lb) |
| Length | 6.7m (22ft) |
| Height | Unknown |
| Meaning of name | 'Beautiful one' in the Mongolian language |

Although not specified in the original description, the translation of *Saichania*'s name ('beautiful one') likely refers to the striking preservation of the fossil skull. The find, first described in 1977, included two complete skulls. *Saichania* is a close relative of *Ankylosaurus*. It was well suited to the hot, dry climate in which it lived 80 million years ago, with nasal passages that may have moisturized the dry air it breathed in from its harsh environment.

**WHERE IN THE WORLD?**

*Saichania* was found in Mongolia.

**FOSSIL EVIDENCE**
The skeleton of *Saichania* was preserved by a sandstorm that killed the dinosaur but ensured that its body remained more or less intact. Erosion destroyed the rear section of *Saichania*, but further partial skeletons discovered in Mongolia have filled in the information gaps that were left. *Saichania* was very well protected by its thick armour. Ridged plates covered its body and tail. *Saichania* had a powerful club at the end of its tail, similar to that of an *Ankylosaurus*.

DINOSAUR

LATE CRETACEOUS

**HEAD**
The bony plates covering *Saichania*'s head gave the impression that it was perpetually snarling.

**MISSING PARTS**
The picture of a complete *Saichania* came from several skeletons because parts of each of them were destroyed or missing.

**TAIL**
Formed of separate pieces of bone fused together – the tail was an excellent weapon against predators.

# LATE CRETACEOUS

• **ORDER** • Ornithischia • **FAMILY** • Ankylosauridae • **GENUS & SPECIES** • *Saichania chulsanensis*

### DISCOVERIES IN THE DESERT
*Saichania* was found in Mongolia's Gobi Desert in 1971. Mongolia, like North America, is a rich source for dinosaur hunters.

#### BABY DINOSAURS
These baby dinosaurs were lucky to emerge alive from their eggs. Many dinosaur eggs never hatched because sandstorms buried them.

### RICH FINDINGS
The Gobi Desert in Mongolia, where the first dinosaur find was discovered in 1922, has since proved a phenomenally rich source for palaeontologists. For example, in a single week in 2006, 67 dinosaur skeletons were found at a site two days' drive from the Mongolian capital, Ulan Bator. The previous year, 30 dinosaur fossils were found in the same area.

**HOW BIG IS IT?**

**TIMELINE** (millions of years ago)

| 540 | 505 | 438 | 408 | 360 | 280 | 248 | 208 | 146 | 65 | 1.8 to today |

319

LATE CRETACEOUS
# Saichania

# LATE CRETACEOUS

• **ORDER** • Ornithischia • **FAMILY** • Ankylosauridae • **GENUS & SPECIES** • *Saichania chulsanensis*

**DINOSAURS IN MONGOLIA**

In 1993, a team of scientists from the American Museum of Natural History found specimens that helped to picture why so many complete, articulated dinosaur fossils are found in certain areas of the Gobi Desert of Mongolia. In Ukhaa Tolgod ('Brown Hills'), southern Mongolia, it appears that, at times, large amounts of water had soaked into the desert sands, causing massive avalanches to run down the sides of the dunes. The dinosaurs became completely buried before other animals could scavenge them or the weather could erode and destroy them. The speed of the event also accounted for the incredible state of preservation in which the dinosaurs' fossils were found. The desert sands had, of course, always been treacherous and avalanches were only one manifestation of such dangers. One *Saichania chulsanensis* individual was discovered well preserved in sandstone where it had been possibly overtaken by a sandstorm and died millions of years before.

LATE CRETACEOUS

# Struthiosaurus

| VITAL STATISTICS | |
|---|---|
| Fossil Location | Across southern Europe |
| Diet | Herbivorous |
| Pronunciation | Struth-io-SAWR-us |
| Weight | 300kg (661lb) |
| Length | Up to 2m (6ft 6in) |
| Height | Unknown |
| Meaning of name | 'Ostrich lizard', because of the alleged resemblance of the back of its skull to that of a bird |

As dinosaurs go, *Struthiosaurus* was on the small side. It was, in fact, one of the smallest dinosaurs yet discovered. *Struthiosaurus* was first described in 1871 by the German palaeontologist Emanuel Bunzel. Further *Struthiosaurus* finds were made in 1915 in Transylvania, in Romania, and more were uncovered in Languedoc in southern France in 2003. The curious thing about the finds in southern Europe is that all dinosaurs from this region have turned out to be dwarf species. This includes a sauropod, a hadrosaurid and an iguanodontid.

**WHERE IN THE WORLD?**

*Struthiosaurus* was found in France, Hungary, Austria and Romania.

## FOSSIL EVIDENCE

A fair amount of fossil evidence exists for *Struthiosaurus*, but palaeontologists have not agreed about how to interpret it. In 1871, Emanuel Bunzel classified it as a separate order of reptiles known as Ornithocephala, meaning 'bird heads'. In 1915, Baron Franz von Nopsca, who worked on the Transylvanian finds, classed *Struthiosaurus* as one of the smallest armoured dinosaurs with a bird-like head. But in 1994, two other palaeontologists examining the *Struthiosaurus* fossils concluded that they were those of a young nodosaur, a relation of *Ankylosaurus*.

DINOSAUR

LATE CRETACEOUS

**Armour**
*Struthiosaurus* had kinds of armour on its back and tail, including large spine plates and long shoulder spines.

**Skull**
*Struthiosaurus* acquired the curious name 'ostrich lizard' because the back of its head looked like the skull of a bird.

# LATE CRETACEOUS

• **ORDER** • Ornithischia • **FAMILY** • Nodosauridae •
• **GENUS & SPECIES** • Several species within the genus *Struthiosaurus*

## SMALL SPECIES

*Struthiosaurus* was a small dinosaur that evolved on a chain of islands that would later be part of the continent of Europe. As the resources available for survival on each island were possibly limited, the dinosaurs did not evolve and grow to enormous sizes that would have needed to consume huge quantities of vegetation to survive.

**LEGS**
*Struthiosaurus* had long legs for an Ankylosaur. Its legs were not as sturdy as other Ankylosaurs as they did not have to support such a huge weight.

**HOW BIG IS IT?**

**TIMELINE (millions of years ago)**

| 540 | 505 | 438 | 408 | 360 | 280 | 248 | 208 | 146 | 65 | 1.8 to today |

LATE CRETACEOUS

# Xiphactinus

• ORDER • Pachycormiformes • FAMILY • Ichthyodectidae • GENUS & SPECIES • *Xiphactinus audax*

| VITAL STATISTICS | |
|---|---|
| FOSSIL LOCATION | North America |
| DIET | Carnivorous |
| PRONUNCIATION | Zy-FACT-in-us |
| WEIGHT | at least 227kg (500lb) |
| LENGTH | 4.3–5.1m (14–17ft) |
| HEIGHT | Unknown |
| MEANING OF NAME | 'Sword ray' for the sword-like fin rays of its pectoral fins |

An efficient predator, Xiphactinus could swim at very high speeds. It may have stunned prey with a smack of its forked tail. It may have leaped above the waves to dislodge parasites.

**MOUTH**
Xiphactinus' upturned jaw could open wide enough to let this fish swallow prey the size of a human adult whole.

### WHERE IN THE WORLD?

*Xiphactinus* swam in the Western Interior Sea, above what is now the middle of North America.

## FOSSIL EVIDENCE

At least a dozen specimens of *Xiphactinus* have been found containing the remains of prey in its stomach. Inside one is a perfectly preserved ichthyodectid, *Gillicus arcatus*. One assumption is that it struggled after having been swallowed whole and ruptured one of the organs of its predator, causing it to die shortly afterwards. One fossil of *Xiphactinus* with a *Gillicus* in its gust is displayed at the Sternberg Museum of Natural History in Kansas, USA. In 2002, an incomplete skull was found in the Czech Republic, and may be a new species of *Xiphactinus*.

| PREHISTORIC ANIMAL |
| LATE CRETACEOUS |

**HOW BIG IS IT?**

**BODY**
A dark blue-black and silver belly meant that *Xiphactinus* may have been dark above and light below — a common type of camouflage for ocean fish.

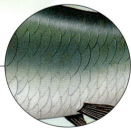

**TIMELINE (millions of years ago)**

540 | 505 | 438 | 408 | 360 | 280 | 248 | 208 | 146 | 65 | 1.8 to today

# Deinosuchus

**LATE CRETACEOUS**

• ORDER • Crocodilia • FAMILY • (Superfamily) Alligatoroidea •
• GENUS & SPECIES • Several species within the genus *Deinosuchus*

### VITAL STATISTICS

| | |
|---|---|
| FOSSIL LOCATION | North America |
| DIET | Carnivorous |
| PRONUNCIATION | DIE-no-SOO-kus |
| WEIGHT | 2–3 tonnes (2.2–3.3 tons) |
| LENGTH | 10–12m (33–40ft) |
| HEIGHT | Unknown |
| MEANING OF NAME | 'Terrible crocodile' |

This was one of the largest of the prehistoric crocodiles and is remarkably similar to the crocs of today. It was a big, armoured reptile with powerful jaw.

### WHERE IN THE WORLD?

*Deinosuchus* was found in various US states from Texas to New Jersey, and across the border of Mexico.

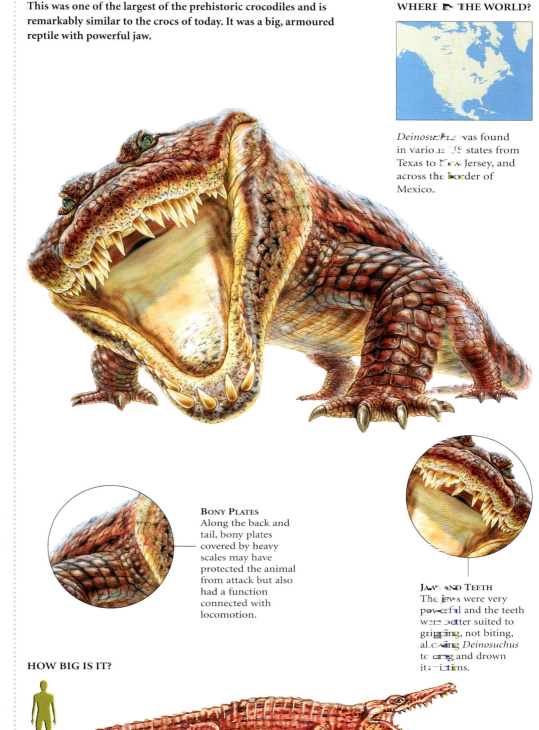

**BONY PLATES**
Along the back and tail, bony plates covered by heavy scales may have protected the animal from attack but also had a function connected with locomotion.

**JAW AND TEETH**
The jaws were very powerful and the teeth were better suited to gripping, not biting, allowing *Deinosuchus* to drag and drown its victims.

### FOSSIL EVIDENCE

Only parts of *Deinosuchus* have been found, the most useful being a complete skull that is 2m (6ft 6in) long, from which its size has been estimated. The jaw was powered by strong muscles and equipped with blunt teeth. It would have used these to drag prey that had strayed too near the edge of the water under the waves, possibly knocking them breathless with the 'death roll' twist. It could have preyed on the plentiful hadrosaurs of the time, and also eaten fish.

### HOW BIG IS IT?

PREHISTORIC ANIMAL
LATE CRETACEOUS

**TIMELINE** (millions of years ago)

| 540 | 505 | 438 | 408 | 360 | 280 | 248 | 208 | 146 | 65 | 1.8 to today |

LATE CRETACEOUS

# Euoplocephalus

• **ORDER** • Ornithischia • **FAMILY** • Ankylosauridae • **GENUS & SPECIES** • *Euoplocephalus tutus*

| VITAL STATISTICS | |
|---|---|
| Fossil Location | North America |
| Diet | Herbivorous |
| Pronunciation | YOU-oh-plo-SEF-ah-lus |
| Weight | 2–3.3. tonnes (1.8–3 tons) |
| Length | 6m (20ft) |
| Height | Unknown |
| Meaning of name | 'Well-armoured head' after the armoured plates on its skull |

Its entire head and body were covered with armour that was studded with spikes, and Euoplocephalus even had bony shutters that slid over its eyes for protection. Horns projected from the back of its skull.

### WHERE IN THE WORLD?

*Euoplocephalus* was found in North America, specifically Alberta, Canada, and Montana, USA.

**ARMOUR PLATING**
The armour plating developed in narrow bands, allowing *Euoplocephalus* to be surprisingly agile.

### FOSSIL EVIDENCE

*Euoplocephalus* has been found in several locations, making it one of the most common ankylosaurs found in North America. Only isolated specimens have been found, suggesting that it lived a solitary life, although there is fossil evidence that some of its relatives lived in herds. The armour plating running across its back offered *Euoplocephalus* good protection. Indeed, it was probably vulnerable only if flipped over on its back: an analysis of dinosaur bones in Alberta, Canada, reveal no bite marks on *Euoplocephalus* or any of its armoured relatives.

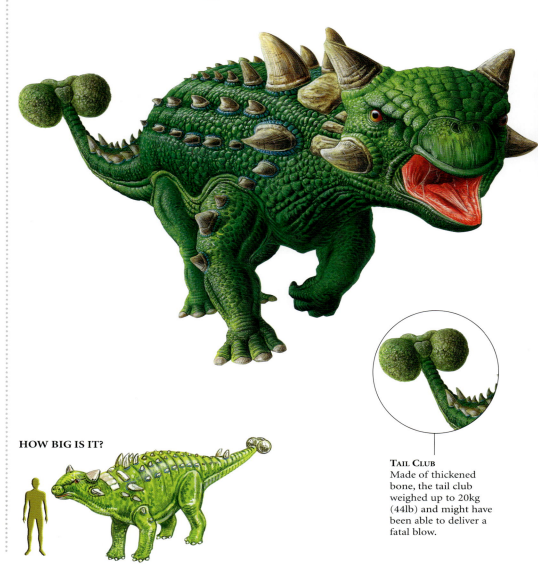

**HOW BIG IS IT?**

**TAIL CLUB**
Made of thickened bone, the tail club weighed up to 20kg (44lb) and might have been able to deliver a fatal blow.

DINOSAUR
LATE CRETACEOUS

**TIMELINE (millions of years ago)**

| 540 | 505 | 438 | 408 | 360 | 280 | 248 | 208 | 146 | 65 | 1.8 to today |

# LATE CRETACEOUS

# Gallimimus

• ORDER • Saurischia • FAMILY • Ornithomimidae • GENUS & SPECIES • *Gallimimus ballutus*

*Gallimimus* was a large, ostrich-like dinosaur whose long legs and hollow bones possibly allowed it to run faster than almost any other animal of its time, including the carnivores that hunted it.

### VITAL STATISTICS

| | |
|---|---|
| Fossil Location | Mongolia |
| Diet | Omnivorous |
| Pronunciation | Gal-i-MY-mus |
| Weight | 440kg (970lb) |
| Length | 5m (17ft) |
| Height | 3.4m (11ft) |
| Meaning of name | 'Chicken mimic' because of the similarity of its neck structure to that of chickens |

### WHERE IN THE WORLD?

Specimens have been found in the Bashin Tsav region of southeastern Mongolia.

**EYES**
Eyes on the side of its head suggest that *Gallimimus* would have had an all-round view, but little depth of vision.

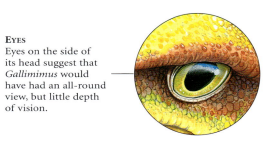

### FOSSIL EVIDENCE

Various individuals have been found, both adult and juvenile, some of them in large bonebeds that suggest *Gallimimus* moved in herds. Long limbs with slender foot bones and short toes indicate that it could run well, possibly as fast as today's ostriches, which reach 70kmh (43mph). Mystery surrounds its diet. The toothless beak could not cope with tough meat, and the claws seem ill-equipped for grasping. It may have eaten small insects, leaves and berries and possibly dinosaur eggs scooped from the ground with its claws or long shovel-like bill.

**HOW BIG IS IT?**

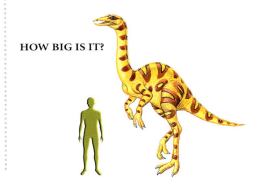

**HANDS**
The three-fingered hands were poorly-designed for grasping and might have been used for scratching and digging at the ground.

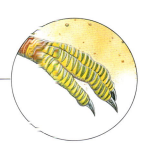

DINOSAUR
LATE CRETACEOUS

**TIMELINE (millions of years ago)**

| 540 | 505 | 438 | 408 | 360 | 280 | 248 | 208 | 146 | 65 | 1.8 to today |

LATE CRETACEOUS

# Hesperornis

• ORDER • Hesperornithiformes •
• FAMILY • Hesperornithidae • GENUS & SPECIES • Several species within the genus *Hesperornis*

## VITAL STATISTICS

| | |
|---|---|
| FOSSIL LOCATION | North America |
| DIET | Carnivorous |
| PRONUNCIATION | hess-puh-RAWR-nihs |
| WEIGHT | Unknown |
| LENGTH | 1.5m (5ft) |
| HEIGHT | Unknown |
| MEANING OF NAME | 'Western bird' because it was found in the western US |

### WHERE IN THE WORLD?

*Hesperornis* hunted in the North American Inland Sea, the Turgai Strait and the prehistoric North Sea.

**Hesperornis was a huge bird that could not fly or walk on land, but was a fast swimmer and accomplished diver. It was one of the greatest predators of the ocean.**

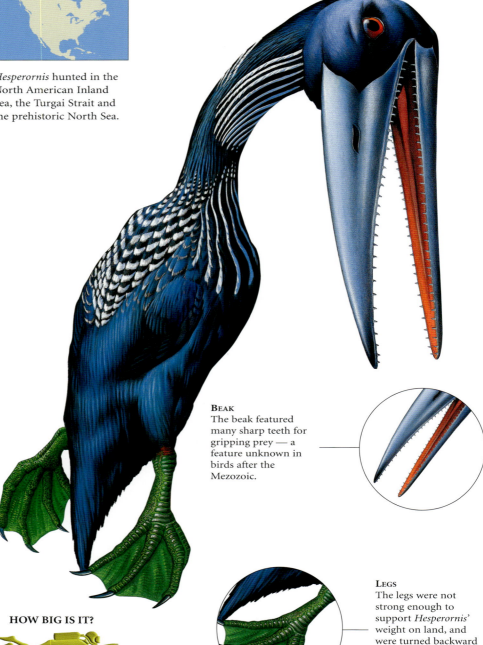

### FOSSIL EVIDENCE

This was one of the largest birds of the Age of Dinosaurs. On land, *Hesperornis* could only dig with its legs and push itself along on its belly like a sea turtle, making it very vulnerable. It may have nested on isolated islands, or stayed in the sea and given birth to live young. In water it could dive and swim, powered by backward-facing legs with webbed or lobed feet.

**BEAK**
The beak featured many sharp teeth for gripping prey — a feature unknown in birds after the Mezozoic.

**LEGS**
The legs were not strong enough to support *Hesperornis'* weight on land, and were turned backward to improve their value in swimming.

DINOSAUR
LATE CRETACEOUS

### HOW BIG IS IT?

## TIMELINE (millions of years ago)

| 540 | 505 | 438 | 408 | 360 | 280 | 248 | 208 | 146 | 65 | 1.8 to today |

# Lambeosaurus

LATE CRETACEOUS
• ORDER • Ornithischia •
• FAMILY • Hadrosauridae • GENUS & SPECIES • Several species within the genus *Lambeosaurus*

*Lambeosaurus* was one of the largest of the duck-billed dinosaurs and featured a head crest that varied between species. This crest continues to baffle paleontologists. Was it for display, to make sounds, or for picking up scent?

## VITAL STATISTICS

| | |
|---|---|
| FOSSIL LOCATION | North America |
| DIET | Herbivorous |
| PRONUNCIATION | Lam-bee-oh-SAWR-us |
| WEIGHT | Up to 23 tonnes (25 tons) |
| LENGTH | 9–15m (30–50ft) |
| HEIGHT | 2.1m (7ft) at the hips |
| MEANING OF NAME | 'Lambe's lizard' after Canadian fossil hunter Charles Lambe |

## FOSSIL EVIDENCE

More than 20 fossils have been found and a number of species named, some of which may simply be juveniles rather than new, small species. This variety has led to widely varying estimates of *Lambeosaurus*' typical size. All of the *Lambeosaurus* species had crests. One was shaped like a hatchet buried in the skull; another was a single ridge. The purpose of these crests is unknown. Their hollow structure may have added volume to the animal's cries. Alternatively, they may have been for ritual display or perhaps to show gender.

DINOSAUR
LATE CRETACEOUS

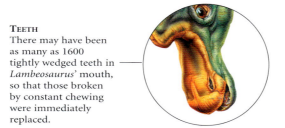

**TEETH**
There may have been as many as 1600 tightly wedged teeth in *Lambeosaurus*' mouth, so that those broken by constant chewing were immediately replaced.

## WHERE IN THE WORLD?

Remains show that *Lambeosaurus* roamed in Alberta, Canada, and in Montana and New Mexico in the USA.

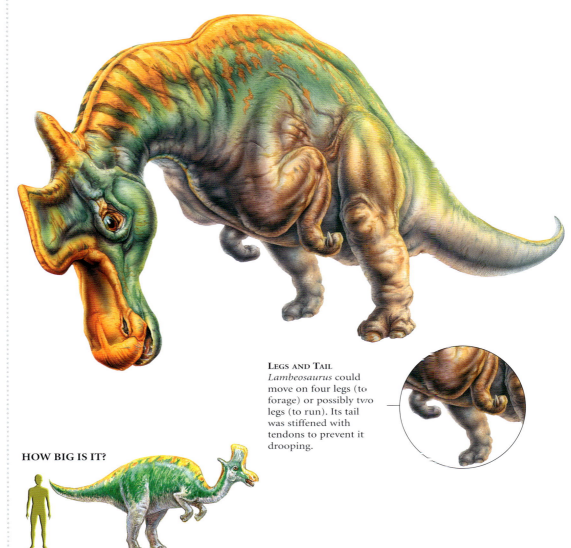

**LEGS AND TAIL**
*Lambeosaurus* could move on four legs (to forage) or possibly two legs (to run). Its tail was stiffened with tendons to prevent it drooping.

**HOW BIG IS IT?**

TIMELINE (millions of years ago)

| 540 | 505 | 438 | 408 | 360 | 280 | 248 | 208 | 146 | 65 | 1.8 to today |

LATE CRETACEOUS

# Libonectes

• **ORDER** • Plesiosauria • **FAMILY** • Elasmosauridae • **GENUS & SPECIES** • *Libonectes morgani* and *L. atlasense*

*Libonectes* was a type of very long-necked plesiosaur known as an elasmosaur, a group of marine animals with four strong, paddle-like flippers that swam in the Late Cretaceous seas.

| VITAL STATISTICS | |
|---|---|
| Fossil Location | USA |
| Diet | Carnivorous |
| Pronunciation | lee-bon-EK-teez |
| Weight | 5–8 tonnes (5.5–9 tons) |
| Length | 7–14m (23–26ft) |
| Height | Unknown |
| Meaning of name | 'Southwest swimmer' because it was found in the southwest US |

## FOSSIL EVIDENCE

One early fossil hunter thought the neck of *Libonectes*' relatives was a tail, as they had never seen a creature shaped like it. Then it was believed that their necks were as flexible as a snake's body. Now it is known that the neck was relatively inflexible. *Libonectes* is thought to have used its swimming proficiency to follow shoals of fish and attack from underneath, snaring them in its cage-like mouth. It swallowed rocks which may have been used to improve its stability in the water.

PREHISTORIC ANIMAL

LATE CRETACEOUS

**NECK**
There were 62 bones in the long neck, which made up almost half of the length of *Libonectes*' body.

**WHERE IN THE WORLD?**

Remains have been found in Texas and Kansas, USA.

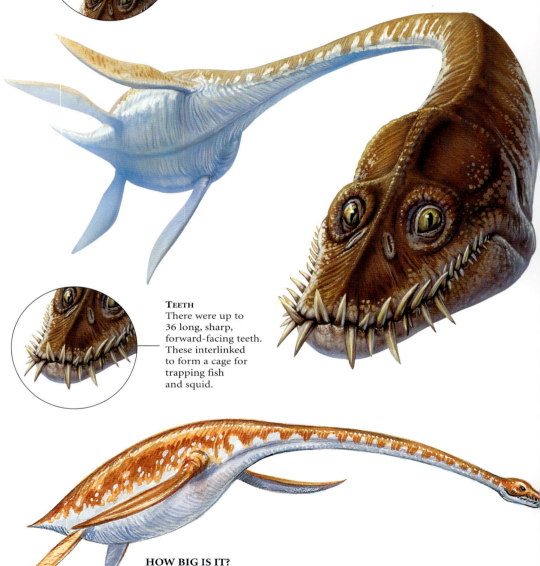

**TEETH**
There were up to 36 long, sharp, forward-facing teeth. These interlinked to form a cage for trapping fish and squid.

**HOW BIG IS IT?**

**TIMELINE** (millions of years ago)

| 540 | 505 | 438 | 408 | 360 | 280 | 248 | 208 | 146 | 65 | 1.8 to today |

# Mononykus

• ORDER • Saurischia • FAMILY • Alvarezsauridae • GENUS & SPECIES • *Mononykus olecranus*

Is it a bird? Is it a bird-like dinosaur? Early opinion was divided because *Mononykus* is so closely related to birds. This was a small, fast, keen-eyed predator that roamed open desert plains.

## VITAL STATISTICS

| | |
|---|---|
| FOSSIL LOCATION | Mongolia |
| DIET | Carnivorous, possibly omnivorous |
| PRONUNCIATION | MON-o-NYE-kus |
| WEIGHT | Unknown |
| LENGTH | 90cm (3ft) |
| HEIGHT | Unknown |
| MEANING OF NAME | 'Single claw' because of the unusual digits on its forelimbs |

### FOSSIL EVIDENCE

Light, hollow bones do not fossilize well, so the remains of *Mononykus* are frustratingly incomplete. The most intriguing feature is the large single claw at the end of each short, muscular arm. It was not likely used for snatching prey or for digging, and has led some palaeontologists to suggest *Mononykus* used it to break open termite mounds in search of food. Its large eyes may have enabled *Mononykus* to hunt at night, and it may have eaten plant food as well as insects and lizards.

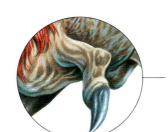

**STERNUM AND FIBULA**
The small, keeled sternum and reduced fibula have led some analysts to suggest *Mononykus* was a primitive bird and not a non-avian dinosaur.

### WHERE IN THE WORLD?

Remains have been found in the Bugin Tsav region of the Gobi Desert in southeastern Mongolia.

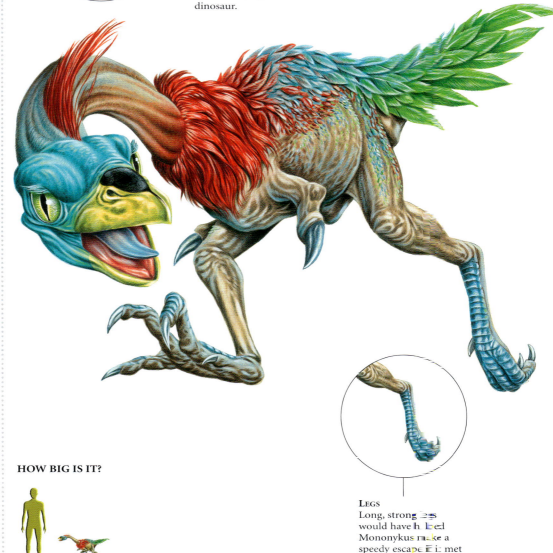

**LEGS**
Long, strong legs would have helped *Mononykus* make a speedy escape if it met a threatening predator.

### HOW BIG IS IT?

**TIMELINE (millions of years ago)**

540 | 505 | 438 | 408 | 360 | 280 | 248 | 208 | 146 | 65 | 1.8 to today

LATE CRETACEOUS

# Mosasaur

• ORDER • Squamata • FAMILY • Mosasauridae • GENUS & SPECIES • Various

Mosasaurs, a type of large marine squamate, were very successful predators. Squamates are the group made up of lizards and snakes, and mosasaurs included the largest squamates of all time. They were large marine animals that evolved paddle-like limbs and flattened tails for swimming after their terrestrial ancestors entered the sea.

## VITAL STATISTICS

| | |
|---|---|
| FOSSIL LOCATION | Worldwide |
| DIET | Carnivorous |
| PRONUNCIATION | MO-za-sore |
| WEIGHT | At least 20 tons for the largest ones |
| LENGTH | Up to 17m (57ft) |
| HEIGHT | Unknown |
| MEANING OF NAME | 'Meuse lizard' because the first one named was discovered near the Meuse River in the Netherlands |

### WHERE IN THE WORLD?

Mosasaur fossils have been found on every continent in marine sediments of the Late Cretaceous. They are especially well known from the Western Interior Seaway deposits of North America.

**FORELIMBS**
These flippers were useful for moving around in the water. Their finger bones, which were inside the flippers, were more numerous than in the human hand.

**TEETH**
Mosasaurs were all flesh eaters. Most had mouths full of sharp conical teeth, which were replaced throughout their lives.

## FOSSIL EVIDENCE

Although most mosasaur localities preserve isolated teeth and bone fragments, mosasaurs are well known from the abundant complete skeletons. They have a very good fossil record because dead marine animals have a better chance of being buried and preserved than do land animals, and they were also very common and global in distribution. They were apparently restricted to the Late Cretaceous Period, but were able to diversify and spread across the globe in that time. They became extinct at the end of the Cretaceous Period along with all of the non-avian dinosaurs and certain other groups.

PREHISTORIC ANIMAL

LATE CRETACEOUS

**HOW BIG IS IT?**

**TIMELINE** (millions of years ago)

| 540 | 505 | 438 | 408 | 360 | 280 | 248 | 208 | 146 | 65 | 1.8 to today |

# Parasaurolophus

**LATE CRETACEOUS**

• ORDER • Ornithischia • FAMILY • Hadrosauridae • GENUS & SPECIES • Several species within the genus *Parasaurolophus*

This hadrosaur had a flamboyant crest curving from the back of its head that may have helped to communicate to others and could have acted as a resonator for its calls.

## VITAL STATISTICS

| | |
|---|---|
| FOSSIL LOCATION | North America |
| DIET | Herbivorous |
| PRONUNCIATION | Pah-ra-SAWR-ro-lo-fous |
| WEIGHT | 2.5 tonnes (2.7 tons) |
| LENGTH | 9.5m (31ft) |
| HEIGHT | 4.9m (15ft) |
| MEANING OF NAME | 'Near-Saurolophus' since it was thought to be a very close relative of that dinosaur |

## FOSSIL EVIDENCE

The cranial crest varies in size between specimens. At its longest it forms a curved hollow tube 1.8m (6ft) long. *Parasaurolophus* has been called the 'trumpet dinosaur' as its crest could have amplified the sounds it made when communicating. The crest may, however, have served mainly for visual recognition. The large eye sockets suggest *Parasaurolophus* had keen vision and may have been active at dusk. Fossilized stomach contents show that it ate land plants, contradicting early suggestions that it lived mainly in water.

**JAW** The jaw was structured to allow the numerous cheek teeth to perform a grinding motion in order to chew.

### WHERE IN THE WORLD?

First unearthed in Alberta, Canada, later findings emerged in New Mexico and Utah, USA.

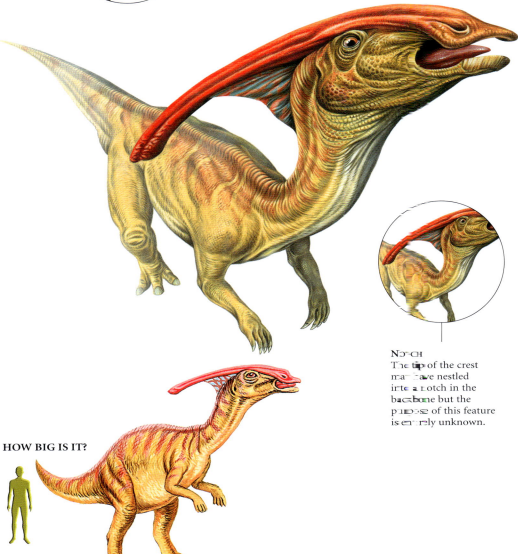

**NOTCH** The tip of the crest may have nestled into a notch in the backbone but the purpose of this feature is entirely unknown.

**HOW BIG IS IT?**

DINOSAUR
LATE CRETACEOUS

**TIMELINE** (millions of years ago)

| 540 | 505 | 438 | 408 | 360 | 280 | 248 | 208 | 146 | 65 | 1.8 to today |

LATE CRETACEOUS

# Quetzalcoatlus

• ORDER • Pterosauria • FAMILY • Azhdarchidae • GENUS & SPECIES • Quetzalcoatlus northropi

| VITAL STATISTICS | |
|---|---|
| Fossil Location | North America |
| Diet | Carnivorous |
| Pronunciation | Kett-zal-coe-AT-luss |
| Weight | 100kg (220lb) |
| Wingspan | 11m (36ft) |
| Height | 7m (22ft) |
| Meaning of name | 'Plumed serpent' from the name of the Aztec and Toltec feathered snake god |

**FOSSIL EVIDENCE**

*Quetzalcoatlus* most probably walked on all fours, but was best suited to flight, holding out its enormous wings to glide on warm air currents and breezes. Its paper-thin, hollow bones supported a light, aerodynamic body with an extremely long neck holding a head topped by a bony crest. It lived inland and probably fished over lakes and river. It may also have scavenged flesh from dead bodies.

PREHISTORIC ANIMAL

LATE CRETACEOUS

This was one of the largest flying animals the world has seen. A pterosaur rather than a dinosaur, it glided on vast wings that may have measured 11m (36ft) from tip to tip.

**NECK**
The neck is estimated at 2.4m (8ft) long. It was kept rigid by tendons and muscles, allowing *Quetzalcoatlus* to stay streamlined in flight.

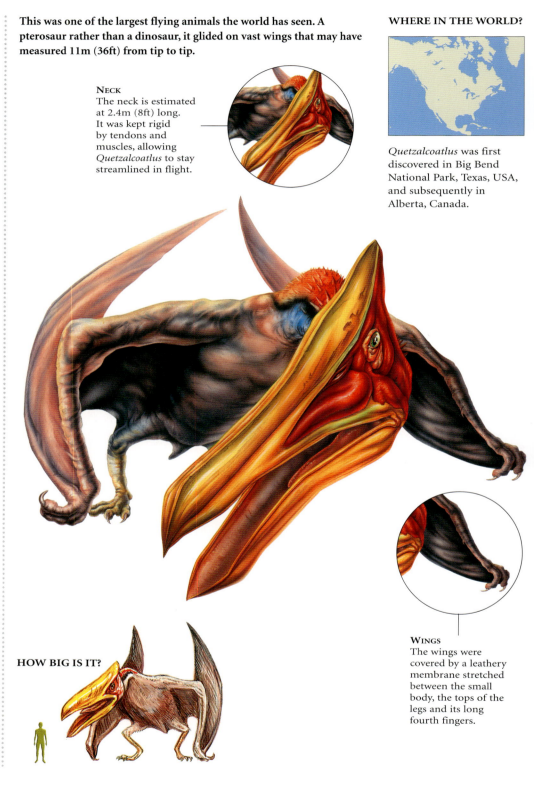

**WHERE IN THE WORLD?**

*Quetzalcoatlus* was first discovered in Big Bend National Park, Texas, USA, and subsequently in Alberta, Canada.

**WINGS**
The wings were covered by a leathery membrane stretched between the small body, the tops of the legs and its long fourth fingers.

**HOW BIG IS IT?**

**TIMELINE (millions of years ago)**

| 540 | 505 | 438 | 408 | 360 | 280 | 248 | 208 | 146 | 65 | 1.8 to today |

LATE CRETACEOUS

# Saltasaurus

• ORDER • Saurischia • FAMILY • Saltasauridae • GENUS & SPECIES • *Saltasaurus loricatus*

## VITAL STATISTICS

| | |
|---|---|
| FOSSIL LOCATION | Argentina |
| DIET | Herbivorous |
| PRONUNCIATION | Salta-SAWR-us |
| WEIGHT | 6–350kg (7 tons) |
| LENGTH | 12m (40ft) |
| HEIGHT | 5m (16ft) |
| MEANING OF NAME | 'Salta lizard' after Salta Province, Argentina, where it was found |

Like many plant-eating dinosaurs, *Saltasaurus* had a very large body and a small head and long neck so it could stretch up to reach leaves or fruits on high branches.

### EGGS
*Saltasaurus* was a very big dinosaur, but eggs of its close relatives found in 1997 in Patagonia, Argentina, were only around 12cm (1 in) long.

## FOSSIL EVIDENCE

*Saltasaurus* was found in 1980. Although it has been discovered in only one place so far, there are a reasonable number of fossils for palaeontologists to study. Salta Province in northwest Argentina has yielded several of the egg-shaped or circular armoured plates that once covered *Saltasaurus'* back and may have protected it from attack by predators. Hundreds of 6.7mm (¼in) bumps grew on the plates. Several part-skeletons found have included vertebrae, leg bones and jawbones.

### ARMOUR
*Saltasaurus* was well protected by its armour, which covered its back and tail and extended up its neck.

### HOW BIG IS IT?

### WHERE IN THE WORLD?

*Saltasaurus* was found in northwest Argentina in South America, in the area around Salta Province.

DINOSAUR
LATE CRETACEOUS

**TIMELINE (millions of years ago)**

| 540 | 505 | 438 | 408 | 360 | 280 | 248 | 208 | 146 | 65 | 1.8 to today |

LATE CRETACEOUS

# Tyrannosaurus

| VITAL STATISTICS | |
|---|---|
| Fossil Location | North America |
| Diet | Carnivorous |
| Pronunciation | Tee-RAN-oh-sawr-us rex |
| Weight | 7000kg (7 tons) |
| Length | 13m (43ft) |
| Height | 4m (13ft) |
| Meaning of name | 'Tyrant lizard' because of its huge size (*rex* means king) |

The fearsome *Tyrannosaurus rex* is probably the most famous dinosaur of them all. Tyrannosaurus rex is usually considered to be the greatest meat-eater of all time, although it was not the largest dinosaur. (Some larger theropods include *Spinosaurus*, *Carcharodontosaurus*, and *Giganotosaurus*.) *Tyrannosaurus* was 13m (43ft) long.

**Neck**
*Tyrannosaurus rex's* short, thick neck had powerful muscles to support the dinosaur's outsized head.

### WHERE IN THE WORLD?

*Tyrannosaurus rex* was found throughout western North America.

**FOSSIL EVIDENCE**
One of the first discoveries of the fossils of *Tyrannosaurus rex* occurred in 1874, when some of its 33cm (13in)-long teeth were unearthed in the US state of Colorado. In 1890, bones from *Tyrannosaurus'* skull were found in Wyoming, followed in 1892 by fragments of its vertebra. Wyoming was also the location where the first part-skeleton of *Tyrannosaurus* was found in 1900.

**Arms**
Palaeontologists have as yet been unable to agree on the function of *Tyrannosaurus'* tiny arms and two-fingered hands.

| DINOSAUR |
|---|
| LATE CRETACEOUS |

LATE CRETACEOUS

• **ORDER** • Saurischia • **FAMILY** • Tyrannosauridae • **GENUS & SPECIES** • *Tyrannosaurus rex*

**COLOSSAL TEETH**
*Tyrannosaurus rex* had bigger teeth than any other carnivorous dinosaur. The teeth in the upper jaw were larger than most of the teeth in the lower jaw, the largest measuring 33cm (13in) from root to sharp, pointed tip. Some of *T. rex*'s other teeth were shaped like blades with chiselled tips. The teeth at the front of the upper jaw were closely packed together.

**LEGS**
It was once thought that *Tyrannosaurus* plodded slowly along on its large legs, but palaeontologists have recently suggested that it may have been a reasonably fast runner.

**BALANCING ABILITY**
*Tyrannosaurus rex's* ability to stand upright required it to perform a balancing act. The tail, which contained up to 40 vertebrae, was big and heavy and provided a counterweight for its torso and enormous head, which could be up to 1.5m (5ft) long. However, many bones in *Tyrannosaurus'* body were hollow to make up for the dinosaur's enormous bulk. Its big, muscle-bound legs were some of the longest seen in any dinosaur in proportion to the size of its body. In addition, *Tyrannosaurus'* skull bones contained tiny air spaces, which made them lighter.

**HOW BIG IS IT?**

**TIMELINE (millions of years ago)**

| 540 | 505 | 438 | 408 | 360 | 280 | 248 | 208 | 143 | 65 | 1.8 to today |

LATE CRETACEOUS
# Tyrannosaurus rex

# LATE CRETACEOUS

• **ORDER** • Saurischia • **FAMILY** • Tyrannosauridae • **GENUS & SPECIES** • *Tyrannosaurus rex*

## HOW LONG DID *TYRANNOSAURUS* LIVE?

For many animals, species survival depends, in part, on having a very large number. This may have been the case for *Tyrannosaurus rex*. Many, if not most, *Tyrannosaurus* failed to survive their first few years. Some palaeontologists have suggested that the mortality rate among juvenile *Tyrannosaurus* was generally quite low. They believe this because relatively few *Tyrannosaurus rex* juveniles have been found in fossilized form, which suggest that juveniles did not die in large numbers. There could be many other reasons for their poor fossil record. At age 14, a young *Tyrannosaurus* began to grow quite dramatically, and though growth began to slow by about age 16, it could gain 6000kg (13000lb) by the time it was approximately 18. By then, palaeontologists reckon that *Tyrannosaurus rex* had reached maturity and was able to start reproducing. But it may have had a mere 6–10 years to do so; average life of expectancy was only about 28 years.

LATE CRETACEOUS

# Carnotaurus

| VITAL STATISTICS | |
|---|---|
| Fossil Location | South America |
| Diet | Carnivorous |
| Pronunciation | Car-no-TAWR-us |
| Weight | 1730kg (620lb) |
| Length | 7.5m (25ft) |
| Height | 2.7m (9ft) |
| Meaning of name | 'Meat-eating bull' because of its flesh-eating habit and bull-like horns |

**Carnotaurus lived in the Cretaceous Period about 90 million years ago. At present, *Carnotaurus sastrei* is the only known species of this strange-looking dinosaur. Its head resembled that of a bulldog and the horns on top of its head looked like those of a bull. *Carnotaurus*' arms were unusually short and it had extremely small four-fingered hands. Unlike most other dinosaurs, *Carnotaurus*' eyes faced a little way forwards, so it probably possessed a limited form of binocular vision. This meant that it could use both eyes to achieve some vision in depth.**

**Horns**
The bull-like horns were probably used during mating rituals or to head-butt rivals and drive them away.

**FOSSIL EVIDENCE**
A single almost complete skeleton of *Carnotaurus* was discovered and named in 1985. The description includes an account of the skin along *Carnotaurus*' entire right side, which, unlike some similar coelurosaurian theropods, seem to have had no feathers. Instead, the skin of *Carnotaurus* featured rows of swellings that became larger the closer they were to the dinosaur's spine.

DINOSAUR

LATE CRETACEOUS

340

LATE CRETACEOUS

• **ORDER** • Saurischia • **FAMILY** • Abelisauridae • **GENUS & SPECIES** • *Carnotaurus sastrei*

**TEETH**
The rows of close-packed teeth in its upper and lower jaws show that *Carnotaurus* was a champion meat-eater.

### HUMAN AND DINOSAUR COEXISTENCE
Humans and non-avian dinosaurs never inhabited Earth at the same time, but had they done so, many dinosaurs would have towered over humans. *Carnotaurus* was not nearly as hefty as some other dinosaurs, such as *Tyrannosaurus rex*, but even they were much faster on their feet than humans, and they were a great deal taller.

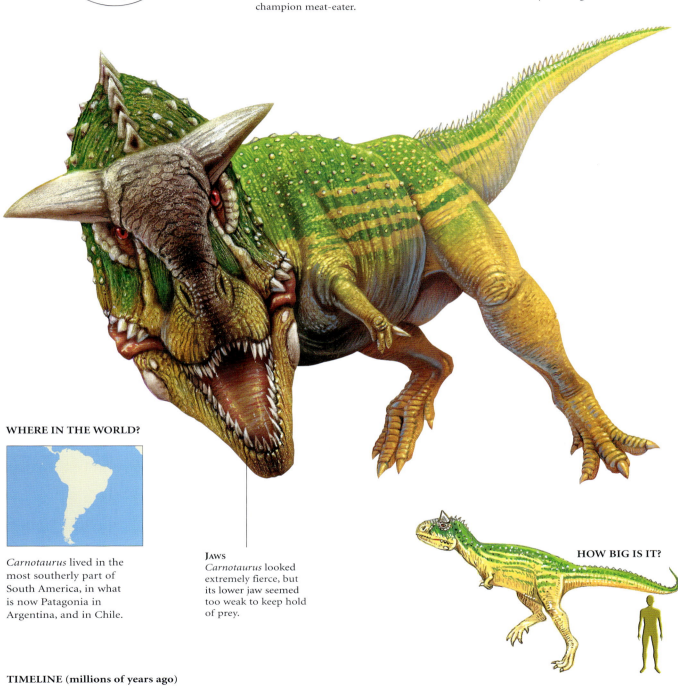

**WHERE IN THE WORLD?**

*Carnotaurus* lived in the most southerly part of South America, in what is now Patagonia in Argentina, and in Chile.

**JAWS**
*Carnotaurus* looked extremely fierce, but its lower jaw seemed too weak to keep hold of prey.

**HOW BIG IS IT?**

**TIMELINE (millions of years ago)**

| 540 | 505 | 438 | 408 | 360 | 280 | 248 | 208 | 146 | 65 | 1.8 to today |

LATE CRETACEOUS
# Carnotaurus

# LATE CRETACEOUS

• **ORDER** • Saurischia • **FAMILY** • Abelisauridae • **GENUS & SPECIES** • *Carnotaurus sastrei*

**WEIRDEST DINOSAUR EVER?**

South America has proved to be the home of some very unusual dinosaurs, of which *Carnotaurus* is a prime example. Some think it could be a candidate for the title 'Weirdest Dinosaur Ever'. *Carnotaurus* was an abelisaurid and belonged to a group of theropods with deep skulls and snouts that carried a mouthful of large, blade-shaped teeth. The abelisaurid, which gives its name to the group *Abelisaurus* (Greek for 'Abel's lizard'), was named after its discoverer, Roberto Abel, one-time director of the Museo de Cipolletti, Argentina. Despite its deep skull, *Carnotaurus*' lower jaw was thin and weak, and the upper jaw was unusually short, giving *Carnotaurus* its distinctly snub-nosed profile. By contrast, *Carnotaurus*' neck was unusually long for a Theropod, and it also had strange arms. Palaeontologists have long been puzzled as to how the dinosaur could have used them, if it used them at all. Like so many dinosaurs, *Carnotaurus*' arms were tiny in the context of its overall size. The upper arm seemed strong enough, but the two forearms were extremely short and the whole arrangement appeared very stunted, as did *Carnotaurus*' hands. One of its four fingers was little more than a backwards-facing spike and the palms faced outwards.

LATE CRETACEOUS

# Deinocheirus

| VITAL STATISTICS | |
|---|---|
| Fossil Location | Mongolia |
| Diet | Probably carnivorous |
| Pronunciation | Dye-no-KYE-rus |
| Weight | Unknown |
| Length | Unknown |
| Height | Unknown |
| Meaning of name | 'Terrible hand' for its huge arms and claws |

*Deinocheirus'* species name *mirificus*, which is the Latin word for 'unusual' or 'peculiar', pinpoints the mystery surrounding this dinosaur. The difficulty has been exacerbated by the fact that the only remains of *Deinocheirus* so far discovered consist of two arms – each 2.6m (8ft 7in) long and featuring a scythe-like claw – and the parts of a few ribs and the dinosaur's backbone. Although *Deinocheirus* was a theropod it may have been a plant-eater, using its long arms to reach up and pull down leaves, fruit and other food from the branches overhead.

### WHERE IN THE WORLD?

Remains have been found in the Gobi Desert of southern Mongolia in central Asia.

**FOREARMS**
Long and powerful arms may have allowed *Deinocheirus* to reach leaves on trees for it to feast on.

**LEGS**
To support the heavy frame that is suggested by the huge arms, Deinocheirus would have huge and strong legs.

**FOSSIL EVIDENCE**
Comparatively little has been found of *Deinocheirus* so far, yet the two arms that were discovered tell us a fair amount about this dinosaur. The arms end in claws that, like all vertebrate claws, have sheaths of keratin or 'horn'. This would have covered the outside of the bone and may have measured up to 1m (3ft 7in) long. Most palaeontologists believe that *Deinocheirus* was a carnivore rather than a herbivore, and used its large, sharp claws to kill its prey and tear it apart.

DINOSAUR

LATE CRETACEOUS

LATE CRETACEOUS

• **ORDER** • Saurischia • **FAMILY** • Deinocheridae • **GENUS & SPECIES** • *Deinocheirus mirificus*

## BIRD-LIKE DINOSAUR

Some palaeontologists have suggested that *Deinocheirus* was a bird-like dinosaur, somewhat like an ostrich. This accounts for its scientific description ornithomimosaur, which breaks down as bird (ornitho-), mimic (-mimo) and lizard (-saur.) Another use for *Deinocheirus*' immense dagger-like claws may have been to rip apart ants' nests before devouring their contents, using its strong beak to scoop up any fleeing ants.

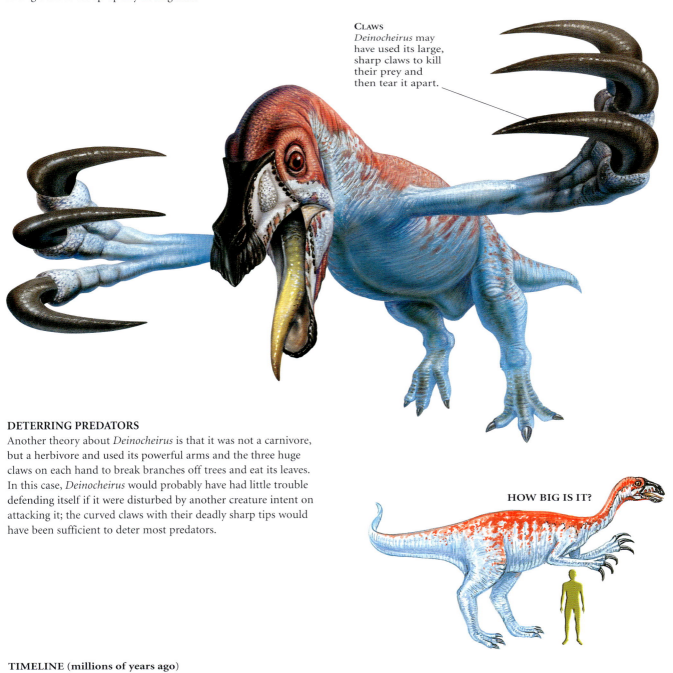

**CLAWS**
*Deinocheirus* may have used its large, sharp claws to kill their prey and then tear it apart.

## DETERRING PREDATORS

Another theory about *Deinocheirus* is that it was not a carnivore, but a herbivore and used its powerful arms and the three huge claws on each hand to break branches off trees and eat its leaves. In this case, *Deinocheirus* would probably have had little trouble defending itself if it were disturbed by another creature intent on attacking it; the curved claws with their deadly sharp tips would have been sufficient to deter most predators.

**HOW BIG IS IT?**

**TIMELINE (millions of years ago)**

| 540 | 505 | 438 | 408 | 360 | 280 | 248 | 208 | 146 | 65 | 1.8 to today |

LATE CRETACEOUS

# Edmontonia

| VITAL STATISTICS | |
|---|---|
| Fossil Location | North America |
| Diet | Herbivorous |
| Pronunciation | Edmon-TONE-ee-uh |
| Weight | 3500kg (1587lb) |
| Length | 7m (23ft) |
| Height | 2m (6ft) |
| Meaning of name | Named after the Edmonton Rock Formation in Canada where it was found |

*Edmontonia* acquired its name in 1928, four years after it was discovered in the Edmonton Formation in Alberta, Canada. As a herbivore it was unlikely to have been as fierce as a carnivore, yet *Edmontonia* carried large, wicked-looking spikes that jutted out from its sides. These were probably used to defend *Edmontonia* territory and keep rivals away from its mate. They may also have served as protection against attack. Another self-defence strategy was probably for *Edmontonia* to crouch low to the ground so that its unarmoured underside was not exposed to its enemies.

**WHERE IN THE WORLD?**

*Edmontonia* was found in Edmonton in Alberta, Canada, and in Montana, South Dakota and Texas in the USA.

**FOSSIL EVIDENCE**
*Edmontonia* was a nodosaurid ankylosaur, that is, an armoured dinosaur, but it lacked a bony club on its thick tail. It was very bulky and got around on four thick legs and wide five-toed feet. Its armour consisted of a mass of plates and spikes that covered its back and tail. The fossils found in the Edmonton Formation belonged to the species *E. longiceps*. They were found 11km (7 miles) west of the village of Morrin in central Alberta.

DINOSAUR

LATE CRETACEOUS

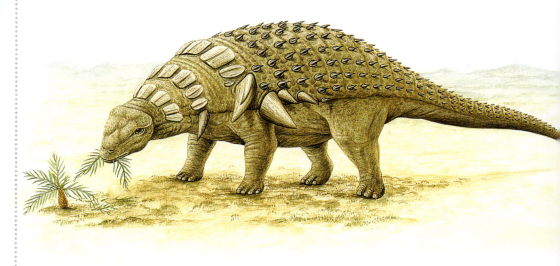

**Legs**
Short, stubby legs made it easy for *Edmontonia* to reach low-lying plants, or to crouch down quickly if under attack.

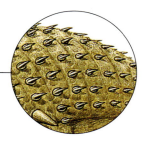

**Spiky Armour**
*Edmontonia's* back was thickly covered in dorsal armour, edged with intimidating spikes. The rest of its body was not similarly protected.

LATE CRETACEOUS

• **ORDER** • Ornithischia • **FAMILY** • Nodosauridae • **GENUS & SPECIES** • *Edmontonia rugosidens, E. longiceps*

### FEEDING LOW TO THE GROUND
So many fossils of *Edmontonia* have been found that it has been relatively easy for palaeontologists to reconstruct the whole dinosaur. It likely lived in the woodlands of prehistoric North America, which provided plenty for it to eat. *Edmontonia*'s physique, with its short neck and stubby legs, was seemingly adapted for feeding on ferns, cycads and other low-lying plants.

**SPIKES**
*Edmontonia*'s armour included spikes that radiated from the sides of its body, especially near its neck, the top of which was also protected by a shield of large plates.

### PROCESSING PLANTS
A dinosaur's teeth can indicate a great deal about the food it ate and how it was broken down and digested. *Edmontonia* had to deal with tough plants. Although it had teeth in its cheeks, these were too small and its jaw too weak to do all the work of chewing them up. Some scientists believe that this was done instead by internal fermentation chambers. Here, chemicals broke the plants down so that they could be digested.

**HOW BIG IS IT?**

**TIMELINE (millions of years ago)**

LATE CRETACEOUS

# Edmontonia

### Edmontonia
This was one of the most common armoured dinosaurs in Alberta. Large and well protected, it had small teeth and weak jaws. Unlike some of its close relatives, it lacked a tail club and had spines only around its neck and shoulders.

# LATE CRETACEOUS

• **ORDER** • Ornithischia • **FAMILY** • Nodosauridae • **GENUS & SPECIES** • *Edmontonia rugosidens, E. longiceps*

## EDMONTONIA AND THE END OF THE NON-AVIAN DINOSAURS

*Edmontonia*, which first appeared in the Late Cretaceous Period some 76 million years ago, existed for about eight million years before the Age of the Dinosaurs ended. There are many theories about this extinction, some suggesting swift death by sudden disaster, and others favouring a more gradual demise. One of these less dramatic, but nevertheless fatal, events seems to have occurred while *Edmontonia* was still alive. It was, of course, a herbivore reliant on a constant supply of vegetation for its diet. But evidence of a disruption in its food supply has been found in rings left in the petrified wood of trees that grew in the forests in *Edmontonia's* time. These rings show that there was a lessening in rainfall and a rise in temperature when the extinction of the non-avian dinosaurs was imminent. These changes probably led to drought, the destruction of plant life and with that an acute shortage of food. Deprived of water and sustenance, the last of the non-avian dinosaurs may have died of thirst and starvation. Palaeontologists have discovered large numbers of ankylosaurs, like *Edmontonia*, buried in sand or mud intact with their body armour still perfectly preserved.

LATE CRETACEOUS

# Maiasaura

| VITAL STATISTICS | |
|---|---|
| Fossil Location | North America |
| Diet | Herbivorous |
| Pronunciation | My-uh-sore-uh |
| Weight | 3900kg (1770lb) |
| Length | 9m (30ft) |
| Height | 2.25m (7ft 5in) |
| Meaning of name | 'Good mother lizard' because of its presumed parenting skills |

The discovery of *Maiasaura* in Montana provided palaeontologists with an exciting first. It was the first adult, non-avian dinosaur whose fossils were found together with its unhatched eggs and their nests. Fossils of the just-hatched baby *Maiasaura* measured 30cm (1ft) long. One of the palaeontologists who named *Maiasaura* in 1979 was Jack R. Horner (he subsequently acted as adviser for the film *Jurassic Park*). *Maiasaura* very likely lived in herds, sometimes containing as many as 10,000 individual dinosaurs. This helps explain why the *Maiasaura* site on Egg Mountain, Montana, contained thousands of fossils.

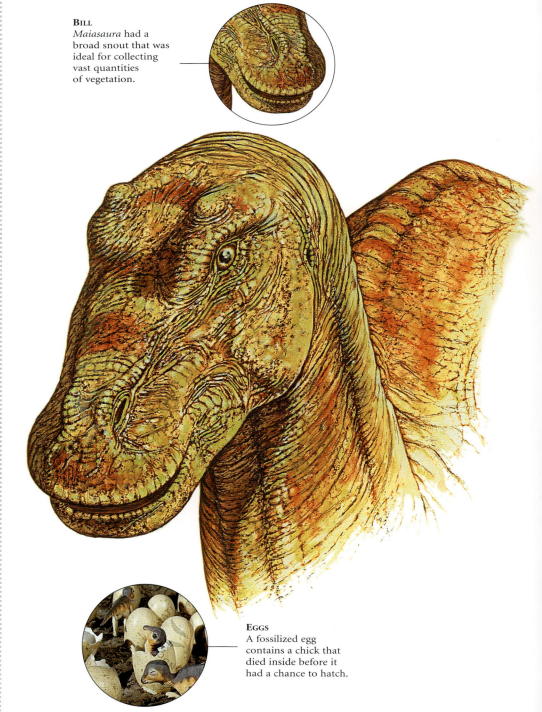

**Bill**
*Maiasaura* had a broad snout that was ideal for collecting vast quantities of vegetation.

**FOSSIL EVIDENCE**
There is no shortage of fossil evidence for the Cretaceous-age dinosaur *Maiasaura*, which lived around 80 million years ago; so it is one of the best known of all non-avian dinosaurs. The fossils found range from eggs the size of grapefruit laid about 25 at a time, through to embryos and hatchlings, juveniles and adults. A huge bonebed representing as many as 10,000 individuals of *Maiasaura* has been found in Montana, USA.

**Eggs**
A fossilized egg contains a chick that died inside before it had a chance to hatch.

DINOSAUR

LATE CRETACEOUS

# LATE CRETACEOUS

• **ORDER** • Ornithischia • **FAMILY** • Hadrosauridae • **GENUS & SPECIES** • *Maiasaura peeblesorum*

### WHERE IN THE WORLD?

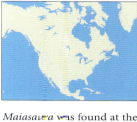

*Maiasaura* was found at the Egg Mountain site near Choteau, western Montana, in the USA.

**TAIL**
*Maiasaura* had few defences against attack except for its heavy tail and the safety in numbers offered by herd life.

### DINO IN SPACE

*Maiasaura*, or at least parts the species *M. peeblesorum*, represented the first dinosaur to go into space; a piece of a baby *Maiasaura* bone and an eggshell were placed on board the US spaceship that carried out an eight-day mission to Spacelab 2 in 1985. In the same year, *Maiasaura* became the official dinosaur of the state of Montana.

**HANDS AND FEET**
*Maiasaura's* hands had four fingers and its feet had hoof-shaped claws.

### HOW BIG IS IT?

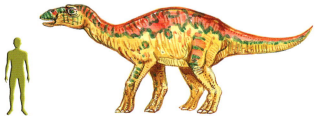

### PRAIRIES, SWAMPS AND MOUNTAINS

The area where *Maiasaura* fossils have been discovered now lies close to the Rocky Mountains that dominate the west of Canada and the USA. The scene was very different when *Maiasaura* lived there. At that time, in the Late Cretaceous Period, *Maiasaura's* home territory was on the coast of a vast ocean that once covered what are now the prairies of North America. As the Rockies rose, the sea receded and the future prairies turned into a vast swamp.

**TIMELINE (millions of years ago)**

| 540 | 505 | 438 | 408 | 360 | 280 | 248 | 208 | 143 | 65 | 1.8 to today |

LATE CRETACEOUS
# Maiasaura

# LATE CRETACEOUS

• **ORDER** • Ornithischia • **FAMILY** • Hadrosauridae • **GENUS & SPECIES** • *Maiasaura peeblesorum*

### EGG MOUNTAIN, MONTANA

Egg Mountain, a veritable dinosaur treasure house, lies 19km (12 miles) west of Choteau in the US state of Montana. It has produced more evidence and information about the dinosaurs of the Cretaceous Period, which occurred 143–65 million years ago, than any other site of its kind in the world. The rich findings first emerged with the discovery of *Maiasaura* in 1978 and the proliferation of fossilized eggs that were found there. After that, find after find was made, including some of the first non-avian dinosaur embryos, a vast dinosaur bonebed of some 2 sq km (0.5 sq mile) and, in addition to *Maiasaura* and *Troodon*, the fossils (including eggs) of *Orodromeus*, *Albertosaurus*, *Ankylosaurus*, *Montanoceratops* and several other creatures. Investigations have also established that different species of dinosaur came to Egg Mountain every year to lay, tend and hatch their eggs. Round-shaped nests of fossilized eggs up to 2m (7ft) wide were discovered at the top of mounds built from mud or earth, together with pieces of eggshell left behind after the dinosaur chicks hatched. Different patterns on the eggshells of different dinosaurs enabled palaeontologists to identify which of these creatures laid them.

LATE CRETACEOUS

# Oviraptor

| VITAL STATISTICS | |
|---|---|
| Fossil Location | Mongolia, China |
| Diet | Possibly omnivorous |
| Pronunciation | Ovi-RAP-tore |
| Weight | 20kg (44lb) |
| Length | 2m (6.6ft) |
| Height | 0.8m (2ft 6in) |
| Meaning of name | 'Egg thief' because the first specimen was found with eggs it was assumed to be stealing |

*Oviraptor*, which lived some 80 million years ago and had a crest on its head and a toothless beak. The crest was probably used for display in mating rituals. The first *Oviraptor* was discovered in the Gobi Desert of Mongolia in 1924, when a fossil was found on top of a clutch of eggs. From this, it was presumed that the *Oviraptor* had eaten the embryos inside. Later, it appeared that the dinosaur was the egg layer, and was not necessarily an egg stealer after all.

### WHERE IN THE WORLD?

*Oviraptor* was found in the Djadokhta Formation of Mongolia and the Neimongol Autonomous Region of Bayan Mandahu, China.

**Beak**
*Oviraptor* had a toothless beak. Further back in the mouth there was a pair of tooth-like prongs.

**Evolving Imagery**
This illustration shows a featherless, egg-stealing *Oviraptor* (with a skull based on the crushed original find) consistent with what scientists thought decades ago. Better fossils, including perfect skulls, significantly changed our view of the animal.

**FOSSIL EVIDENCE**
Although its skeleton was exceptionally bird-like and was very likely feathered, *Oviraptor* was a non-avian dinosaur. Its hands had three fingers with sharp claws more than 7cm (3in) long. It also three-toed feet and a long, stiff tail. *Oviraptor*'s jaws were toothless and its diet is unknown, but since it was a theropod, it was likely at least partly carnivorous.

DINOSAUR

LATE CRETACEOUS

# LATE CRETACEOUS

• **ORDER** • Saurischia • **FAMILY** • Oviraptoridae • **GENUS & SPECIES** • *Oviraptor philoceratops*, *O. mongoliensis*

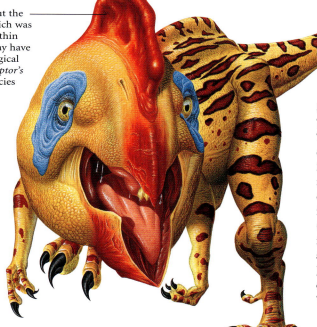

**HEAD CREST**
There are many theories about the crest on *Oviraptor's* head, which was hollow and lined with a very thin sheet of bone. Although it may have had some unknown physiological function, it's likely that *Oviraptor's* crest was used mainly for species recognition.

**STAR OF FILM AND TV**
Today, *Oviraptor* has become a film and TV star, but its image has changed from egg thief to something more positive. In 2000, when the Walt Disney Pictures movie entitled *Dinosaur* was released, a computer-generated *Oviraptor* was shown furtively stealing an *Iguanodon* egg. *Oviraptor* was later found to possibly be innocent of such types of theft. In 2002, it appeared in a TV mini-series based on James Gurney's *Dinotopia* in the more caring guise of Ovinutrix, which means 'egg nurse'.

**BEAK**
Mother *Oviraptor* may have used her beak to turn eggs to ensure they would be uniformly warm.

**EGG NURSE**
*Oviraptor* adults protected their eggs by sitting over them as they lie in the nest.

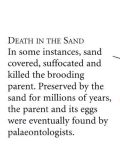

**DEATH IN THE SAND**
In some instances, sand covered, suffocated and killed the brooding parent. Preserved by the sand for millions of years, the parent and its eggs were eventually found by palaeontologists.

**HOW BIG IS IT?**

**TIMELINE (millions of years ago)**

| 540 | 505 | 438 | 408 | 360 | 280 | 248 | 208 | 146 | 65 | 1.8 to today |

LATE CRETACEOUS

# Oviraptor

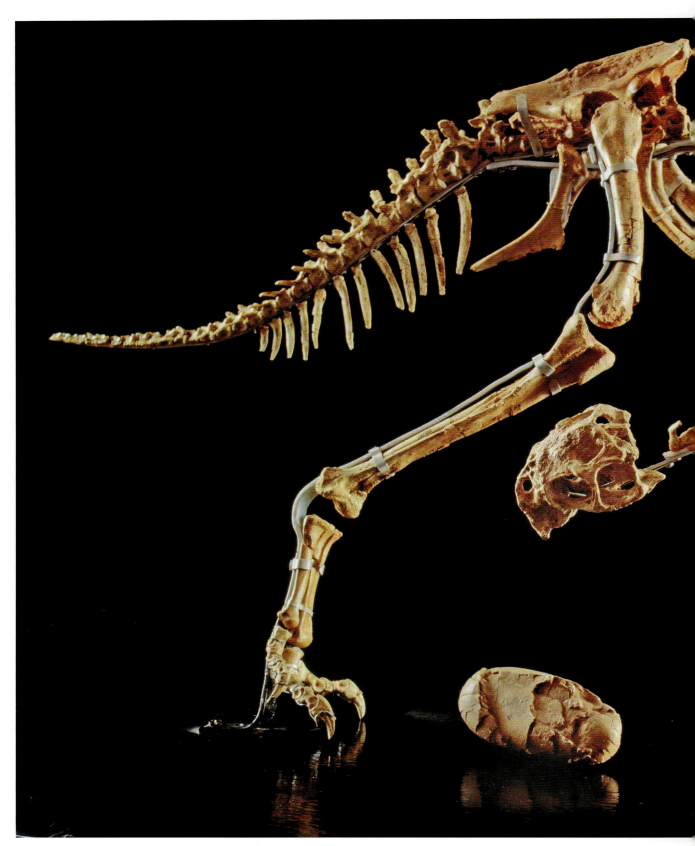

# LATE CRETACEOUS

• **ORDER** • Saurischia • **FAMILY** • Oviraptoridae • **GENUS & SPECIES** • *Oviraptor philoceratops, O. mongoliensis*

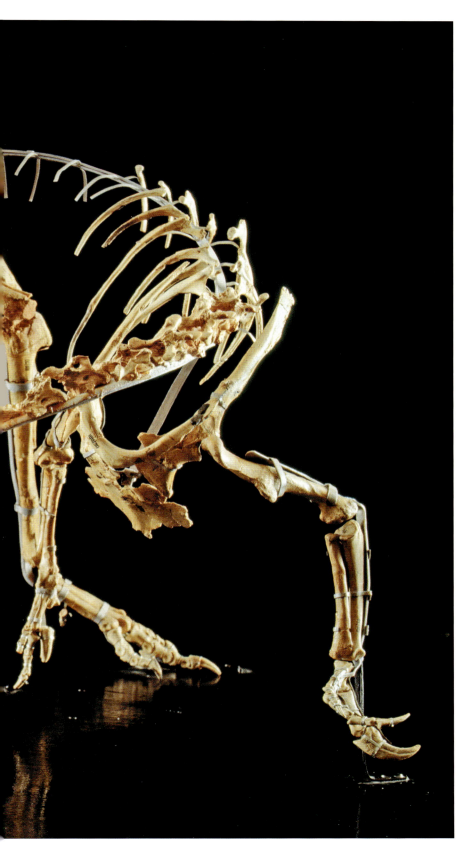

**A GIANT OVIRAPTORID IN CHINA**

In 2005, the fossil of a giant bird-like dinosaur thought to belong to the oviraptorid group was unearthed in China. News of the find, however, did not break until 2007. The location was the Erlian Basin in the Chinese region of Inner Mongolia, where the giant was discovered quite unintentionally. A team from the Institute of Vertebrate Palaeontology and Palaeoanthropology in Beijing, headed by the distingished Chinese fossil-finder Xu Xing, was demonstrating to journalists how previous fossils had been found when they happened upon the 1m (3ft)-long leg bone of a young adult. On examination, the giant dinosaur was found to be as large as some tyrannosaurs, which made it more than six times taller than the 'standard' oviraptorid. Named, appropriately enough, the *Gigantoraptor*, the find was reckoned to be 8m (26ft) long, 5m (16ft) tall and to weigh 1400kg (1.38 ton). Although some palaeontologists have expressed surprise that an oviraptorid could be so enormous, Philip Currie of the University of Alberta, in Canada, has suggested that animals tended to become larger as evolution progressed because their size made it easier for them to obtain food, attract mates and fight off predators.

LATE CRETACEOUS

# Pachycephalosaurus

| VITAL STATISTICS | |
|---|---|
| Fossil Location | North America |
| Diet | Herbivorous |
| Pronunciation | Pakky-SE-fallo-sawr-us |
| Weight | 430kg (950lb) |
| Length | 4.6m (15ft) |
| Height | 4.3m (17ft 6in) |
| Meaning of name | 'Thick-headed lizard' because of its thickened skull roof |

*Pachycephalosaurus* was not a particularly large dinosaur, but it had a huge head with a very thick dome-shaped skull that was up to 25cm (10in) thick. It owed its strange appearance to the row of bony knobs that encircled the top of its head and its snout. Pachycephalosaurs used their thick skulls to kill or hurt attackers.

**WHERE IN THE WORLD?**

*Pachycephalosaurus* remains have been found in Montana, South Dakota and Wyoming in the US.

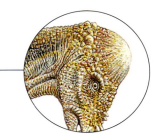

**SKULL**
The skull was up to 25cm (10in) thick. This was used to kill or hurt attackers.

**FOSSIL EVIDENCE**
*Pachycephalosaurus* fossils were first discovered in 1938 on a ranch near Ekalaka, Montana, and named in 1943. Complete fossils of this largest of thick-headed dinosaurs are scarce, but fragments of their super-thick skulls have proved plentiful. One unusual feature was its large, round forward-facing eye sockets. This suggests that *Pachycephalosaurus* possessed binocular vision – a valuable asset in a dangerous world where herbivores were often at the mercy of marauding carnivores.

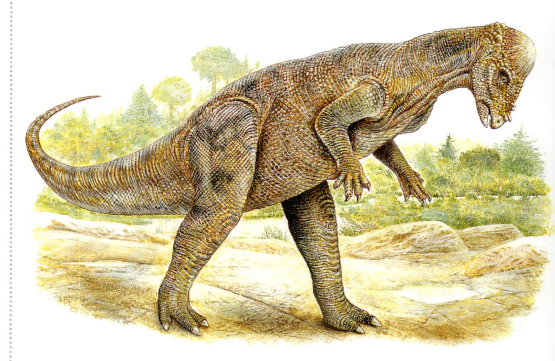

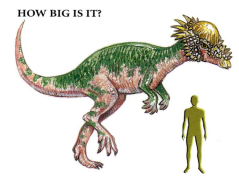

**HOW BIG IS IT?**

| DINOSAUR |
| LATE CRETACEOUS |

358

LATE CRETACEOUS

• ORDER • Ornithischia • FAMILY • Pachycephalosauridae • GENUS & SPECIES • *Pachycephalosaurus wyomingensis*

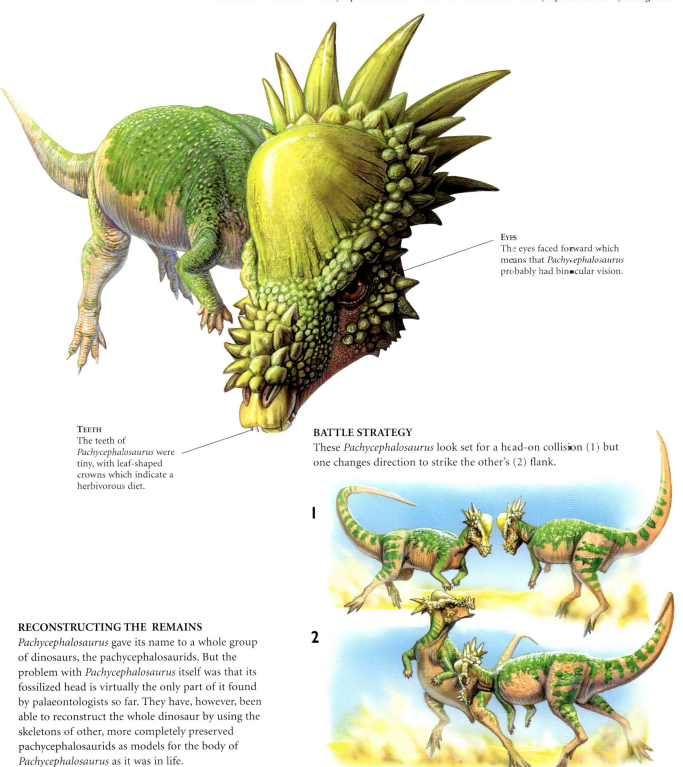

**EYES**
The eyes faced forward which means that *Pachycephalosaurus* probably had binocular vision.

**TEETH**
The teeth of *Pachycephalosaurus* were tiny, with leaf-shaped crowns which indicate a herbivorous diet.

**BATTLE STRATEGY**
These *Pachycephalosaurus* look set for a head-on collision (1) but one changes direction to strike the other's (2) flank.

**RECONSTRUCTING THE REMAINS**
*Pachycephalosaurus* gave its name to a whole group of dinosaurs, the pachycephalosaurids. But the problem with *Pachycephalosaurus* itself was that its fossilized head is virtually the only part of it found by palaeontologists so far. They have, however, been able to reconstruct the whole dinosaur by using the skeletons of other, more completely preserved pachycephalosaurids as models for the body of *Pachycephalosaurus* as it was in life.

**TIMELINE (millions of years ago)**

| 540 | 505 | 438 | 408 | 360 | 280 | 248 | 208 | 146 | 65 | 1.8 to today |

LATE CRETACEOUS
# Pachycephalosaurus

LATE CRETACEOUS

• **ORDER** • Ornithischia • **FAMILY** • Pachycephalosauridae • **GENUS & SPECIES** • *Pachycephalosaurus wyomingensis*

**A GREAT FOSSIL FIND**
Retrieving dinosaur fossils is not always just a matter of digging them up, dusting them off and examining them to discover their secrets. Fossilization can imprison prehistoric remains in ways that may defeat palaeontologists' best efforts, or at least give them a very hard time before they succeed in unlocking the remains. A case in point was a *Dracorex hogwartsia* skull apparently encased in a solid block of stone. Some believe that *Dracorex* is a juvenile *Pachycephalosaurus*. The block weighed nearly 91kg (200lb) and was found in 2003 by three amateur fossil hunters in the state of Iowa, USA. They managed to get the block to St Luke's Regional Medical Centre in Sioux City, Iowa, where they hoped it could be electronically dissected. The Centre radiologist and a veterinary surgeon agreed to passed the block through a computer tomography scanner. Such scanners use digital geometry processing to obtain a three-dimensional image of any object inside a stone block. To the delight of the fossil hunters, the scan successfully revealed the *Dracorex* skull, confirming that they had made a great palaeontological find.

LATE CRETACEOUS

# Protoceratops

| VITAL STATISTICS | |
|---|---|
| Fossil Location | Mongolia and China |
| Diet | Herbivorous |
| Pronunciation | Proh-toh-SEH-ratops |
| Weight | 181kg (400lb) |
| Length | 1.8m (6ft) |
| Height | 0.6m (2ft) |
| Meaning of name | 'First horn face' because it was first thought to be ancestral to later horned dinosaurs |

*Protoceratops* had a large bony frill around its neck. It is often stated that the frill was meant to protect *Protoceratops'* neck and strengthen the jaw muscles, but its fragile structure may have precluded those functions. The muscles had plenty of work to do because, like most other herbivores, *Protoceratops* consumed masses of vegetation. Although it was fairly long, measuring up to 1.8m (6ft), *Protoceratops* was comparatively low to the ground, only 0.6m (2ft) in height. Its body was shaped like a barrel, and it had a short tail, short legs with five-toed feet and bird-like hips.

**WHERE IN THE WORLD?**

*Protoceratops* has been found in Gansu and in the Bayan Mandahu Formation of Mongolia.

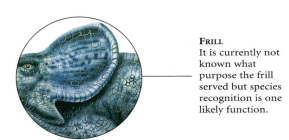

**Frill**
It is currently not known what purpose the frill served but species recognition is one likely function.

**FOSSIL EVIDENCE**
The first *Protoceratops* find was made in 1922 in the Gobi Desert of Mongolia by a photographer, J.B. Shackleford. Shackleford belonged to a U.S. expedition exploring the desert in search of the ancestors of humans. The *Protoceratops* fossils were very well preserved. In 1971, a dramatic find was made of a double fossil showing a *Protoceratops* fighting with a *Velociraptor*. It seems that the two dinosaurs died together, possibly in a mudslide, while engaged in combat.

DINOSAUR

LATE CRETACEOUS

LATE CRETACEOUS

• **ORDER** • Ornithischia • **FAMILY** • Protoceratopsidae • **GENUS & SPECIES** • *Protoceratops andrewsi, P. hellenikorhinus*

## FANCY FRILLS
*Protoceratops*' head frills proved durable enough to survive millions of years after its death and yet retain much of its original form and the curious eagle-like look its bony mask gave to its face. In *Protoceratops andrewsi* there were two forms of frills. One type was more extensive than the other and suggests that the two forms represent males and females.

## HERDING ANIMAL
*Protoceratops* may have been a herding animal, and palaeontologists have referred to it as 'the sheep of the Cretaceous' because its fossils are so common in the Gobi Desert.

### HOW BIG IS IT?

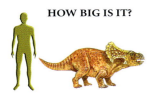

**TIMELINE (millions of years ago)**

| 540 | 505 | 438 | 408 | 360 | 280 | 248 | 208 | 146 | 65 | 1.8 to today |

LATE CRETACEOUS

# Protoceratops

LATE CRETACEOUS

• **ORDER** • Ornithischia • **FAMILY** • Protoceratopsidae • **GENUS & SPECIES** • *Protoceratops andrewsi*, *P. hellenikorhinus*

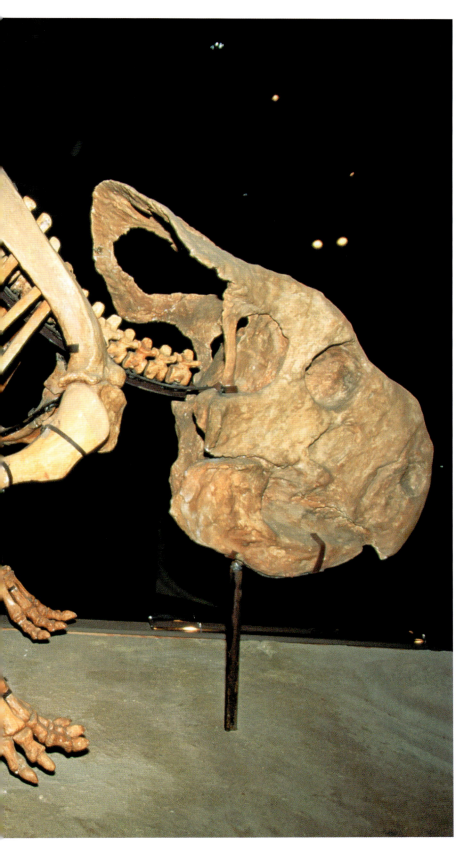

**GRIFFIN OR PROTOCERATOPS?**
The griffin was a mythical creature that was supposed to have an eagle's head and wings. It was also said to have the body of a lion, and laid its eggs in nests on the ground. However, more than 2600 years ago, in 675 BCE, Ancient Greek writers recorded the discovery of griffins as if they were real-life creatures. At this time, stories about griffins reached Ancient Greece after contact was made with the Scythian nomads, who originated in southeastern Europe. These nomads had been mining for gold in the Tian Shan and Altai mountains of Central Asia. During their diggings in the hills and red sandstone formations, the Scythians may have come across fossilized skeletons and other remains. The creatures they described to the Ancient Greeks closely fitted descriptions of *Protoceratops* made by palaeontologists hundreds of years later. The Scythians may have created the griffin myth in order to explain the remains of an animal they eventually considered guardians of gold mines. However, it seems that what the Ancient Greeks were describing and what the Scythians actually found were the fossilized remains of *Protoceratops*. In an interesting twist to the tale, the areas in China and Mongolia where *Protoceratops* finds have since proliferated also contain run-offs of gold from nearby mountains.

LATE CRETACEOUS

# Pteranodon

| VITAL STATISTICS | |
|---|---|
| FOSSIL LOCATION | USA |
| DIET | Fish |
| PRONUNCIATION | Teh-RAH-noh-don |
| WEIGHT | 15.8kg (35lb) |
| LENGTH | Up to 9m (30ft) |
| HEIGHT | 1.8m (6ft) |
| MEANING OF NAME | 'Winged and toothless' |

*Pteranodon* is one of the largest pterosaurs or flying reptiles ever and lived some 89.3 million years ago in what is now the central region of the USA. *Pteranodon*, one of the last pterosaurs, was once considered to be the largest creature that could fly. Except for its enormous size, *Pteranodon* resembled present-day birds in that its beak contained no teeth and its bones were hollow. Its huge wingspan is similar to that of a small aircraft, which made it a veritable giant of the air. *Pteranodon's* long pointed beak made it ideal for plunging into the sea to snatch up the fish on which it lived.

**WHERE IN THE WORLD?**

*Pteranodon* remains have been found in central USA, in Kansas, Alabama, Nebraska, Wyoming and South Dakota.

**HUGE FEMALE**
What seem to to be the female of *Pteranodon* were two-thirds of the size of the male, but they were still huge, with a wingspan measuring 6m (20ft) across.

**FOSSIL EVIDENCE**
*Pteranodon* fossils were collected at the Smoky Hill Chalk in the north of the Niobrara Formation in western Kansas as early as 1870. *Pteranodon sternbergi*, collected in 1952, had an upright skull crest less familiar than *P. longiceps*. *P. sternbergi* is recognized as the probable ancestor of a later species of *Pteranodon*, *P. longiceps* was named in 1876 from an almost complete fossil that had a wingspan of 7m (23ft). Fossilized fish bones found in the specimen's stomachs proved these were fish-eaters.

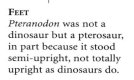

**FEET**
*Pteranodon* was not a dinosaur but a pterosaur, in part because it stood semi-upright, not totally upright as dinosaurs do.

PREHISTORIC ANIMAL

EARLY CRETACEOUS

LATE CRETACEOUS

• ORDER • Pterosauria • FAMILY • Pteranodontidae • GENUS & SPECIES • *Pteranodon longiceps, P. sternbergi*

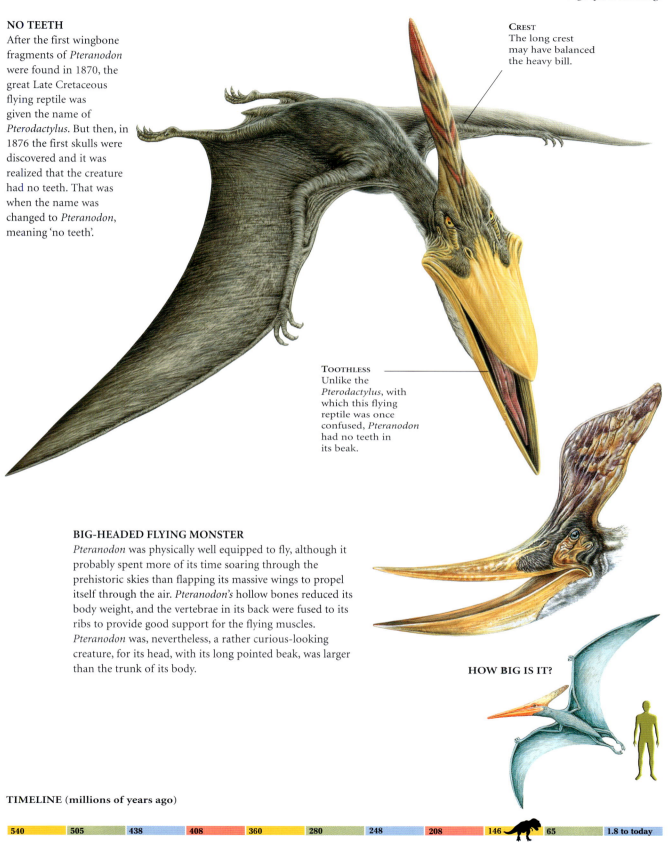

**C**REST
The long crest may have balanced the heavy bill.

## NO TEETH

After the first wingbone fragments of *Pteranodon* were found in 1870, the great Late Cretaceous flying reptile was given the name of *Pterodactylus*. But then, in 1876 the first skulls were discovered and it was realized that the creature had no teeth. That was when the name was changed to *Pteranodon*, meaning 'no teeth'.

**T**OOTHLESS
Unlike the *Pterodactylus*, with which this flying reptile was once confused, *Pteranodon* had no teeth in its beak.

## BIG-HEADED FLYING MONSTER

*Pteranodon* was physically well equipped to fly, although it probably spent more of its time soaring through the prehistoric skies than flapping its massive wings to propel itself through the air. *Pteranodon*'s hollow bones reduced its body weight, and the vertebrae in its back were fused to its ribs to provide good support for the flying muscles. *Pteranodon* was, nevertheless, a rather curious-looking creature, for its head, with its long pointed beak, was larger than the trunk of its body.

**HOW BIG IS IT?**

**TIMELINE** (millions of years ago)

| 540 | 505 | 438 | 408 | 360 | 280 | 248 | 208 | 146 | 65 | 1.8 to today |

LATE CRETACEOUS
# Pteranodon

# LATE CRETACEOUS

• **ORDER** • Pterosauria • **FAMILY** • Pteranodontidae • **GENUS & SPECIES** • *Pteranodon longiceps, P. sternbergi*

**MYSTERIOUS CREST**

*Pteranodon* had two notable features: first it had a very short tail – missing was the long ruddered tail of earlier pterosaurs, and it carried a very long bony crest that, in some of these flying reptiles, doubled the length of its head. Together, *Pteranodon's* head and crest measured around 1.8m (6ft). No one is certain what the crest was for, although there have been several suggestions. One is that the crest balanced the long beak – an important consideration for *Pteranodon's* stability in flight. Another is that the crest was used as a kind of rudder to aid in landing. The crest might even have acted as a brake to stop the *Pteranodon* coming in to land too quickly and sustaining injury. Some *Pteranodon* carried no crest at all, so it may have been a characteristic of males, possibly for display in mating rituals or simply for identification. Whatever role the crest played, *Pteranodon's* expertise in the air may not have been all that impressive. Despite a wingspan exceeding 8m (27ft), it may have done more gliding than flying.

LATE CRETACEOUS

# Tarbosaurus

| VITAL STATISTICS | |
|---|---|
| Fossil Location | Mongolia, China |
| Diet | Carnivorous |
| Pronunciation | Tar-boh-SAWR-us |
| Weight | 4900kg (10800lb) |
| Length | 10m (33ft) |
| Height | 5m (16ft) |
| Meaning of name | 'Terrifying lizard' |

*Tarbosaurus*, a smaller relative of the famous *Tyrannosaurus rex*, lived on the humid floodplains of Mongolia and China around 75 million years ago. As a relative of *Tyrannosaurus rex* it lived up to the predatory family characteristic, and may have taken on dinosaurs much larger than itself, such as the hadrosaurs. Like *Tyrannosaurus*, *Tarbosaurus* moved around on two legs, and it was a major predator. However, *Tarbosaurus'* arms and two-fingered hands were the smallest found in any dinosaur of the Tyrannosaurid family to which it belonged.

**WHERE IN THE WORLD?**

*Tarbosaurus* was found in the Nemegt Formation in Mongolia's Gobi Desert and the Suba Shi Formation in the Xingjiang Autonomous Region of China.

**FOSSIL EVIDENCE**
In 1946, the Soviet Union and Mongolia teamed up to send an expedition into the Gobi Desert, where it found a large skull together with some back bones belonging to a member of the theropod group. Three more skulls were discovered there in 1948 and 1949. These finds were classed in 1965 as different stages of the same species, *Tarbosaurus bataar* ('Tarbosaurus the hero'). Subsequently, Mongolia's Nemegt Formation became of great interest to expeditions from Poland, Japan and Canada, as well as from Mongolia itself. So far, more than 30 *Tarbosaurus* fossils have been located, together with 15 skulls.

| DINOSAUR |
|---|
| LATE CRETACEOUS |

**TEETH**
*Tarbosaurus* had up to 64 sharp teeth in its jaws, the largest of which, situated in its upper jawbone, were 8.5cm (3in) long.

**CLAWS**
*Tarbosaurus* had three sharp, curved 1.7cm (4.5in)-long claws on each of its back feet and two on each of its fore feet.

# LATE CRETACEOUS

• **ORDER** • Saurischia • **FAMILY** • Tyrannosauridae • **GENUS & SPECIES** • *Tarbosaurus bataar*

## ONE OF A KIND
The name *Tarbosaurus* was created to designate a new tyrannosaur nearly as large as *Tyrannosaurus* and very similar in detail. Some scientists believe these details are trivial and thus not enough to warrant a unique genus. They consider *Tarbosaurus bataar* to be a species within the genus *Tyrannosaurus*, i.e., *Tyrannosaurus bataar*. This may have some merit since strictly North American *Tyrannosaurus* and strictly Asian *Tarbosaurus* lived in places that were connected by a land bridge during the Late Cretaceous.

## MORE FOSSIL EVIDENCE NEEDED
A 2003 study identified the tyrannosaur *Alioramus* as *Tarbosaurus*' closest known relative. If correct, this relationship would cast doubt on the *Tarbosaurus/Tyrannosaurus* link, lending support to the idea that separate tyrannosaur lineages evolved in North America and Asia. But some scientists think that the single known specimen of *Alioramus* is clearly separate from *Tarbosaurus*. More fossils are needed to resolve this issue.

### SMALL HANDS
Like all later tyrannosaurs, Tarbosaurus had short arms with only 2-clawed fingers on its hands. Although some scientists calculate great strength for tyrannosaur arms, others believe they were vestigial and of lessening importance to the group.

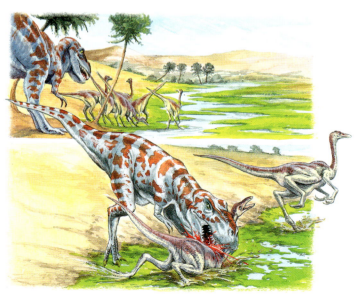

## HOW BIG IS IT?

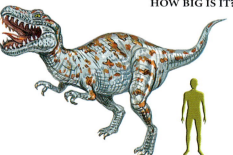

**TIMELINE (millions of years ago)**

| 540 | 505 | 438 | 408 | 360 | 280 | 248 | 208 | 146 | 65 | 1.8 to today |

LATE CRETACEOUS

# Tarbosaurus

# LATE CRETACEOUS

• **ORDER** • Saurischia • **FAMILY** • Tyrannosauridae • **GENUS & SPECIES** • *Tarbosaurus bataar*

**NEW FIND IN THE GOBI DESERT**

Palaeontology is an ongoing process and new discoveries arise all the time. In 2006, a joint venture between members of the Centre of Palaeontology at the Mongolian Academy of Sciences and experts from the Hayashibara Company, a biotechnology firm based in Okayama, Japan, made an exciting find. The team discovered a nearly intact skeleton of a *Tarbosaurus* contained within a block of sandstone. The skeleton belonged to a young *Tarbosaurus* and has been hailed as one of the best-preserved fossils of its kind ever discovered. The only parts missing were the neck bones and bones at the tip of the dinosaur's tail. The find was notable, as the skeletons of young dinosaurs are often found in poor condition, eroded by weather or destroyed by predators. The *Tarbosaurus*, the gender of which is as yet unknown, measures 2m (6ft 6in) long, one-sixth the estimated size of an adult. It was approximately five years old when it died some 70 million years ago.

LATE CRETACEOUS

# Triceratops

| VITAL STATISTICS | |
|---|---|
| Fossil Location | North America |
| Diet | Herbivorous |
| Pronunciation | Try-SER-a-tops |
| Weight | 12,000kg (26455lb) |
| Length | 9m (29ft 6in) |
| Height | 3m (10ft) |
| Meaning of name | 'Three-horned face' |

*Triceratops* is one of the most famous dinosaurs. In the ensuing years, hundreds of *Triceratops* skulls and other remains have been found, with the Hell Creek Formation in Montana providing a particularly rich yield. *Triceratops* skulls were very rugged, which helps explains why they have survived so well for 72 million years.

**WHERE IN THE WORLD?**

*Triceratops* has been found in Colorado, Wyoming, Montana and South Dakota in the US and in Alberta and Saskatchewan in Canada.

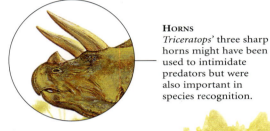

**HORNS**
*Triceratops*' three sharp horns might have been used to intimidate predators but were also important in species recognition.

**FOSSIL EVIDENCE**
The first *Triceratops* find of 1877 was sent to Othniel Charles Marsh, the renowned palaeontologist. He thought it came from the Pliocene period, 5.3–1.8 million years ago. Marsh also reckoned, just as inaccurately, that the remains had been a type of bison that he called *Bison alticornis*. He changed his mind in 1888, however, after examining two more skulls. These convinced him that what he had thought was a bison was actually a dinosaur.

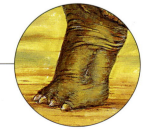

**LEGS**
*Triceratops* had huge chunky legs that supported its heavy bulk.

DINOSAUR

LATE CRETACEOUS

LATE CRETACEOUS

• **ORDER** • Ornithischia • **FAMILY** • Ceratopsidae • **GENUS & SPECIES** • *Triceratops horridus, T. prorsus*

### IDENTIFYING MALES AND FEMALES
In 1986, more than a century after *Triceratops* was first discovered, the American palaeontologist Thomas M. Lehman suggested that male and female examples of this dinosaur could be identified by their skulls and horns. According to Lehman, the males had larger skulls and more upright horns. They were taller than the female *Triceratops*, which had shorter horns that leaned forwards.

*Triceratops* was not among the largest herbivores, but it was tough and was likely a formidable opponent when threatened or attacked.

### TEETH IN RESERVE
The teeth of *Triceratops* lined its mouth in batteries. Up to 40 columns of teeth grew on each side of its jaw, with around five teeth stacked in each column. This meant that *Triceratops* possessed some 432 teeth in all, and if any of them were damaged or broken, another would grow in its place. This arrangement of teeth would have made it easy for *Triceratops* to eat large quantities of tough, fibrous plants such as cycads, palms or ferns.

### HOW BIG IS IT?

**TIMELINE (millions of years ago)**

LATE CRETACEOUS

# Triceratops

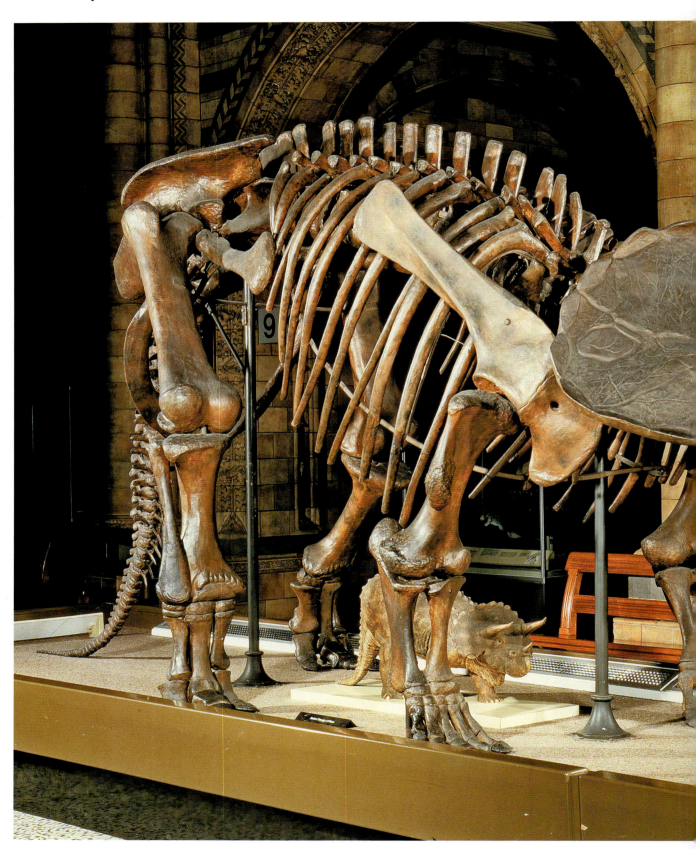

# LATE CRETACEOUS

• **ORDER** • Ornithischia • **FAMILY** • Ceratopsidae • **GENUS & SPECIES** • *Triceratops horridus*, *T. prorsus*

**PROLIFIC FINDINGS**

*Triceratops* fossils are among the most frequently found non-avian dinosaurs known. This is probably testimony to their great abundance in the Late Cretaceous since, even though bone beds of Triceratops are uncommon, most finds are of single individuals. Despite this, *Triceratops* finds have been so prolific that one palaeontologist, Bruce Erickson of the Science Museum of Minnesota, observed around 200 examples of *Triceratops prorsus*, which was first discovered in 1890 by Charles Othniel Marsh in the Hell Creek Formation in Montana. Another palaeontologist, Barnum Brown, did even better. He claimed to have seen more than 500 skulls in the same area. In the west of North America, hundreds of fragments of *Triceratops* teeth, horns, frills and other pieces of skull have been discovered, leading to the conclusion that *Triceratops* was one of the most common herbivores of the Late Cretaceous. Some palaeontologists go further and class this dinosaur as the most dominant of its time. American palaeontologist Robert Bakker reckoned in 1986 that *Triceratops* accounted for more than 80 per cent of all large dinosaurs at the end of the Cretaceous Period.

LATE CRETACEOUS

# Troodon

| VITAL STATISTICS | |
|---|---|
| Fossil Location | North America |
| Diet | Carnivorous |
| Pronunciation | TROH-don |
| Weight | 60kg (130lb) |
| Length | 2m (6ft 6in) |
| Height | 1m (39in) |
| Meaning of name | 'Wounding tooth' |

*Troodon*, one of the first dinosaurs to be found in North America, was discovered in 1855. It was comparatively small and lived around 75 million years ago in what is now the USA and Canada. To judge by its long, narrow limbs, it was a quick mover. Some palaeontologists believe it was an omnivore that ate plants but also hunted insects and other creatures smaller than itself, such as lizards. *Troodon* has been classed as one of the most intelligent non-avian dinosaurs because its brain was among the largest relative to its body mass.

**HEAD**
*Troodon* was a bird-like dinosaur with a head that bore some resemblance to the head of the present-day ostrich.

**FOSSIL EVIDENCE**
*Troodon* fossils were discovered in North America in places many hundreds of kilometers distant from each other and separated by up to 10 million years. This makes it very unlikely that they belonged to a single species of *Troodon*, although it remains unknown how many species there were. A *Troodon* tooth was the first fossil found (in 1901), followed in 1932 by a foot, a hand and vertebrae. A dramatic feature of the foot was the enlarged claw attached to the second toe.

**CLAWS**
*Troodon* may have been comparatively small, but the large, curved claws on its hands and feet made it a very dangerous predator.

DINOSAUR

LATE CRETACEOUS

LATE CRETACEOUS

• ORDER • Saurischia • FAMILY • Troodontidae • GENUS & SPECIES • *Troodon formosus*

## WHERE IN THE WORLD?

*Troodon* has been found in Montana, Alaska and Wyoming in the USA, and in Alberta in Canada.

### LARGE BRAIN AND BINOCULAR VISION

Some palaeontologists believe *Troodon* was among the most intelligent of the dinosaurs, with a large brain relative to its body mass and binocular vision that was better than most other creatures of its kind possessed. Its vision was important if, as some palaeontologists believe, *Troodon* was a predator living off small animals because it gave *Troodon* the facility of seeing in depth.

### OMNIVOROUS DIET

Was *Troodon* exclusively a herbivore? It certainly had the kind of teeth herbivores needed to cut and shred plants and leaves, and the strong, clawed fingers and toes needed to pull down and keep hold of branches while eating. But these features could also have been used for shredding and tearing flesh. It has therefore been suggested that *Troodon* was in fact an omnivore. Its claws and mouthful of sharp, hooked fangs also seem to support this idea.

## HOW BIG IS IT?

**TIMELINE (millions of years ago)**

| 540 | 505 | 438 | 408 | 360 | 280 | 248 | 208 | 146 | 65 | 1.8 to today |

379

LATE CRETACEOUS

# Velociraptor

| VITAL STATISTICS | |
|---|---|
| Fossil Location | Mongolia, Chinese Inner Mongolia |
| Diet | Carnivorous |
| Pronunciation | Vel-ossi-RAP-tor |
| Weight | 15kg (33lb) |
| Length | 2m (7ft) |
| Height | 0.5m (1ft 7in) at the hip |
| Meaning of name | 'Swift thief' |

*Velociraptor* had feathers, a long tail and claws on all its fingers and toes. The claws were probably used to kill *Velociraptor's* prey and tear its flesh. It appears to have lived in an arid environment full of sand dunes but very little water. *Protoceratops* likely provided a meal for *Velociraptor*. In 1971, a find that came to be known as the 'Fighting Dinosaurs' was discovered, in which *Protoceratops* and *Velociraptor* were found locked together in deadly combat.

**WHERE IN THE WORLD?**

*Velociraptor* was found at Omnogovi and Tugrugeen Shireh in Mongolia, and in Chinese Inner Mongolia.

**FOSSIL EVIDENCE**
Around 12 fossil skeletons of *Velociraptor* have been discovered, more than any other member of the bird-like dromaeosaurid family. *Velociraptor's* teeth were made for slicing meat, with up to 28 of them spaced along each side of its jaw. The teeth were serrated front and back, a feature that helped *Velociraptor* slice the flesh of its prey. *Velociraptor's* hands with their three strong claws were constructed in the same way as the wing bones of modern birds.

**Claws**
Sickle-shaped claws on its feet measured more than 6.5cm (2.5in) each.

DINOSAUR

LATE CRETACEOUS

LATE CRETACEOUS

• ORDER • Saurischia • FAMILY • Dromaeosauridae • GENUS & SPECIES • *Velociraptor mongoliensis, V. osmolskae*

## MISSED DISCOVERIES

During the Cold War, palaeontologists from the West were banned from Communist-ruled Mongolia and were unable to explore its rich sources of dinosaur finds. Therefore, Western palaeontologists missed out on important discoveries including the so-called 'Fighting Dinosaurs'. Western palaeontologists were allowed to return to Mongolia in 1990, with the end of the Cold War.

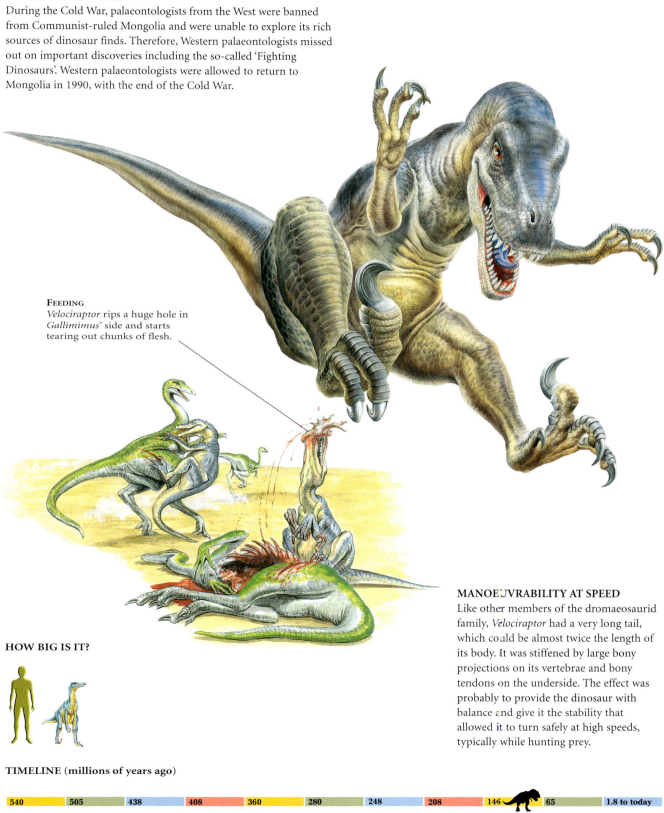

**FEEDING**
*Velociraptor* rips a huge hole in *Gallimimus*' side and starts tearing out chunks of flesh.

## MANOEUVRABILITY AT SPEED

Like other members of the dromaeosaurid family, *Velociraptor* had a very long tail, which could be almost twice the length of its body. It was stiffened by large bony projections on its vertebrae and bony tendons on the underside. The effect was probably to provide the dinosaur with balance and give it the stability that allowed it to turn safely at high speeds, typically while hunting prey.

**HOW BIG IS IT?**

**TIMELINE (millions of years ago)**

| 540 | 505 | 438 | 408 | 360 | 280 | 248 | 208 | 146 | 65 | 1.8 to today |

LATE CRETACEOUS

# Velociraptor

LATE CRETACEOUS

• **ORDER** • Saurischia • **FAMILY** • Dromaeosauridae • **GENUS & SPECIES** • *Velociraptor mongoliensis, V. osmolskae*

**EVIDENCE OF FEATHERS**

*Velociraptor* was first discovered in Mongolia in 1922, long before birds were conclusively recognized as dinosaurs. When this idea began to take hold in the 1960s, scientists began to suppose that *Velociraptor*, since was it was so close to birds, was feathered. it was presumed that it was unable to fly. It wasn't until 2007 that fossil evidence was uncovered that proved this supposition. In September of that year, a study made on the bones of a *Velociraptor* forearm found in Mongolia revealed rows of small bumps. These were soon identified as quill knobs, which served as anchors for feathers. From this discovery the conclusion was drawn that *Velociraptor* wore feathers. When this revelation was added to the many other similarities between *Velociraptor* and modern birds, it became even clear that despite the interval of millions of years that lay between them, they had a great deal in common. As Mark Norell, curator in charge of fossil reptiles, amphibians and birds at the American Museum of Natural History, put it: 'the more we learn about these animals, the more we find that there is basically no difference between birds and their closely related dinosaur ancestors like *Velociraptor*. Both have wishbones, brooded their nests, possessed hollow bones and were covered in feathers. If animals like *Velociraptor* were still alive today, our first impression would be that they were just very unusual-looking birds.'

LATE CRETACEOUS

# Styracosaurus

• **ORDER** • Ornithischia • **FAMILY** • Ceratopsidae • **GENUS & SPECIES** • *Styracosaurus albertensis*

| VITAL STATISTICS | |
|---|---|
| Fossil Location | North America |
| Diet | Herbivorous |
| Pronunciation | Styra-co-SAWR-us |
| Weight | 3000kg (6613lb) |
| Length | 5.5m (18ft) |
| Height | 1.8m (6ft) |
| Meaning of name | 'Spiked lizard' |

**WHERE IN THE WORLD?**

*Styracosaurus* remains were found in Montana and Wyoming in the USA and at the Judith River Group in Alberta, Canada.

*Styracosaurus'* terrifying appearance probably deterred attackers but was also likely used for species recognition. Its head was adorned with several long, barbarous-looking spikes and its nose had a fearsome protruding horn.

**NOSE HORN**
The long horn on the centre of *Styracosaurus'* nose measured 57cm (22in) in length in the skull but was even longer in the living animal because of the keratin sheath that it would have had over the bone.

**FOSSIL EVIDENCE**
*Styracosaurus* was a large, horned and frilled dinosaur with a body resembling that of a rhinoceros. The first *Styracosaurus* fossils (parts of a skull and skeleton) were discovered in 1913 in Alberta, Canada, in the area now named Dinosaur Park Formation. Later finds, such as the lower jaws and the remainder of the skeleton, were made in 1935. In 1915, a US expedition uncovered a nearly complete skeleton with part of its skull, also in the Dinosaur Park. Further discoveries were made in 2006.

**BEAK**
*Styracosaurus'* sharply pointed beak, used for snapping off vegetation.

DINOSAUR

LATE CRETACEOUS

**TIMELINE** (millions of years ago)

| 540 | 505 | 438 | 408 | 360 | 280 | 248 | 208 | 146 | 65 | 1.8 to today |

NEOGENE

# Borhyaena

• ORDER • Sparassodonta • FAMILY • Borhyaenidae • GENUS & SPECIES • *Borhyaena macrodenta, B. tuberata*

### VITAL STATISTICS

| | |
|---|---|
| Fossil Location | South America |
| Diet | Carnivorous |
| Pronunciation | Bore-high-EE-nuh |
| Weight | 100kg (220lb) |
| Length | 1.5m (5ft) |
| Height | Unknown |
| Meaning of name | 'Devouring hyena' because of its presumed ravenous nature and similarity to hyenas |

*Borhyaena* was a marsupial – in other words, a mammal that carried its young in a pouch. It was heavily built and therefore probably could not run particularly fast on its flat feet and short legs.

### WHERE IN THE WORLD?

*Borhyaena* was found in what is now Argentina in South America.

### FOSSIL EVIDENCE

Fossils of the hyena-like *Borhyaena* were found in the early rocks laid down in the Miocene era in what is now Argentina. The *Borhyaena* skull was large and its teeth, as befitted a ferocious carnivore, were heavy, broad and capable of making an efficient job of crushing its prey, although its fangs were not particularly long. *Borhyaena* was not likely a swift mover. It may have ambushed prey from behind cover, grasping its victims in its claws.

PREHISTORIC ANIMAL

NEOGENE

**Feet**
The four claws on each of *Borhyaena*'s feet made it a formidable predator.

**Teeth**
The teeth were heavy and broad and possibly strong enough to crush bone.

### HOW BIG IS IT?

### TIMELINE (millions of years ago)

| 540 | 505 | 438 | 408 | 360 | 280 | 248 | 208 | 146 | 65 | 1.8 to today |

PALEOGENE

# Gastornis

| VITAL STATISTICS | |
|---|---|
| Fossil Location | Western and central Europe, North America |
| Diet | Possibly omnivorous |
| Pronunciation | gas-TOR-nis |
| Weight | 170kg (385lb) |
| Length | Unknown |
| Height | 2.13m (7ft) |
| Meaning of name | 'Gaston's bird' after Gaston Planté, who had discovered the first fossils near Paris |

**WHERE IN THE WORLD?**

*Gastornis* was found in France, Belgium, Germany in Europe and in North America.

*Gastornis* lived some 60 million years ago. Despite its slight resemblance to the modern parrot, if an exceedingly large one, it has no known close relatives among the birds alive today. The environment in which it lived was mainly marked by thick forests with a varying climate that was moist or partly arid and tropical, or subtropical at different times.

## FOSSIL EVIDENCE

*Gastornis* was named in 1855, after Gaston Planté, a French physicist, who made the first fossil discovery in the Argile Plastique Formation at Meudon, near Paris. Around 1870, a famous American palaeontologist, Edward Drinker Cope, found more fossils and named them Diatryma. *Gastornis's* fossilized footprints indicate that it had very large feet. One footprint found at Montmorency, near Paris, was 40cm (16in) long. Another, found in the Green River Valley near Black Diamond, Washington State, USA, measured 27cm (11in) wide and 32cm (13in) long.

**Beak**
*Gastornis's* talons and sharp beak made it look fearsome, but it is not certain that it was a predator.

**Wings**
*Gastornis's* tiny wings could not possibly have lifted this large and heavy bird into the air to fly.

DINOSAUR

PALEOGENE

386

PALEOGENE

• ORDER • Gastornithiformes • FAMILY • Gastornithidae • GENUS & SPECIES • Several species within the genus *Gastornis*

## TWO BIRDS AS ONE
The fossils that Gaston Planté discovered in 1855 were fragmentary, so Gastornis was initially classified as a crane-like bird. More fossils were discovered some 15 years later, and in 1884, the American ornithologist Elliott Coues suggested that Gastornis and Diatryma must be one and the same creature. However, a century passed before this idea was widely accepted.

## BIG BIRD
If *Gastornis* was a carnivore, then the prehistoric horse, known as *Hyracotheum*, could have been one of its victims. *Gastornis's* huge size and superior height would have enabled it to dominate the 60cm (2ft)-long horse, and its mighty talons could easily have held it down.

**LEGS**
*Gastornis* was a hefty bird, with thick, strong legs and deadly claws.

**HOW BIG IS IT?**

**TIMELINE (millions of years ago)**

| 540 | 505 | 438 | 408 | 360 | 280 | 248 | 208 | 146 | 65 | 1.8 to today |

387

PALEOGENE
# Gastornis

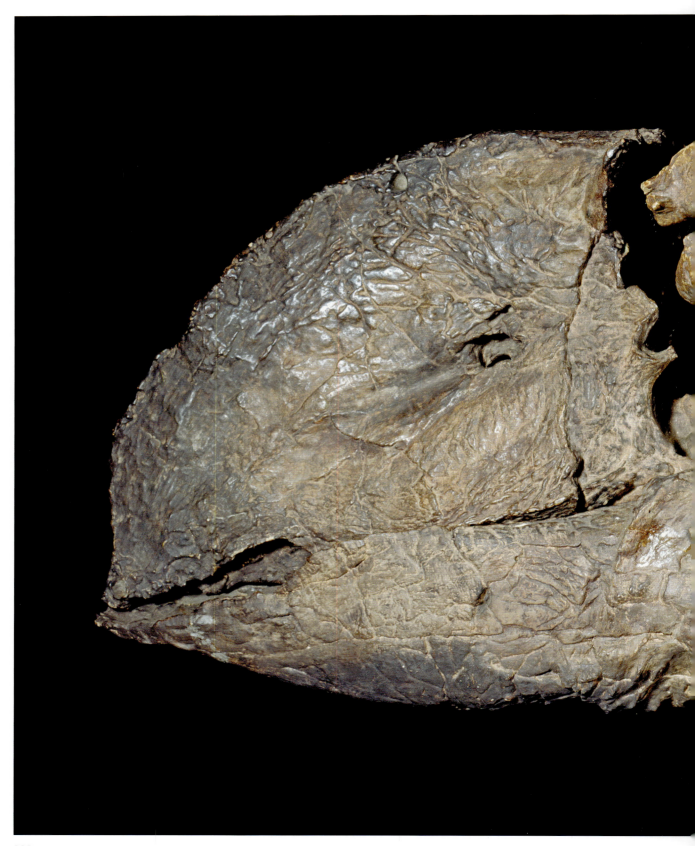

PALEOGENE

• **ORDER** • Gastornithiformes • **FAMILY** • Gastornithidae • **GENUS & SPECIES** • Several species within the genus *Gastornis*

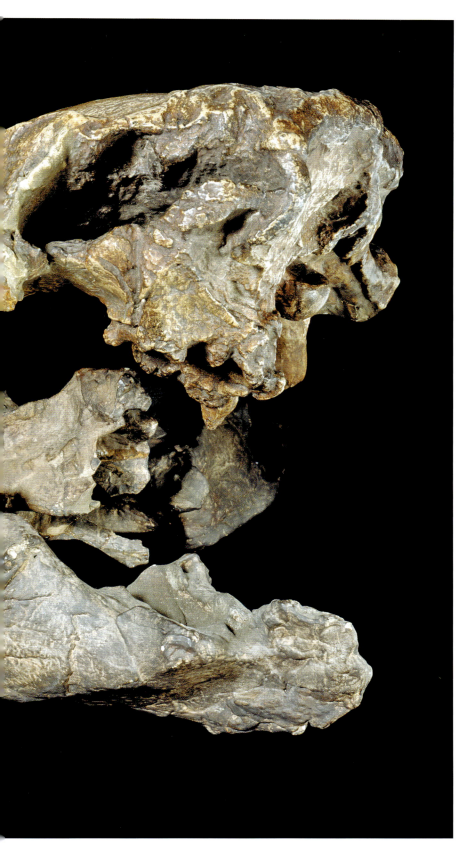

**DIETARY CONTROVERSY**

Although *Gastornis* is frequently classed as an omnivore, palaeontologists have not always agreed about what food it ate and whether it was a predator and carnivore. Its large, powerful legs and ferocious-looking claws seemed to some to suggest that *Gastornis* was a hunter and meat-eater, but others doubt that *Gastornis* would have been fast or agile enough to pursue, catch and hold down its prey as carnivores normally did. It has been suggested that *Gastornis*'s huge, beak is not strong enough nor designed well for a 'predatory habit'. Were its sharply pointed upper and lower ends used to stab and tear the flesh of a victim, or was *Gastornis* a herbivore, using its beak to crush seeds and strip vegetation off plants and trees? The alternative to these theories is that *Gastornis* was an omnivore – one that confined its meat-eating to smaller creatures, vertebrate or invertebrate, that could be easily pursued and caught.

NEOGENE

# Carcharocles

• ORDER • Lamniformes • FAMILY • Lamnidae • GENUS & SPECIES • Carcharocles megalodon

| VITAL STATISTICS | |
|---|---|
| FOSSIL LOCATION | Worldwide |
| DIET | Carnivorous |
| PRONUNCIATION | car-CAR-o-kleez MEG-ah-lo-don |
| WEIGHT | 45,360kg (50 tons) |
| LENGTH | 18m (60ft) |
| HEIGHT | Unknown |
| MEANING OF NAME | 'Jagged clamp' or 'Jagged and famous' |

### WHERE IN THE WORLD?

Teeth of *C. megalodon* (sometimes loosely referred to as just *Megalodon*) have been found worldwide. They have even been dredged up from the South Pacific sea floor.

Imagine a prehistoric shark so huge that it preyed on whales; it was as long as a school bus and had teeth as big as an adult man's hand.

*Carcharocles megalodon* was a huge shark with giant, serrated teeth. In general design, it was much like the modern Great White Shark *Carcharodon carcharias*, but more robust. For a long time, it was thought to be a species of giant Great White, but recent work suggests that it is more closely related to mako sharks.

### FOSSIL EVIDENCE

Giant teeth of prehistoric sharks have been known for hundreds of years. Shark teeth are probably the most common vertebrate fossils because sharks have a long, diverse history, are (and were) extremely abundant, and produce (and lose) tens of thousands of hard, durable teeth during their lifetimes. But little else of the skeletons of these animals is known. This is because sharks have cartilaginous skeletons that don't often preserve well. However, in rare cases, jaws and vertebrae have been found and, along with the teeth, these indicate that *Carcharocles megalodon* possibly grew to 18m (60ft) in length.

**TEETH**
The largest known teeth of this shark reached 20cm (8in) in length, but there were also much smaller ones in its mouth.

**VERTEBRAE**
Sometimes shark vertebrae mineralize enough during life that they can more readily become fossils. Known *C. megalodon* vertebrae are enormous.

### HOW BIG IS IT?

**TIMELINE** (millions of years ago)

| 540 | 505 | 438 | 408 | 360 | 280 | 248 | 208 | 146 | 65 | 1.8 to today |

# Platybelodon

NEOGENE
• ORDER • Proboscidea • FAMILY • Gomphotheriidae
• GENUS & SPECIES • Several species within the genus *Platybelodon*

## VITAL STATISTICS

| | |
|---|---|
| FOSSIL LOCATION | Worldwide |
| DIET | Herbivorous |
| PRONUNCIATION | Platty-BELL-oh-don |
| WEIGHT | 4000kg (3.9 tons) |
| LENGTH | 6m (20ft) |
| HEIGHT | 2.8m (9ft) at the shoulder |
| MEANING OF NAME | Possibly 'Broad point tooth' (or broad tusk) for its flat lower tusks |

### WHERE IN THE WORLD?

*Platybelodon* fossils have been found worldwide.

*Platybelodon*, which lived around 15 million years ago, was related to the modern elephant. Living in swampy savannahs or prairies, *Platybelodon* likely ate soft leaves and tree bark.

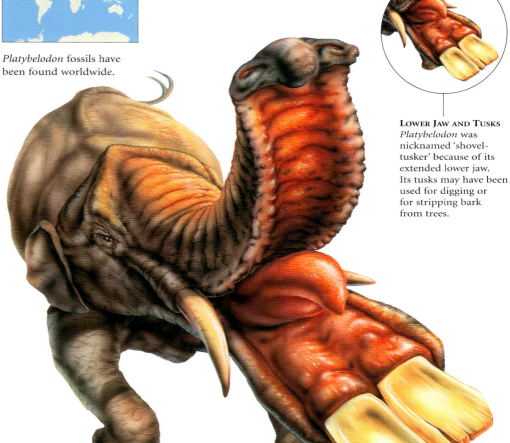

**LOWER JAW AND TUSKS**
*Platybelodon* was nicknamed 'shovel-tusker' because of its extended lower jaw. Its tusks may have been used for digging or for stripping bark from trees.

## FOSSIL EVIDENCE

*Platybelodon* may have crossed the landbridge between northwest North America and eastern Siberia, leaving its fossils in both areas. Its huge lower jaw had two flat teeth at the front. In its cheeks were teeth for grinding up leaves before swallowing them. *Platybelodon* also had two sharp, downward-facing tusks. These might have been used to strip the bark off trees for food, to dig for water when the ground was dry, or as weapons.

**TEETH**
*Platybelodon's* teeth and the wear patterns on their teeth and tusks suggest it was a versatile browser.

**HOW BIG IS IT?**

PREHISTORIC ANIMAL
NEOGENE

**TIMELINE (millions of years ago)**

| 540 | 505 | 438 | 408 | 360 | 280 | 248 | 208 | 146 | 65 | 1.8 to today |

NEOGENE

# Thylacosmilus

• **ORDER** • Sparassodonta • **FAMILY** • Thylacosmilidae • **GENUS & SPECIES** • *Thylacosmilus atrox, T. lentis*

| VITAL STATISTICS | |
|---|---|
| Fossil Location | South America |
| Diet | Carnivorous |
| Pronunciation | thigh-LAK-o-SMY-lus |
| Weight | 90kg (200lb) |
| Length | 2.5m (8ft) |
| Height | 0.6m (2ft) |
| Meaning of name | 'Pouch sabre' (pouched mammal with sabre teeth) |

The marsupial answer to the sabre-toothed cats, *Thylacosmilus* filled the same carnivorous niche in South America.

*Thylacosmilus'* fangs look very much like those of a sabre-toothed cat, but so does the general body plan of the animal. This is because the two animals led very similar predatory lives. One outward difference would have been the marsupium, or brood pouch, used to raise the young of the species.

### WHERE IN THE WORLD?

Most *Thylacosmilus* fossils have been found in Argentina; a few have been found at other locations in South America.

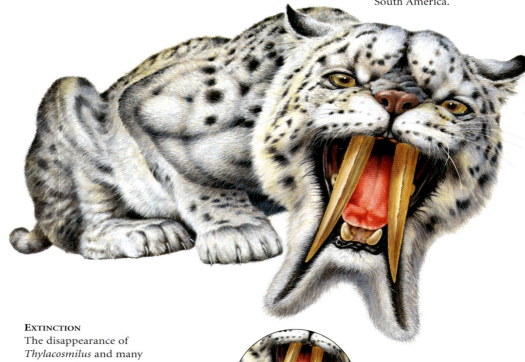

### FOSSIL EVIDENCE

Most of what is known about *Thylacosmilus* is gleaned from two partial skeletons found in Pliocene deposits in Argentina. But other even more incomplete specimens indicate that it lived from at least the late Miocene and on into the Pleistocene until about two million years ago. Land carnivores are generally rarer than herbivores and fewer specimens of *Thylacosmilus* are expected in the fossil record than their prey. The hunting style of *Thylacosmilus* probably differed significantly from its cat analogs, as evidenced by the non-retractable nature of their claws. All cats, except the cheetah, have retractable claws; this helps keep them sharp.

### EXTINCTION
The disappearance of *Thylacosmilus* and many other South American marsupials roughly coincides with the Pleistocene exchange of faunas between the Americas.

### HOW BIG IS IT?

### FANGS
The sabres of some mammalian carnivores are usually seen as stabbing weapons, but those of *Thylacosmilus* might not have been strong enough for that function.

PREHISTORIC ANIMAL

NEOGENE

**TIMELINE (millions of years ago)**

| 540 | 505 | 438 | 408 | 360 | 280 | 248 | 208 | 146 | 65 | 1.8 to today |

TERTIARY (EOCENE)
• ORDER • Pantodonta • FAMILY • Coryphodontidae
• GENUS & SPECIES • Several species within the genus *Coryphodon*

# Coryphodon

## VITAL STATISTICS

| | |
|---|---|
| FOSSIL LOCATION | Europe, North America |
| DIET | Herbivorous |
| PRONUNCIATION | Cor-ee-FOE-don |
| WEIGHT | 500kg (1102lb) |
| LENGTH | 2.25m (7ft 6in) |
| HEIGHT | 1m (3ft 4in) at the shoulder |
| MEANING OF NAME | Possibly 'Point tooth' for the pointed tooth ridges |

### WHERE IN THE WORLD?

*Coryphodon* was found in Europe, and in North Dakota in the USA.

*Coryphodon* inhabited swamps and marshes 55 million years ago. It is the largest mammal so far known from its time, with long forelimbs and short back legs, which were needed to support its weight.

**TUSKS**
*Coryphodon's* tiny tusks, a feature likely of the male of the species, were used to pull up plants in the marshes.

## FOSSIL EVIDENCE

*Coryphodon* was a primitive mammal that lived in the early Eocene epoch some 55 million years ago. Like the modern hippopotamus (to which it is not closely related), *Coryphodon* seems to have led a semi-aquatic lifestyle in a marsh habitat. It varied in size between the modern tapir and rhinoceros. *Coryphodon* was not one of the more intelligent early mammals: palaeontologists have estimated that its brain-to-bodyweight ratio was 90g (1lb 3oz) to 500kg (1102lb).

**FEET**
Each foot ended in five toes that resemble those of the modern-day elephant. Each toe ended in a small hoof.

### HOW BIG IS IT?

PREHISTORIC ANIMAL

TERTIARY (EOCENE)

**TIMELINE (millions of years ago)**

| 540 | 505 | 438 | 408 | 360 | 280 | 248 | 208 | 116 | 65 | 1.8 to today |

TERTIARY (EOCENE)

# Hyracotherium

• **ORDER** • Perissodactyla • **FAMILY** • Palaeotheriidae • **GENUS & SPECIES** • *Hyracotherium leporinum*

| VITAL STATISTICS | |
|---|---|
| FOSSIL LOCATION | Europe, North America |
| DIET | Herbivorous |
| PRONUNCIATION | High-rah-co-THEER-ium |
| WEIGHT | 6.8kg (15lb) |
| LENGTH | 60cm (2ft) |
| HEIGHT | 23cm (9in) at the shoulder |
| MEANING OF NAME | 'Hyrax beast' because of its supposed resemblance with hyraxes |

### WHERE IN THE WORLD?

Fossils have been found in England and in Utah in the USA.

*Hyracotherium* which lived around 50 million years ago, was once considered the earliest known horse but it is now considered a paleothere – a group closely related to horses.

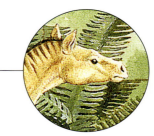

**FACE**
*Hyracotherium*'s face resembled that of modern Arab horses. It also had a diastema (space) between its front and back teeth.

### FOSSIL EVIDENCE

The first *Hyracotherium* fossils were found in England in 1841 by the palaeontologist Richard Owen. The discovery was not a complete skeleton, and Owen named it 'a hyrax-like beast.' (The hyrax is a small herbivorous mammal). In 1876, the American palaeontologist Othniel C. Marsh found a complete skeleton, which he named *Eohippus* (meaning *'dawn horse'*). However, *Hyracotherium* is the proper name. The skeleton had four toes, with hooves on each of the front feet and three toes with hooves on each back foot. *Hyracotherium* had a long skull and 44 teeth.

PREHISTORIC ANIMAL

TERTIARY (EOCENE)

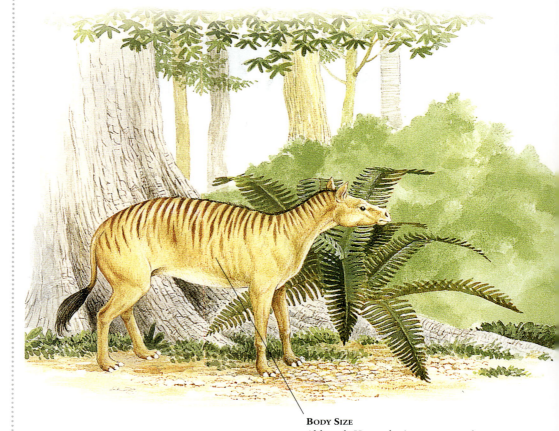

**BODY SIZE**
Although *Hyracotherium* was an early relative of much larger animals including the rhinoceros, it was only the size of a small dog.

### HOW BIG IS IT?

**TIMELINE (millions of years ago)**

| 540 | 505 | 438 | 408 | 360 | 280 | 248 | 208 | 146 | 65 | 1.8 to today |

TERTIARY (EOCENE)

# Mesonyx

• **ORDER** • Mesonychia • **FAMILY** • Mesonychidae • **GENUS & SPECIES** • *Mesonyx obtusidens, M. uintensis*

*Mesonyx* was a wolf-like mammal that probably lived by the sea around 45 million years ago.

## VITAL STATISTICS

| | |
|---|---|
| FOSSIL LOCATION | North America, east Asia |
| DIET | Carnivorous |
| PRONUNCIATION | Mez-ON-icks |
| WEIGHT | Unknown |
| LENGTH | 2.5m (8ft) |
| HEIGHT | Unknown |
| MEANING OF NAME | 'Middle claw' |

### WHERE IN THE WORLD?

*Mesonyx* was found in Wyoming and northern Utah in the USA, and in east Asia.

**PREDATOR ORIGINS**
*Mesonyx* was a mesonychid, a family of predators sometimes also referred to by its older name Acreodi, which first appeared some 50 million years ago.

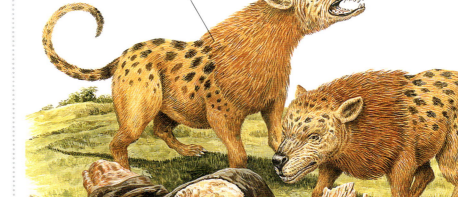

## FOSSIL EVIDENCE

Palaeontologists discovered *Mesonyx uintensis,* one of the two officially recognized species, in Wyoming in the USA. Their find dated from the Upper Eocene era of northern Utah, 45 million years ago. From the fossils they unearthed, they calculated that *Mesonyx's* face measured 20.6cm (8in) in length and that its skull was 43cm (17in) long. This was certainly the skull of a predator; *Mesonyx* had a large saggital crest above its braincase, which anchored powerful jaw muscles that ensured a strong bite.

**FEET**
*Mesonyx* probably hunted hoofed plant-eaters and was likely fast on its feet.

PREHISTORIC ANIMAL

TERTIARY (EOCENE)

### HOW BIG IS IT?

## TIMELINE (millions of years ago)

| 540 | 505 | 438 | 408 | 360 | 280 | 248 | 208 | 146 | 65 | 1.8 to today |

TERTIARY (EOCENE)

# Pristichampsus

• ORDER • Crocodilia • FAMILY • Pristichampsidae
• GENUS & SPECIES • Several species within the genus *Pristichampsus*

## VITAL STATISTICS

| | |
|---|---|
| FOSSIL LOCATION | Europe, Asia, North America |
| DIET | Carnivorous |
| PRONUNCIATION | Pristi-CHAMP-sus |
| WEIGHT | Unknown |
| LENGTH | 3m (10ft) |
| HEIGHT | Unknown |
| MEANING OF NAME | Probably 'Sawfish tooth' for the resemblance of its teeth to those of sawfish *Pristis* |

*Pristichampsus* was a crocodile with an armoured body, teeth with serrated edges all round, an unarmoured tail and long legs suitable for running quickly in pursuit of prey.

### HOOVES
Unlike its more aquatic relatives, *Pristichampsus* had hooves rather than claws – another adaptation that points to a terrestrial habit.

### WHERE IN THE WORLD?

*Pristichampsus* was found in France and Germany, eastern Kazakhstan, in Hengdon Basin in China, and in Wyoming and western Texas in the USA.

## FOSSIL EVIDENCE
Prehistoric crocodiles like *Pristichampsus* were more widespread across the world than crocodiles are now. As the fossil record shows, *Pristichampsus* had longer limbs than modern crocodiles and, although modern crocodiles do sometimes gallop, *Pristichampsus* seems to have been better suited to it. But the fossil record could be misleading. *Pristichampsus*' teeth were compressed from side to side so that they were at first mistaken for the teeth of theropods, and, like some of them, had managed to survive the mass extinction that took place at the end of the Cretaceous Period.

PREHISTORIC ANIMAL

TERTIARY (EOCENE)

### DID YOU KNOW?
The big carnivorous prehistoric crocodiles like *Pristichampsus* could find plenty of small mammals like *Hyracotherium* to prey on.

### HOW BIG IS IT?

## TIMELINE (millions of years ago)

| 540 | 505 | 438 | 408 | 360 | 280 | 248 | 208 | 146 | 65 | 1.8 to today |

TERTIARY (EOCENE)

# Uintatherium

• ORDER • Dinocerata • FAMILY • Uintatheriidae • GENUS & SPECIES • *Uintatherium anceps*

## VITAL STATISTICS

| | |
|---|---|
| Fossil Location | USA |
| Diet | Herbivorous |
| Pronunciation | You-in-ta-THER-ium |
| Weight | 2721kg (6000lb) |
| Length | 4.5m (15ft) |
| Height | 1.5m (5ft) |
| Meaning of name | 'Uinta beast' for the Uinta Basin in Utah, where the fossils were found |

*Uintatherium* was a mammal closely resembling the modern rhinoceros. It lived 45 million years ago and its family, the Uintatheriidae, had skulls unlike any other group of mammals.

**Ossicones**
*Uintatherium* had three pairs of bony ossicones on its head. The function of these is unknown; they could have been for defence or for sexual display just like the ossicones that giraffes have.

### WHERE IN THE WORLD?

Fossils have been found in Wyoming, USA.

## FOSSIL EVIDENCE

*Uintatherium* fossils have been discovered near Fort Bridger in Wyoming, USA. Six probable males were found, with skulls that featured six prominent bony knobs in the front. The considerable weight of the skull was due to its extremely thick walls. As with the modern elephant, the weight was lessened somewhat by cavities within the bone. Some scientists believe that *Uintatherium* rarely strayed far from rivers or lakes, where it fed on marsh plants and other water vegetation.

**Fangs**
The purpose of the huge canines is unknown but likely included threat or species recognition.

### HOW BIG IS IT?

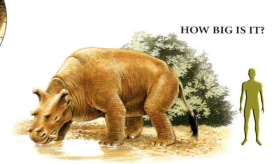

PREHISTORIC ANIMAL

TERTIARY (EOCENE)

**TIMELINE (millions of years ago)**

| 540 | 505 | 438 | 408 | 360 | 280 | 248 | 208 | 143 | 65 | 1.8 to today |

TERTIARY (EOCENE)

# Andrewsarchus

| VITAL STATISTICS | |
|---|---|
| Fossil Location | Mongolia |
| Diet | Possibly omnivorous |
| Pronunciation | AN-droo-SAR-kus |
| Length | 83cm (2.7ft) (skull) |
| Length | 4m (13 ft). (body) |
| Height | Unknown |
| Meaning of name | 'Andrews' chief', named for expedition leader Roy Chapman Andrews |

The huge, enigmatic skull of *Andrewsarchus* has left palaeontologists wondering whether it was the largest mammalian land carnivore of all time or an omnivore. *Andrewsarchus* had a very big skull and enormous jaw muscles for a powerful bite. But it also had relatively small eyes and blunt teeth, which suggest that it might not have been an active hunter. If its body was similar to its relatives, it walked on hooved feet rather than having the claws expected for a carnivore.

**WHERE IN THE WORLD?**

Fossils have been found in Irdin Mahna (also spelled Erdeni-Mandal), in the Gobi Desert, Mongolia.

**FOSSIL EVIDENCE**
A single skull lacking lower jaws is all that is known of *Andrewsarchus*. It was discovered in the early 1920s, and subsequent expeditions to the area of the find have failed to reveal any more specimens of this animal. The skull is well preserved though, and nearly complete, showing huge blunt teeth and wide attachments for the jaw muscles. If *Andrewsarchus* had similar proportions to its more completely known relatives, it may have been 4m (13ft) long and 1.8m (6ft) high at the shoulders.

**Teeth**
The immensely powerful bite and blunt teeth suggest *Andrewsarchus* was able to crush large bones and so might have been a scavenger.

| PREHISTORIC ANIMAL |
|---|
| TERTIARY (EOCENE) |

# TERTIARY (EOCENE)

•**ORDER** • Mesonychia • **FAMILY** • Triisodontidae • **GENUS & SPECIES** • *Andrewsarchus mongoliensis*

**EYES**
The eyes were fairly small and set low on the skull, close to the rear of the tooth row.

**DID YOU KNOW?**
*Andrewsarchus* was once considered to be a close relative to the ancestor of whales. More recent work has shown that they are not as closely related as was first assumed.

**HOW BIG IS IT?**

**TIMELINE (millions of years ago)**

| 540 | 505 | 438 | 408 | 360 | 280 | 248 | 208 | 146 | 65 | 1.8 to today |

399

TERTIARY (EOCENE)

# Basilosaurus

| VITAL STATISTICS | |
|---|---|
| Fossil Location | USA, Egypt, Pakistan |
| Diet | Carnivorous |
| Pronunciation | Baz-illo-SAWR-us |
| Weight | 6300kg (6.1 tons) |
| Length | 45m (150ft) |
| Height | Unknown |
| Meaning of name | 'King lizard' because it was first assumed to be a marine reptile |

*Basilosaurus* was a marine mammal that inhabited the prehistoric ocean some 40 million years ago. Its bones were first discovered in the early nineteenth century in Alabama in the USA. However, it was not widely known until 1845, when the palaeontologist Albert Koch reconstructed a giant 35m (115ft)-long skeleton and called it a sea serpent. Later, it was realized that the bones were found to have come from five different individuals, and that the so-called sea serpent was a fake. Later, a very well-preserved species of *Basilosaurus* was found in the Wadi Al-Hitan (Arabic for 'Whale Valley') in Egypt, and remains of another species were discovered in Pakistan.

**Body Shape**
Despite its colossal size, the sleek, streamlined body of *Basilosaurus* slid through the water with grace and ease.

**Teeth**
The teeth were pointed at the front of the jaw and saw-edged at the rear.

### FOSSIL EVIDENCE

The extreme length of *Basilosaurus* came about because of an increase in the number and through the lengthening of its vertebrae, making it the 'nearest a whale ever came to a snake', as it was once described. *Basilosaurus* could move like an eel, although up and down rather than side to side. Its vertebrae seem to have been hollow and were probably filled with fluid. Recently, fossils of *Basilosaurus*'s tiny hind legs were found but they clearly showed that they were not supposed to move about on land.

| PREHISTORIC ANIMAL |
|---|
| TERTIARY (EOCENE) |

TERTIARY (EOCENE)

• ORDER • Cetacea • FAMILY • Basilosauridae • GENUS & SPECIES • Several species within the genus *Basilosaurus*

## WHERE IN THE WORLD?

Remains have been found in Alabama in the USA, in Egypt's Zeuglodon Valley, and in Pakistan.

## EVIDENCE OF LIMBS

*Basilosaurus* carried indications of its ancestry in its 60cm (2ft)-long hind limbs, which evidently had little use in enabling such an enormous creature to move through the water. *Basilosaurus* had been land creatures and these tiny limbs were the remnants of the legs that they used to walk on.

## DID YOU KNOW?

*Basilosaurus* is the state fossil of Mississippi and Alabama in the southern USA, where palaeontologists first discovered its remains.

## FOSSIL FURNITURE

In around 1834, so many fossils of one *Basilosaurus* species, *B. cetoides*, were being found in the US states of Louisiana and Alabama that local people were turning them into items of furniture. An anatomist, Dr Richard Harlan, acquired some of these fossils and realized that material of great scientific value was being destroyed. Harlan named the fossils *Basilosaurus* and a fellow anatomist, Sir Richard Owen, later identified them as a mammal.

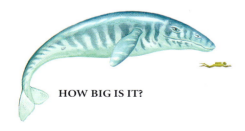

**HOW BIG IS IT?**

**TIMELINE (millions of years ago)**

| 540 | 505 | 438 | 408 | 360 | 280 | 248 | 208 | 146 | 65 | 1.8 to today |

TERTIARY (EOCENE)

# Basilosaurus

# TERTIARY (EOCENE)

• **ORDER** • Cetacea • **FAMILY** • Basilosauridae • **GENUS & SPECIES** • Several species within the genus *Basilosaurus*

**SPURIOUS SEA SERPENT**

In the early days of palaeontology, it was comparatively easy to fool the public, and even other palaeontologists, into believing that an impressive new prehistoric creature had been discovered. This is what happened in 1845, when Albert Koch, who called himself 'Doctor', learned that giant bones had been unearthed in the southern US state of Alabama. Koch travelled there to view the remains. He decided to create an enormous skeleton and put it on show. Eventually, Koch constructed a massive skeleton 35m (115ft) in length, which he described as a sea serpent that was longer, larger and taller than any dinosaur known in the mid-nineteenth century. Today we know that such dimensions would have made the fictitious sea serpent comparable to the 35m (115ft)-long titanosaurian *Argentinosaurus*, which was first described in 1993. However, Koch's sea serpent became a great success when put on display in New York City and it was subsequently shown across Europe. It was, of course, a composite fake – real fossils of several animals reconstructed falsely. Later, it was discovered that the skeleton came from five individuals, not all of which were *Basilosaurus*. Koch's sea serpent eventually perished in the great fire of Chicago in 1871.

TERTIARY (EOCENE)

# Brontotherium

| VITAL STATISTICS | |
|---|---|
| Fossil Location | North America |
| Diet | Herbivorous |
| Pronunciation | Bron-toth-EE-rium |
| Weight | 1.8 tonnes (1.76 tons) |
| Length | Unknown |
| Height | 2.5m (8ft) at the shoulder |
| Meaning of name | 'Thunder beast' for its great size and to honor the legends of the Sioux Native Americans on whose lands it was first found |

**FOSSIL EVIDENCE**

*Brontotherium* is well known from many well preserved skeletons. Their most striking feature is the flattened, forked nasal horns whose purpose is unknown. Due to the huge variability in skulls, many species have been assigned to this genus, but it seems that there are much fewer species than originally suspected. In fact, some recent research proposes that *Brontotherium* should actually be put in the brontothere genus *Megacerops*.

PREHISTORIC ANIMAL

TERTIARY (EOCENE)

*Brontotherium* closely resembled the modern rhinoceros and, as the meaning of its name suggests, it probably shook the ground when on the move. Probably the first discoverers of *Brontotherium* in the USA were Native American Sioux who found their skeletons after rainstorms had washed away soil. The Sioux believed that the noise these creatures made occurred when they were running across the clouds, which was why they called *Brontotherium* 'thunder horse.' Palaeontologists have suggested that the *Brontotherium* specimens that the Sioux found had died en masse as a result of volcanic eruptions.

**Skull**
*Brontotherium*'s huge skull was held up by powerful neck muscles, which were supported by the very the long spine.

**HOW BIG IS IT?**

**LOCAL DANGERS**

*Brontotherium* roamed across a large part of North America around 56 millions years ago, but it was a dangerous place to be. In addition to rainstorms and drought, there was another, even more destructive, peril. Frequent volcanic eruptions occurred as the Rocky Mountains were being formed, and these had the power to destroy entire herds of *Brontotherium* at once.

# TERTIARY (EOCENE)

• **ORDER** • Perissodactyla • **FAMILY** • Brontotheriidae • **GENUS & SPECIES** • Several species within the genus *Brontotherium*

## WHERE IN THE WORLD?

Fossils have been found in South Dakota and Nebraska in the USA.

**BODY**
*Brontotherium* had a beefy physique similar to a modern rhinoceros, yet it was more closely related to the horse.

## CHANGING EARTH

When *Brontotherium* was alive, in the Eocene, plant life on Earth was changing as profoundly as the geography. The first grasslands appeared, and in response, grazing animals developed with the type of teeth and digestive system that were adapted to handle this new abundance of food. In addition, numerous new species of shrubs and trees arrived. The warm Eocene climate brought on the development of a great range of trees, especially deciduous ones. Flowering plants also appeared.

**BROKEN BONES**
According to fossil evidence, *Brontotherium* skeletons found with broken ribs may have been losers in a fight over a mate.

## TIMELINE (millions of years ago)

| 540 | 505 | 438 | 408 | 360 | 280 | 248 | 208 | 146 | 65 | 1.8 to today |

TERTIARY (EOCENE)
# Brontotherium

# TERTIARY (EOCENE)

• **ORDER** • Perissodactyla • **FAMILY** • Brontotheriidae • **GENUS & SPECIES** • Several species within the genus *Brontotherium*

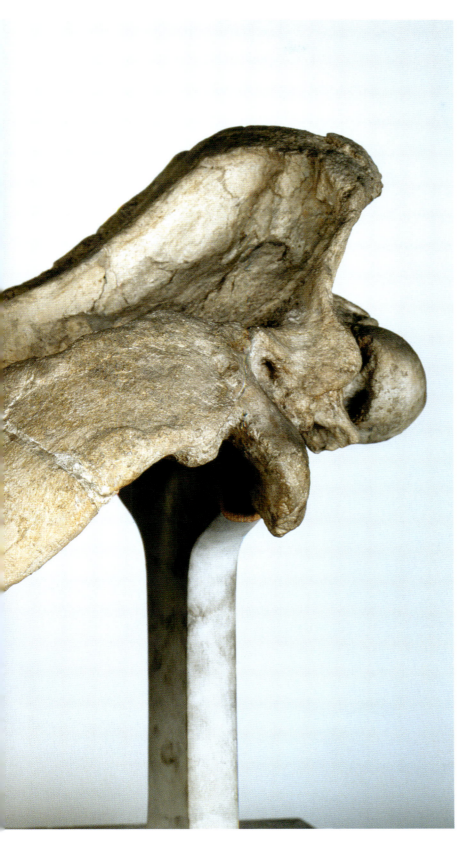

**EOCENE CHANGES**

The Eocene was a period that saw many rapid changes in the geography of Earth. *Brontotherium* lived in the latter part of the Eocene, which lasted until around 38 million years ago. Around this time, its North American habitat, the Great Valley of California, sank below the Pacific Ocean together with a large area of the Atlantic and the coastal plain of the Gulf of Mexico. The area of submerged coastal plain stretched from New Jersey to Texas, into Mississippi River Valley and northwards as far as southern Illinois. A vast segment of *Brontotherium*'s home territory therefore disappeared at this time. Elsewhere, the Eocene Epoch saw the Norwegian–Greenland Sea opening up. In southern Europe, North Africa and southwest Asia, most of the land was inundated by the Mediterranean Sea. The Eocene climate was pleasantly warm. The 'new' Eocene mammals were the ancestors of animals familiar today, including the rhinoceros, tapir, camel, pig, rat and other rodents, the monkey and the whale.

TERTIARY (OLIGOCENE)

# Arsinoitherium

- **ORDER** • Embrithopoda **FAMILY** • Arsinoitheriidae
- **GENUS & SPECIES** • Several species within the genus *Arsinoitherium*

*Arsinoitherium*, which resembled the modern rhinoceros, lived in the tropical rainforest at the edge of swamps some 36 million years ago. Strongly built and hefty, *Arsinoitherium's* size generally made it safe from predators.

## VITAL STATISTICS

| | |
|---|---|
| FOSSIL LOCATION | Africa and Middle East |
| DIET | Herbivorous |
| PRONUNCIATION | ar-sin-OH-ih-THEE-ree-um |
| WEIGHT | Unknown |
| LENGTH | 3m (10ft) |
| HEIGHT | 1.8m (5ft 10in) at the shoulder |
| MEANING OF NAME | Name derived from third-century CE Egyptian Queen, Arsinoe II, whose palace at Fayyum was close to the site where its fossils were discovered |

**LEGS AND FEET**
*Arsinoitherium* spent most of its time in the water. Its broad flat feet and long legs were better adapted for wading and swimming than walking.

**HORNS**
Two huge horns made of slid bone protruded like knives from just above the nose.

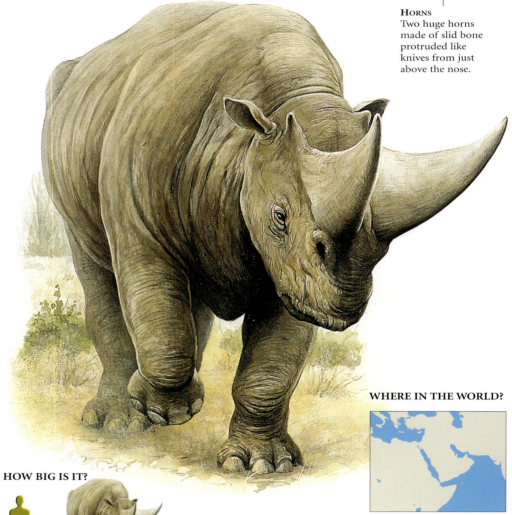

## FOSSIL EVIDENCE

Complete *Arsinoitherium* skeletons have been found only in Egypt. Fragments of the jaws of its relatives were discovered in southeastern Europe and Mongolia and appear to be of an earlier date than the Egyptian finds. The fossils found in Ethiopia in 2003 were some 27 million years old. *Arsinoitherium's* most remarkable features were the two huge horns that stuck up just above its nose. By contrast, the two knob-shaped horns directly behind them were tiny. *Arsinoitherium* seems to have spent most of the day feeding to maintain its hefty physique.

PREHISTORIC ANIMAL

TERTIARY (OLIGOCENE)

**HOW BIG IS IT?**

**WHERE IN THE WORLD?**

Fossils have been found in Fayyum in Egypt, Mongolia, Turkey and Ethiopia.

**TIMELINE (millions of years ago)**

| 540 | 505 | 438 | 408 | 360 | 280 | 248 | 208 | 146 | 65 | 1.8 to today |

TERTIARY (OLIGOCENE)

# Mammalodon

• ORDER • Cetacea • FAMILY • Mammalodontidae • GENUS & SPECIES • *Mammalodon colliveri*

*Mammalodon colliveri* was the first toothed baleen whale ever discovered when English palaeontologist George Pritchard spotted one of its skull bones on Jan Juc Beach, in Torquay, Victoria, in Australia.

### VITAL STATISTICS

| | |
|---|---|
| FOSSIL LOCATION | Australia |
| DIET | Carnivorous |
| PRONUNCIATION | Mamm-AL-oh-don |
| WEIGHT | Unknown |
| LENGTH | 2.5m (8ft) |
| HEIGHT | Unknown |
| MEANING OF NAME | 'Mammal tooth' because its teeth apparently showed its mammalian heritage |

### FOSSIL EVIDENCE

*Mammalodon*, which was discovered in 1932, was an early type of baleen whale. Unlike present-day baleen whales, its mouth had teeth, but also may have featured baleen plates for filtering its food from mouthfuls of water. Its face was short and its jaw contained only one or two incisor teeth. These were widely spaced, but it has not yet been determined whether or not *Mammalodon* could filter small marine animals through its whalebones. *Janjucetus*, a longer-toothed contemporary of *Mammalodon*, was found on the same Australian beach (Jan Juc) in the late 1990s.

PREHISTORIC ANIMAL
TERTIARY (OLIGOCENE)

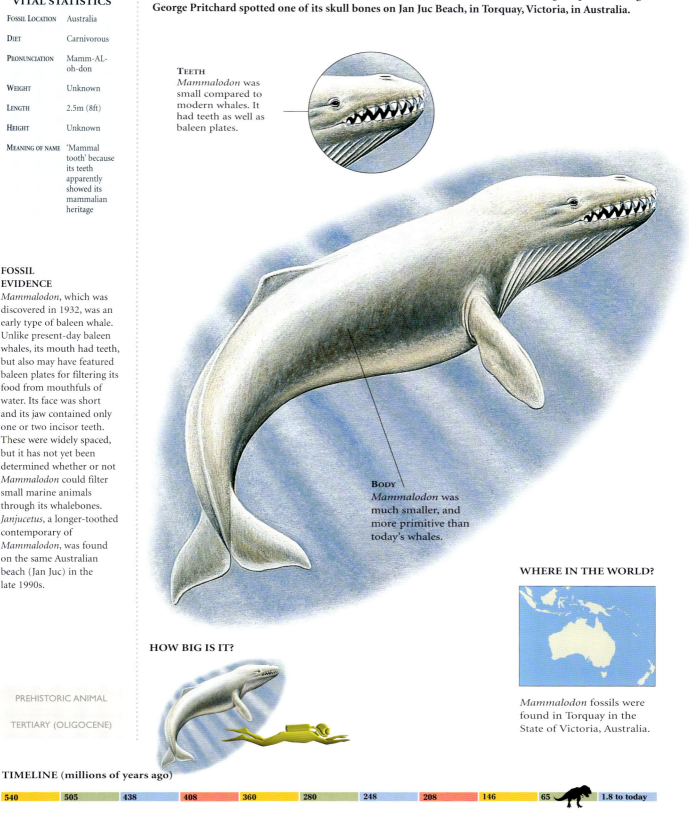

**TEETH**
*Mammalodon* was small compared to modern whales. It had teeth as well as baleen plates.

**BODY**
*Mammalodon* was much smaller, and more primitive than today's whales.

**HOW BIG IS IT?**

### WHERE IN THE WORLD?

*Mammalodon* fossils were found in Torquay in the State of Victoria, Australia.

**TIMELINE (millions of years ago)**

| 540 | 505 | 438 | 408 | 360 | 280 | 248 | 208 | 146 | 65 | 1.8 to today |

TERTIARY (OLIGOCENE)

# Pyrotherium
• **ORDER** • Pyrotheria • **FAMILY** • Pyrotheriidae • **GENUS & SPECIES** • Several species within the genus *Pyrotherium*

| VITAL STATISTICS | |
|---|---|
| Fossil Location | Argentina, Bolivia |
| Diet | Herbivorous |
| Pronunciation | Pie-ro-THEE-rium |
| Weight | Unknown |
| Length | 3m (10ft) |
| Height | 1.5m (5ft) |
| Meaning of name | 'Fire beast' because it was first found in an ancient volcanic ashfall |

*Pyrotherium* lived in Argentina around 34 million years ago. It was named Fire Beast because the site where it was excavated was covered by ash from an ancient volcanic eruption.

### DID YOU KNOW?
*Pyrotherium*'s outward similarity to elephants is due to convergence rather than close kinship.

### WHERE IN THE WORLD?

Fossils have been found in Bolivia and Argentina in South America.

### FOSSIL EVIDENCE
*Pyrotherium* was an ungulate (an animal with hooves) and resembled the modern elephant except for its trunk, which was probably shorter. Like the elephant, *Pyrotherium* had thick, sturdy legs to support its considerable weight. *Pyrotherium* endured from the Early Oligocene in Argentina to the Late Oligocene when it left its fossils at Salla in Bolivia. When *Pyrotherium* lived in the early Oligocene, some of the first grasslands, elephants and the first horses were on the scene.

PREHISTORIC ANIMAL

TERTIARY (OLIGOCENE)

### HOW BIG IS IT?

**Jaw**
There were two flat, forwards-facing tusks on the upper jaw, and two on the lower jaw.

**TIMELINE (millions of years ago)**

| 540 | 505 | 438 | 408 | 360 | 280 | 248 | 208 | 146 | 65 | 1.8 to today |

TERTIARY (OLIGOCENE-MIOCENE)

# Palaeocastor

• **ORDER** • Rodentia • **FAMILY** • Castoridae • **GENUS & SPECIES** • Several species within the genus *Palaeocastor*

*Palaeocastor* was a burrowing animal that fossil evidence suggests lived in family groups. Their burrows were corkscrew-shaped. They excavated them not with their claws but with their incisors.

### VITAL STATISTICS

| | |
|---|---|
| FOSSIL LOCATION | USA |
| DIET | Herbivorous |
| PRONUNCIATION | Pal-aye-oh-CASS-tor |
| WEIGHT | Unknown |
| LENGTH | 20cm (8in) |
| HEIGHT | Unknown |
| MEANING OF NAME | 'Prehistoric beaver' |

**TEETH**
At first scientists thought that *Palaeocaster* dug burrows with its feet, but later analysis showed that it used its strong incisors.

**TAIL**
Since it was terrestrial, *Palaeocaster's* tail was not flatteneed like the swimming tail of modern beavers.

### FOSSIL EVIDENCE

The fossilized burrows of *Palaeocastor*, known as *Daemonelix* or (devil's corkscrews), were found in 1891 in Nebraska, although the mammal itself had first been described in 1869. Fossil evidence suggests that *Palaeocastor* may have lived in family groups like modern beavers, though they were mainly burrowing, not semi-aquatic, mammals. In 1977, surface markings on a fossilized burrow enabled palaeontologists to understand how it made its burrows. The burrows had been a mystery until the fossilized body of a prehistoric beaver was discovered in one of them. Until then, it had been thought that they were the fossilized roots of a plant.

### HOW BIG IS IT?

### WHERE IN THE WORLD?

Remains have been found in North and South Dakota and in Harrison, Nebraska, USA.

PREHISTORIC ANIMAL

TERTIARY (MIOCENE)

**TIMELINE (millions of years ago)**

| 540 | 505 | 438 | 408 | 360 | 280 | 248 | 208 | 146 | 65 | 1.8 to today |

TERTIARY (MIOCENE)

# Amebelodon

• ORDER • Proboscidea • FAMILY • Gomphotheriidae • GENUS & SPECIES • Several species within the genus *Amebelodon*

| VITAL STATISTICS | |
|---|---|
| FOSSIL LOCATION | USA, eastern China, North Africa |
| DIET | Herbivorous |
| PRONUNCIATION | Am-eh-BELL-oh-don |
| WEIGHT | 10,000kg (22,000lb) |
| LENGTH | Unknown |
| HEIGHT | 2.4m (8ft) |
| MEANING OF NAME | Possibly 'Together point tooth' (or 'fused tusk') in reference to its shovel-like lower tusks |

## FOSSIL EVIDENCE

*Amebelodons* belonged to a group called gomphotheres. Although *Amebelodon* was nicknamed a 'shovel-tusker' and used its lower tusks to shovel food into its mouth, it has been suggested that the tusks had more uses than this. For example, they could have been used to strip the bark off of trees, a function that some palaeontologists think is confirmed by the wear patterns they carry. In addition, the short flap-shaped trunk portrayed in many *Amebelodon* illustrations may be in error—some scientists now think that it had a long trunk like a modern elephant's.

*Amebelodon* was the relative of the mammoth and the modern elephant. Arising first in North America, *Amebelodon* migrated to China across the landbridge now covered by the Bering Sea.

**JAW**
*Amebelodon's* shovel-like lower jaw carried two huge teeth.

### WHERE IN THE WORLD?

Fossils have been discovered in North America, China and North Africa.

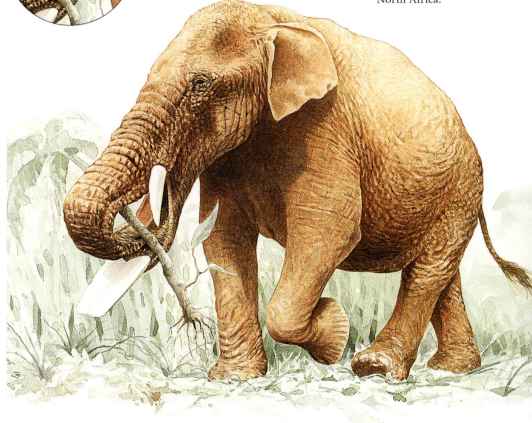

**HOW BIG IS IT?**

PREHISTORIC ANIMAL

TERTIARY (MIOCENE)

**TIMELINE (millions of years ago)**

| 540 | 505 | 438 | 408 | 360 | 280 | 248 | 208 | 146 | 65 | 1.8 to today |

TERTIARY (MIOCENE)

# Daeodon

• ORDER • Artiodactyla • FAMILY • Enteledontidae • GENUS & SPECIES • *Daeodon shoshonensis*

## VITAL STATISTICS

| | |
|---|---|
| FOSSIL LOCATION | Agate Springs Quarry, Nebraska, USA |
| DIET | Probably omnivorous |
| PRONUNCIATION | DAY-oh-don |
| WEIGHT | 907kg (2000lb) |
| LENGTH | 3.4m (11ft) |
| HEIGHT | 2.4m (8ft) at the shoulder |
| MEANING OF NAME | 'Destructive tooth' |

The gigantic hog-like *Daeodon* was a tusked mammal, with enormous jaws that could probably crush bones.

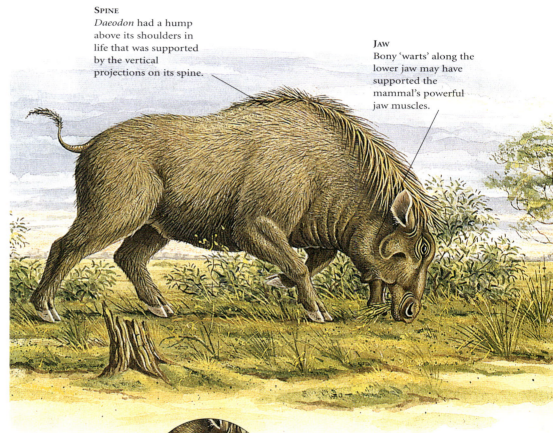

**SPINE**
*Daeodon* had a hump above its shoulders in life that was supported by the vertical projections on its spine.

**JAW**
Bony 'warts' along the lower jaw may have supported the mammal's powerful jaw muscles.

**TEETH**
The robust teeth of *Daeodon* are at first reminiscent of carnivores but certain details suggest omnivory– its feeding behaviour may have included scavenging carcasses.

## FOSSIL EVIDENCE

The massive bonebed at Agate Springs Quarry in Nebraska shows where dozens of *Daeodon* died together, most probably during a drought. Bones of the mammal *Moropus* have been unearthed which seem to carry the tooth marks of the predatory *Daeodon*, which had the teeth of a savage hunter, although it also ate plants. Some palaeontologists believe the 'warts' on *Daeodon*'s skull may have attached to the mammal's powerful jaw muscles.

### WHERE IN THE WORLD?

Fossils have been discovered in North America and East Asia.

### HOW BIG IS IT?

PREHISTORIC ANIMAL
TERTIARY (MIOCENE)

**TIMELINE** (millions of years ago)

| 540 | 505 | 438 | 408 | 360 | 280 | 248 | 208 | 146 | 65 | 1.8 to today |

TERTIARY (MIOCENE)

# Deinogalerix

• ORDER • Erinaceomorpha • FAMILY • Erinaceidae • GENUS & SPECIES • Several species within the genus *Deinogalerix*

*Deinogalerix*, which lived in what is now Italy some 11.2 million years ago, belonged to the gymnures, or moon-rat, family. But it was actually more like a hairy giant hedgehog on long legs.

## VITAL STATISTICS

| | |
|---|---|
| FOSSIL LOCATION | Gervasio Quarry, Foggia Province, Apulia, southern Italy (Gargano Island during the Miocene) |
| DIET | Insectivorous/carnivorous |
| PRONUNCIATION | die-no-ga-LEH-rix |
| WEIGHT | 9kg (20lb) |
| LENGTH | 60cm (24in) |
| HEIGHT | Unknown |
| MEANING OF NAME | Probably 'Terrible *Galerix*' because it was larger than, but similar to, the other fossil hedgehog *Galerix*, whose name, in turn, probably means 'Wearer of a skull cap' |

### FOSSIL EVIDENCE

First described in 1972, fossil finds show that the skull of *Deinogalerix*, with its thin, cone-shaped face, measured 20cm (8in) long, making up to one-third of its body length. This may have been an example of so-called island gigantism, a natural process by which creatures isolated on islands sometimes grow in size compared to their mainland counterparts. Unlike the modern hedgehog that it otherwise resembles, *Deinogalerix* had a hairy skin rather than quills.

PREHISTORIC ANIMAL

TERTIARY (MIOCENE)

**SKULL**
One of the larger Deinogalerix species, *D. koenigswaldi*, had a skull 20cm (8in) long and a body 60cm (24in) in length.

**CLAWS**
*Deinogalerix* likely used its fierce claws to catch its varied prey items, which could have included beetles, dragonflies and crickets, as well as snails and lizards.

## HOW BIG IS IT?

### WHERE IN THE WORLD?

Fossils have been found in Italy.

**TIMELINE (millions of years ago)**

540 | 505 | 438 | 408 | 360 | 280 | 248 | 208 | 146 | 65 | 1.8 to today

TERTIARY (MIOCENE)

# Homalodotherium
• ORDER • Notoungulata • FAMILY • Homalodotheriidae • GENUS & SPECIES • *Homalodotherium cunninghami*

*Homalodotherium*, discovered in 1869, was among the biggest of all the land mammals uncovered at Argentina's Santa Cruz Formation. It was admirably suited to the herbivorous life.

## VITAL STATISTICS

| | |
|---|---|
| Fossil Location | Santa Cruz Formation, Tarija Province, Argentina |
| Diet | Herbivorous |
| Pronunciation | Hoe-mallo-doh-THEH-rium |
| Weight | Unknown |
| Length | 2m (6ft 7in) |
| Height | Unknown |
| Meaning of name | Possibly 'Smooth-tooth beast' |

### WHERE IN THE WORLD?

Remains were found at Santa Cruz Formation in Argentina.

**Jaw**
The probable style of eating adopted by *Homalodotherium* – pulling down branches and feeding on leaves – is used today by giant pandas.

### FOSSIL EVIDENCE

Fossils of *Homalodotherium* showed that the mammal had short, stocky back legs and longer, flexible arms. A significant feature of the fossil was an extremely large upper arm bone. This showed prominent ridges where strong muscles were attached in life. This probably gave *Homalodotherium* extra strength for digging roots or pulling on branches. It walked along like a ground sloth, on the outside edge of its feet in a pigeon-toed fashion.

PREHISTORIC ANIMAL

TERTIARY (MIOCENE)

### HOW BIG IS IT?

**Articulated wrist**
The flexible wrist likely enabled *Homalodotherium* to grasp branches and pull them down in order to obtain leaves and fruit.

**TIMELINE (millions of years ago)**

| 540 | 505 | 438 | 408 | 360 | 280 | 248 | 208 | 146 | 65 | 1.8 to today |

TERTIARY (MIOCENE)

# Argentavis

| VITAL STATISTICS | |
|---|---|
| Fossil Location | Argentina |
| Diet | Carnivorous |
| Pronunciation | ar-jen-TAY-viss |
| Weight | 63–80kg (140–180lb) |
| Length | 8m (26ft) |
| Height | 1.5m (5ft) |
| Meaning of name | 'Argentine bird' |

**The largest flying bird even known, *Argentavis* glided high above its territory looking for prey that it could surprise and overcome.**

*Argentavis* was a gigantic Miocene flying bird. It had a wingspan that may have reached 8m (26ft) and it weighed over 63kg (140lb). Compare this with the bird with the largest wingspan today (the Wandering Albatross at 3.6m/12ft) and the heaviest modern flying birds, weighing 18–20kg (40–45lb), and you have an idea of *Argentavis'* vast size. Its humerus bone alone was as long as an adult human arm.

### WHERE IN THE WORLD?

*Argentavis* comes from Argentina. So far, no fossils of this bird have been found anywhere else.

**FOSSIL EVIDENCE**

Flying birds need very lightweight skeletons to facilitate flying. One way they achieve this is to have hollow bones with very thin, and thus fragile, walls. Because of this, bird fossils are very rare and generally consist of isolated bones and fragments. No complete skeletons exist of *Argentavis*, but a composite reconstruction has been made from the scattered bits available. *Argentavis* fossils have been found from the foothills of the Andes to the pampas. This suggests that this bird chose these areas in order to facilitate soaring using the wind off the Andean slopes and the thermals rising from the pampas to get airborne.

**HOW BIG IS IT?**

DINOSAUR

TERTIARY (PLIOCENE)

# TERTIARY (MIOCENE)

• **ORDER** • Ciconiiformes • **FAMILY** • Teratornithidae • **GENUS & SPECIES** • *Argentavis magnificens*

**WINGS**
Gliders need large wings to travel long distances without flapping. The estimated area of *Argentavis*' wings was 7sq m (75sq ft).

**BEAK**
The hook at the end of *Argentavis*' beak is characteristic of a flesh-eating bird. However, other skull structures suggest that it swallowed much of its food whole.

**DID YOU KNOW?**
Argentavis was probably too big to sustain frequent flapping of its wings in flight. Instead, it took advantage of its environment to lift its huge body skywards using wind and rising heat.

**TIMELINE (millions of years ago)**

| 540 | 505 | 438 | 408 | 360 | 280 | 248 | 208 | 146 | 65 | 1.8 |

TERTIARY (MIOCENE)

# Moropus

• ORDER • Perissodactyla • FAMILY • Chalicotheriidae • GENUS & SPECIES • Several species within the genus *Moropus*

*Moropus* lived around 23.5 million years ago in North America. It was a strange-looking animal that resembled a horse with large claws.

## VITAL STATISTICS

| | |
|---|---|
| FOSSIL LOCATION | USA |
| DIET | Herbivorous |
| PRONUNCIATION | More-OH-puss |
| WEIGHT | Unknown |
| LENGTH | Unknown |
| HEIGHT | 2.43m (8ft) at the shoulder |
| MEANING OF NAME | 'Sluggish foot' because it was presumed to be a slow, clumsy mover |

### WHERE IN THE WORLD?

Remains have been found at Agate Springs Quarry, Nebraska, USA.

**NECK**
A ball-and-socket arrangement in its neck vertebrae enabled *Moropus* to hold its head unusually high when eating leaves.

### FOSSIL EVIDENCE

*Moropus* had short, sturdy back legs and longer, flexible forelimbs. It belonged to the group called chalicotheres, the same group of odd-toed mammals that today include horses, rhinoceroses and tapirs. Its molars were broad and low, ideal for browsing on soft, leafy vegetation. It did not have horns or antlers to hamper it from feeding from shrubs or thick trees.

### HOW BIG IS IT?

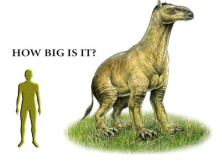

**FEET**
*Morophus* was probably a selective browser and may have used its massive claws to dig for tubers.

| PREHISTORIC ANIMAL |
|---|
| TERTIARY (MIOCENE) |

## TIMELINE (millions of years ago)

| 540 | 505 | 438 | 408 | 360 | 280 | 248 | 208 | 146 | 65 | 1.8 to today |

TERTIARY (MIOCENE)

# Borophagus

• ORDER • Carnivora • FAMILY • Canidae • GENUS & SPECIES • Several species within the genus *Osteoborus*

## VITAL STATISTICS

| | |
|---|---|
| FOSSIL LOCATION | USA |
| DIET | Carnivorous |
| PRONUNCIATION | bor-oh-FAY-gus |
| WEIGHT | Unknown |
| LENGTH | Unknown |
| HEIGHT | 9m (2ft) |
| MEANING OF NAME | 'Devouring eater' because of its presumed ravenousness |

*Borophagus* was a prehistoric dog rather like a hyena that lived in the USA around 6 million years ago. With its mighty jaws, it could probably crack and crush bone.

**NOSE**
Like dogs today, *Borophagus* probably had a keen sense of smell that helped it to locate the bodies of dead animals.

### WHERE IN THE WORLD?

Fossils have been found in the USA at Edson Quarry in Kansas, Ogalalla Formation in Texas, and in Dixie County, Florida.

### FOSSIL EVIDENCE

*Borophagus* was a primitive dog. It was likely a bone-eating scavenger with conical teeth that resembled those of the modern hyena. *Borophagus* was also characterized by a bulging forehead. Wandering the plains of North America, *Borophagus* probably scavenged carcasses.

**HOW BIG IS IT?**

**TEETH**
Broad curved fangs at the front of the mouth were covered with thick enamel.

PREHISTORIC ANIMAL

TERTIARY (MIOCENE)

## TIMELINE (millions of years ago)

| 540 | 505 | 438 | 408 | 360 | 280 | 248 | 208 | 146 | 65 | 1.8 to today |

TERTIARY (MIOCENE)

# Syndyoceras

• ORDER • Artiodactyla • FAMILY • Protoceratidae • GENUS & SPECIES • *Syndyoceras cooki*

*Syndyoceras* looked like a deer, but with a difference. On its head it carried two horns, rather than antlers, but there was a second pair, fused together, at the base of its snout.

## VITAL STATISTICS

| | |
|---|---|
| FOSSIL LOCATION | USA |
| DIET | Herbivorous |
| PRONUNCIATION | Sin-die-AH-she-rass |
| WEIGHT | Unknown |
| LENGTH | 1.5m (5ft) |
| HEIGHT | Unknown |
| MEANING OF NAME | 'Together horn' because its snout horns are fused at the base |

## FOSSIL EVIDENCE

It resembled a deer, but *Syndyoceras* had greater similarity with the prehistoric camels. *Syndyoceras cooki* was first described by the palaeontologist George Barbour in 1905. But it was 1968 before a remarkable collection of fossils was revealed hidden beneath the surface of Wildcat Ridge, Nebraska. Excavation started in earnest in 1999 under the University of Nebraska State Museum's Highway Salvage Palaeontology Program. Among the 46 species discovered on Wildcat Ridge was an extremely rare Syndyoceras. It had teeth similar to present-day cattle and deer.

PREHISTORIC ANIMAL

TERTIARY (MIOCENE)

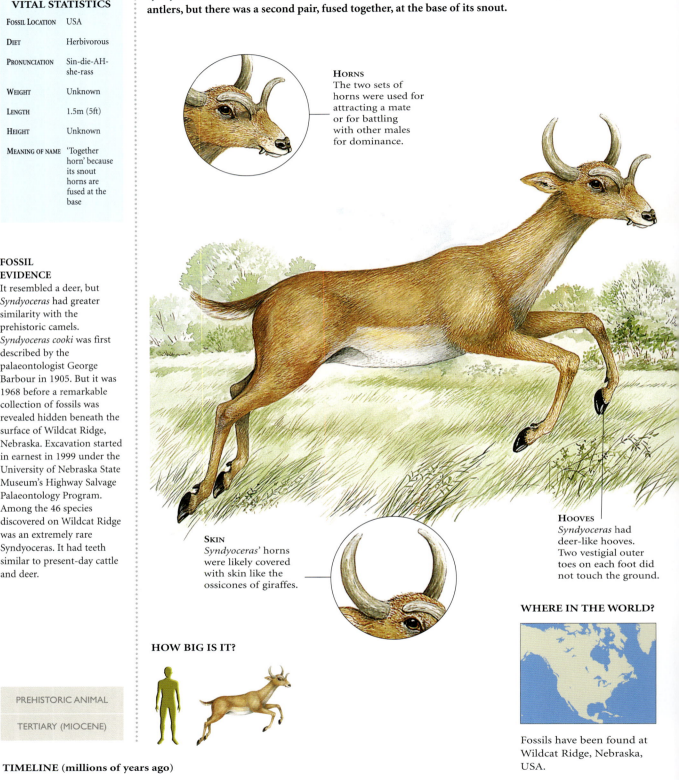

**HORNS**
The two sets of horns were used for attracting a mate or for battling with other males for dominance.

**SKIN**
*Syndyoceras'* horns were likely covered with skin like the ossicones of giraffes.

**HOOVES**
*Syndyoceras* had deer-like hooves. Two vestigial outer toes on each foot did not touch the ground.

**HOW BIG IS IT?**

### WHERE IN THE WORLD?

Fossils have been found at Wildcat Ridge, Nebraska, USA.

**TIMELINE (millions of years ago)**

| 540 | 505 | 438 | 408 | 360 | 280 | 248 | 208 | 146 | 65 | 1.8 to today |

# LATE TERTIARY (PLIOCENE)

# Megatherium

• ORDER • Pilosa • FAMILY • Megatheriidae • GENUS & SPECIES • Several species within the genus *Megatherium*

## VITAL STATISTICS

| | |
|---|---|
| FOSSIL LOCATION | South America |
| DIET | Herbivorous |
| PRONUNCIATION | Megga-thee-rium |
| WEIGHT | 5000kg (5 tons) |
| LENGTH | 5.4m (18ft) |
| HEIGHT | 6m (20ft) when standing on hind legs |
| MEANING OF NAME | 'Great beast' |

Modern sloths are not very large animals, but *Megatherium*, a ground sloth, was one of the largest mammals ever to walk the Earth. Standing on its hind legs, it was roughly twice the height of a present-day African bull elephant.

**CLAWS**
There were three hooked claws on each forelimb and five large claws on both back feet.

### WHERE IN THE WORLD?

*Megatherium* originated in South America before migrating to North America.

## FOSSIL EVIDENCE

The first *Megatherium* fossil was found in Brazil in 1789, and showed that Megatherium had powerful jaws. It probably moved around slowly whether on all fours or on its hind legs. When standing up, *Megatherium* used its short tail to balance its body and was tall enough to reach leaves that grew in the upper branches of trees. *Megatherium* had peg-like teeth and used its powerful cheek muscles to help grind down the vegetation it ate.

**GAIT**
Because the huge curved claws on its toes got in the way, *Megatherium* walked on the sides of its feet.

**HOW BIG IS IT?**

PREHISTORIC ANIMAL

LATE TERTIARY (PLIOCENE)

**TIMELINE** (millions of years ago)

| 540 | 505 | 438 | 408 | 360 | 280 | 248 | 208 | 143 | 65 | 1.8 to today |

LATE TERTIARY (PLIOCENE)

# Sivatherium

• ORDER • Artiodactyla • FAMILY • Giraffidae
• GENUS & SPECIES • Several possible species within the genus *Sivatherium*

*Sivatherium* was an extinct type of giraffe. It resembled the modern okapi, but was very much bigger.

## VITAL STATISTICS

| | |
|---|---|
| Fossil Location | India, southern Asia |
| Diet | Herbivorous |
| Pronunciation | See-va-THEE-rium |
| Weight | Unknown |
| Length | Unknown |
| Height | 2.2m (7ft) at the shoulder |
| Meaning of name | 'Beast of Shiva', after Shiva (regarded as the most important god in the Hindu religion), since it was found in the Siwalik Hills of India |

### WHERE IN THE WORLD?

Remains have been found chiefly in India and around southern Asia.

### FOSSIL EVIDENCE

Its fossils are more commonly found in India. Despite its elk-like appearance, *Sivatherium* was a type of giraffe. Its shoulders were strong and heavily built to support the powerful neck muscles required when lifting its weighty skull. The neck and limbs were relatively short and *Sivatherium* carried long, broad, flat horns on its head. The snout was wide, like that of a moose.

**Ossicones**
*Sivatherium* had two large ossicones (bony lumps) on its head and another, much smaller, pair above its eyes.

**Back and Neck**
Some reconstructions of *Sivatherium* show a sloping back and a longer neck, though not as long as the modern giraffe's.

**HOW BIG IS IT?**

PREHISTORIC ANIMAL
LATE TERTIARY (PLIOCENE)

**TIMELINE (millions of years ago)**

| 540 | 505 | 438 | 408 | 360 | 280 | 248 | 208 | 146 | 65 | 1.8 to today |

PLEISTOCENE

# Colossochelys

• ORDER • Testudines • FAMILY • Testudinidae • GENUS & SPECIES • Colossochelys atlas

## VITAL STATISTICS

| | |
|---|---|
| FOSSIL LOCATION | Western India; Pakistan; possibly southern and eastern Europe; Sulawesi and Timor, Indonesia |
| DIET | Herbivorous |
| PRONUNCIATION | ko-LAHS-oh-KEE-leez |
| WEIGHT | Up to 4000kg (4 tons) |
| LENGTH | 2.7m (9ft) |
| HEIGHT | 1.8m (6ft) |
| MEANING OF NAME | 'Colossal turtle' |

### FOSSIL EVIDENCE

*Colossochelys* was discovered in 1835 by the Scots palaeontologist Dr Hugh Falconer in the Punjab in northwestern India. The discovery site, the Siwalik Hills below the Himalayas, was full of the remains of several other prehistoric species. Most testudinids were (and are) relatively small, so the immense *Colossochelys* was an exception. Like present-day giant tortoises, *Colossochelys* had four huge, strong legs to support its enormous weight. Its shell alone measured some 2m (7ft) long.

### WHERE IN THE WORLD?

India, Pakistan, Indonesia and possibly Europe

*Colossochelys*, which lived up to 11,550 years ago, is also sometimes referred to as *Testudo atlas* (Atlas tortoise). It was one of the largest turtles ever known and probably resembled the Galapagos Islands tortoise of today.

**SHELL**
*Colossochelys* was one of the world's most gigantic turtles. Its shell length outclassed today's biggest, the giant leatherback, by almost 60cm (2ft).

**DEFENCE**
When endangered, *Colossochelys* defended itself, as modern turtles do, by pulling its head and legs back into its shell.

### HOW BIG IS IT?

PREHISTORIC ANIMAL

PLEISTOCENE

**TIMELINE (millions of years ago)**

| 540 | 505 | 438 | 408 | 360 | 280 | 248 | 208 | 146 | 65 | 1.8 to today |

PLEISTOCENE

# Doedicurus

## VITAL STATISTICS

| | |
|---|---|
| Fossil Location | North and South America |
| Diet | Herbivorous |
| Pronunciation | Dee-dik-YOO-rus |
| Weight | 2032kg (1.9 tons) |
| Length | 3.6m (12ft) |
| Height | 1.5m (5ft) |
| Meaning of name | 'Pestle tail' |

*Doedicurus*, a relative of the modern armadillos, lived during the Pleistocene Epoch until the close of the last Ice Age around 11,000 years ago. It carried an enormous domed shell on its back, consisting of scores of bony plates closely fitted together. *Doedicurus*' head had a 'carapace' of its own as a protection against attack. The tail had a covering of bone, ending in a club covered in sharp spikes. Its habitat was woodlands and grassland, which afforded plenty of plants and other vegetation.

### WHERE IN THE WORLD?

Remains have been found in North and South America, particularly the Ensenada Formation in Argentina.

**EYES**
*Doedicurus*' eyes were tiny and had a limited field of vision.

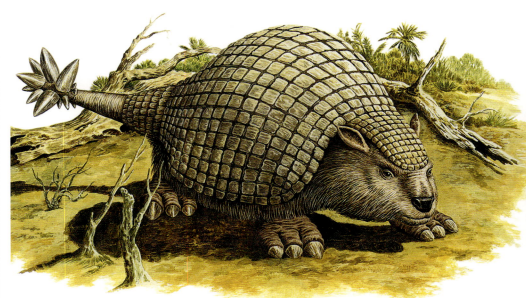

### FOSSIL EVIDENCE

*Doedicurus* fossils are among the most common mammal remains in the Argentine pampas. Unlike the modern armadillo, *Doedicurus*' armour was entirely rigid, and lacked the mobility offered to armadillos by a band of articulated plates. The carapace on *Doedicurus*' back was made up of large, thick osteoderms (plates made of bone). The tail was encased in a protective tube of bone that could have been up to 1.3m (4ft) long. Combined with the horny spines at the end, this was a formidable weapon against predators.

### DENTAL PLAN

The teeth inside *Doedicurus*' powerful jaws were typical of herbivores. It lacked teeth in the front of its mouth, but further back there were teeth adapted for grinding down the types of plants and leaves that grew in the grasslands and woodlands where it lived.

### HOW BIG IS IT?

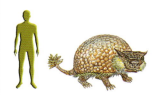

PREHISTORIC ANIMAL

PLEISTOCENE

PLEISTOCENE

• **ORDER** • Cingulata • **FAMILY** • Glyptodontidae • **GENUS & SPECIES** • *Doedicurus clavicaudatus*

## HUMAN PREDATORS
Humanity did not overlap on Earth with most of the prehistoric animals in this encyclopedia. One exception was *Doedicurus clavicaudatus*, a species found in both North and South America. It was still extant in South America some 15,000 years ago, meaning that early humans may well have encountered and hunted it. *Doedicurus'* armour would not have given it total protection in these circumstances, because, unlike other predators, humans wielding bows and arrows knew how to kill their prey from a distance.

**ARMOUR**
*Doedicurus* looked like a formidable proposition for a would-be predator with its heavy back armour and spiked tail club.

**BIG GLYPTODONTID**
*Doedicurus* was the biggest known glyptodontid, the large, heavily armoured relatives of the extinct pampheres (pampas beasts) and modern armadillos.

**TIMELINE (millions of years ago)**

| 540 | 505 | 438 | 408 | 360 | 280 | 248 | 208 | 146 | 65 | 1.8 |

PLEISTOCENE

# Smilodon

| VITAL STATISTICS | |
|---|---|
| Fossil Location | The La Brea Tar Pits, Los Angeles, USA; State of Minas Gerais, Brazil |
| Diet | Carnivorous |
| Pronunciation | SMILE-oh-don |
| Weight | 100–400kg (220–884lb) |
| Length | 6m (19.7ft) |
| Height | 1.2m (4ft) at the shoulder |
| Meaning of name | 'Chisel tooth' |

*Smilodon* was a big cat with a pair of long, sharp curved fangs. It is often popularly referred to as a sabre-toothed tiger, even though it is not closely related to the tiger. It was a very effective killer. To give greater effect to its hunting skills, *Smilodon* could open its mouth wide and position its fangs for the kill at an angle of 120 degrees. Its legs were very strong, and it could probably pull down large prey such as bison, elk, deer, mammoth and mastodons.

**FANGS**
The extraordinarily long fangs could grow to a length of 17cm (7in).

## FOSSIL EVIDENCE
The La Brea Tar Pits in Los Angeles yielded hundreds of *Smilodon* fossils. They probably died there when they became stuck while attempting to kill prey that had made the same mistake and become trapped. This event gave palaeontologists hundreds of complete *Smilodon* skeletons to study. Copious finds were also made in South America in 1841 when a Danish palaeontologist, Peter Wilhelm Lund, found fossils of *Smilodon populator* in caves in the state of Minas Gerais, Brazil.

PREHISTORIC ANIMAL

PLEISTOCENE

**HOW BIG IS IT?**

**WHERE IN THE WORLD?**

Thousands of fossils have been found in tar pits and Rocks from North and South America.

# PLEISTOCENE

• **ORDER** • Carnivora • **FAMILY** • Felidae • **GENUS & SPECIES** • Several species within the genus *Smilodon*

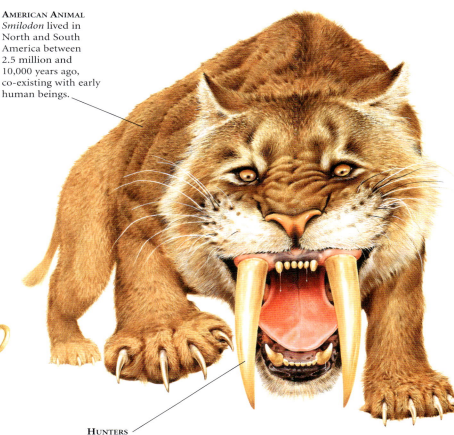

**AMERICAN ANIMAL**
*Smilodon* lived in North and South America between 2.5 million and 10,000 years ago, co-existing with early human beings.

## FAT CATS

The earliest of the three *Smilodon* species officially recognized today was *gracilis*, which lived some 2.5 million years ago. Weighing in at 100kg (221lb), it was the smallest of the species. *Smilodon fatalis* followed around 1.6 million years ago, living in both North America and western South America. *Smilodon fatalis* was a good deal larger and heavier, weighing up to 220kg (486lb). The third species, *Smilodon populator*, arrived 1 million years ago and was the largest of them all at 400kg (884lb).

**HUNTERS**
The fangs of both male and female *Smilodon* were generally of similar length, suggesting that both had roles as hunters.

## SEIZING PREY

Research in 2007 likely revealed how *Smilodon* killed its prey. Using the great strength of its upper body, it seized large victims and grappled with them. It forced them to the ground and held them down while it used its fangs to stab through the large jugular veins. After that, *Smilodon's* prey bled to death very quickly.

**TIMELINE (millions of years ago)**

| 540 | 505 | 438 | 408 | 360 | 280 | 248 | 208 | 146 | 65 | 1.8 |

PLEISTOCENE

# Smilodon

PLEISTOCENE

• **ORDER** • Carnivora • **FAMILY** • Felidae • **GENUS & SPECIES** • Several species within the genus *Smilodon*

**SMILODON FINDINGS**

In 2007, a report published in the proceedings of the American National Academy of Sciences suggested that, despite its ferocious reputation, Smilodon was in reality something of a pussycat. First of all, the report claims, *Smilodon* did not belong to the same fierce species as tigers and lions. *Smilodon* could bite its victims only one-third as effectively as a lion. This was discovered when university palaeontologists used fossils in a digital reconstruction of *Smilodon* and lion skulls. In computerized crash tests, Smilodon's skull and jaw failed to cope as long as the prey remained on its feet and struggled to avoid death. Conversely, in the same circumstances, the lion performed a great deal better and remained in control until the kill was completed. As one of the palaeontologists, Colin McHenry of the University of Newcastle, explained: 'we simulated the forces you might expect if each one was taking large prey.' This was not a matter of superiority or inferiority: the culprit was evolution. *Smilodon*, it appears, had developed a weaker bite because of its small lower jaw, which had evolved to accommodate its long fangs.

PLEISTOCENE

# Woolly Mammoth

| VITAL STATISTICS | |
|---|---|
| Fossil Location | Northern Hemisphere |
| Diet | Herbivorous |
| Pronunciation | WUH-lee MAM-uth |
| Weight | 8000kg (8 tons) |
| Length | 7.6m (25ft) |
| Height | 3m (10ft) |
| Meaning of name | Possibly 'Earth mole' in reference to a mythological, subterraneous animal that emerges near rivers |

The quintessential Ice Age beast, the woolly mammoth is a widely recognized icon of the cold, but fairly recent, geological past. The woolly mammoth is just one of several mammoth species. It was a large relative of elephants that lived in cold climates and wore a heavy coat of hair, as its name implies. Its skull had a more domed roof and its backs sloped down towards the tail, whereas in modern elephants the back is more level.

**DID YOU KNOW?**
Woolly mammoths lived as recently as 4000 years ago on Wrangle Island, north of Siberia. Most other mammoths became extinct 10,000 years ago.

**FOSSIL EVIDENCE**
Woolly mammoths are known from very abundant remains from the Northern Hemisphere. Some of the specimens found in the Siberian permafrost are justly famous for their extraordinary preservation. Freezing preserves the skin, hair, eyes, viscera, muscles, gut contents, blood, parasites and even DNA. Several examples of mummified baby woolly mammoths also exist; a recently discovered example promises to be the best-preserved fossil yet. It will be CT-scanned to reveal the internal organs, which are rarely studied for a long-extinct species. Climate change has caused a lot of melting in Arctic environments, exposing more frozen remains.

PREHISTORIC ANIMAL

PLEISTOCENE

**HOW BIG IS IT?**

**Trunk**
Woolly mammoth trunks had two fleshy 'fingers' at the tip that helped the animals grasp its food.

# PLEISTOCENE

• **ORDER** • Proboscidea • **FAMILY** • Elephantidae • **GENUS & SPECIES** • *Mammuthus primigenius*

### WHERE IN THE WORLD?

Woolly mammoths are known all over the Northern Hemisphere. Colder climates were more widespread in the Pleistocene and their distribution reflects this.

### DID YOU KNOW?
Mammoths were once thought to be giant, subterranean burrowers that died when they accidentally broke through the surface and were exposed. This was a reason once offered to explain the weathering of the mummified remains.

**FEET**
Woolly mammoth feet had wide soles and a spongy pad that cushioned the weight of the animal's enormous body.

### MAMMOTH MEAL
A Woolly mammoth is so huge that it would have been a huge task to butcher the carcass. Humans probably got around this by cooking the animal on the spot rather than attempting to move it.

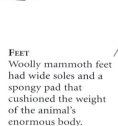

**TIMELINE (millions of years ago)**

PLEISTOCENE

# Woolly Mammoth

PLEISTOCENE

• **ORDER** • Proboscidea • **FAMILY** • Elephantidae • **GENUS & SPECIES** • *Mammuthus primigenius*

### TUNDRA MAMMOTH

The woolly mammoth, also known more prosaically as the tundra mammoth, survived until around 1700BCE. This meant it was one of the animals encountered, and hunted, by early humans. The woolly mammoth was also one of the first animals to have its portrait painted. Prehistoric peoples left its picture behind on the walls of their caves in what is thought by some palaeontologists to have been a ritual designed to ensure successful hunting. A huge elephant-like creature with curved tusks that could be up to 5m (16ft) long, the woolly mammoth was well protected against the intense cold in which it lived in the far north of North America and Eurasia, and the frozen wastes of Siberia. Its shaggy coat was up to 90cm (35in) long, with a fine covering of wool for extra insulation next to its hide. The mammoth's ears were comparatively small at 30cm (12in) long, compared to the modern African elephant's, which measure 180cm (71in). But ears that were smaller meant that less surface area was exposed to the freezing temperatures. Under its skin, the woolly mammoth had a layer of fat up to 8cm (3in) thick, another characteristic that enabled it to retain body heat and keep warm.

PLEISTOCENE

# Homotherium

• ORDER • Carnivora • FAMILY • Felidae • GENUS & SPECIES • Several species within the genus *Homotherium*

*Homotherium* was a so-called sabre-toothed cat that lived between three million and 10,000 years ago.

## VITAL STATISTICS

| | |
|---|---|
| FOSSIL LOCATION | North and South America, Eurasia, Africa |
| DIET | Carnivorous |
| PRONUNCIATION | Ho-mo-THEE-rium |
| WEIGHT | 250kg (552.5lb) |
| LENGTH | 1.6m (5ft) |
| HEIGHT | 1m (3ft) at the shoulder |
| MEANING OF NAME | Either 'Man's Beast' because the first fossils were found in association with human remains and artifacts, or 'Similar Beast' for some unrecorded similarity |

### WHERE IN THE WORLD?

North and South America, Eurasia, Africa.

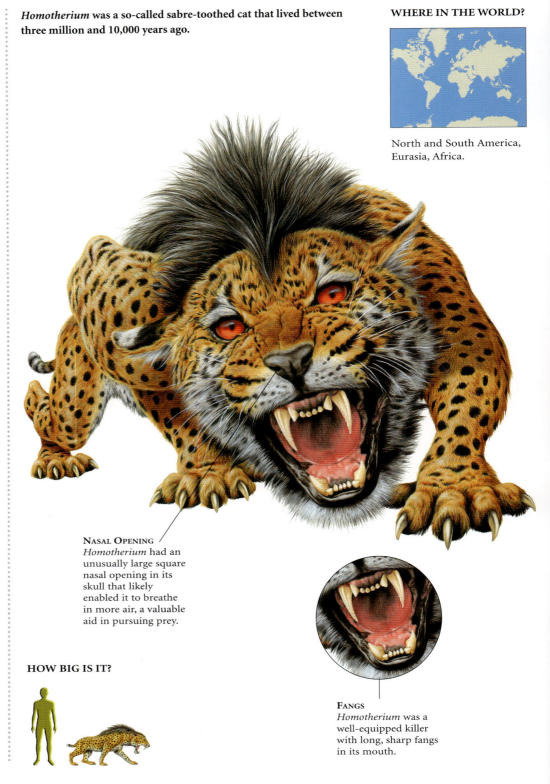

**NASAL OPENING**
*Homotherium* had an unusually large square nasal opening in its skull that likely enabled it to breathe in more air, a valuable aid in pursuing prey.

**FANGS**
*Homotherium* was a well-equipped killer with long, sharp fangs in its mouth.

## FOSSIL EVIDENCE

The fossil record shows that, like the present-day cheetah, *Homotherium* had an unusually large, square nasal opening. Also like the cheetah, its visual cortex was big and likely gave it extremely good sight both by day and night. However, the cat also bore resemblances to the hyena. Its front legs were long but its back legs were squat, so making its back slope downward when it was standing. It may have hunted in packs to take down much larger animals.

PREHISTORIC ANIMAL

PLEISTOCENE

### HOW BIG IS IT?

## TIMELINE (millions of years ago)

540 | 505 | 438 | 408 | 360 | 280 | 248 | 208 | 146 | 65 | 1.8

QUATERNARY (PLEISTOCENE)

# Coelodonta

• ORDER • Perissodactyla • FAMILY • Rhinocerotidae • GENUS & SPECIES • *Coelodonta antiquitatis*

*Coelodonta*, the woolly rhinoceros, first appeared in Europe some 350,000 years ago, about halfway through the last Ice Age. They remained on Earth long enough to be hunted by early humans.

## VITAL STATISTICS

| | |
|---|---|
| FOSSIL LOCATION | Ukraine, England, Belgium, Germany |
| DIET | Herbivorous |
| PRONUNCIATION | See-low-DON-tah |
| WEIGHT | 3000kg (3 tons) |
| LENGTH | 3.7m (11ft) |
| HEIGHT | 2m (6ft 6in) |
| MEANING OF NAME | 'Hollow tooth' |

**FRONT LIP**
*Coelodonta* had a broad front lip that it used to cut down and eat plants.

### WHERE IN THE WORLD?

Fossils have been discovered throughout Europe and Asia.

**FUR COAT**
Heavy, shaggy fur protected *Coelodonta* against the cold temperatures of Ice Age Europe.

**SKULL**
Fossilized *Coelodonta* skulls have measured 76cm (30in) in length.

## FOSSIL EVIDENCE

*Coelodonta* was among the megafauna (large animals) that lived during the last Ice Age, but became extinct around 10,000 years ago. *Coelodonta* first became known through paintings made by prehistoric hunters on the walls of their caves. The first fossil remains, however, were discovered buried in mud at Staruni, Ukraine. This find was a well-preserved female. Fossilized skulls of *Coelodonta* were also found in Germany and Belgium. Others were found frozen in ice or buried in earth saturated with oil.

PREHISTORIC ANIMAL

QUATERNARY

### HOW BIG IS IT?

**TIMELINE (millions of years ago)**

| 540 | 505 | 438 | 408 | 360 | 280 | 248 | 208 | 146 | 65 | 1.8 |

QUATERNARY (PLEISTOCENE)

# Diprotodon

• ORDER • Diprotodontia • FAMILY • Diprotodontidae
• GENUS & SPECIES • Several species within the genus *Diprotodon*

| VITAL STATISTICS | |
|---|---|
| FOSSIL LOCATION | Australia |
| DIET | Herbivorous |
| PRONUNCIATION | Dip-roh-TOH-don |
| WEIGHT | 2786kg (6157lb) |
| LENGTH | 3m (10ft) |
| HEIGHT | 1.7m (6ft) at the shoulder |
| MEANING OF NAME | 'Two forward teeth' |

*Diprotodon*, whose present-day relatives are wombats and koala bears, lived in the Australian forests, woodlands and grasslands throughout most of the Pleistocene Epoch. They first appeared 1.6 million years ago.

**CARING FOR YOUNG**
Like many animals today, *Diprotodon* took great care of their young and taught them how to survive.

**WHERE IN THE WORLD?**

Remains have been found in New South Wales, Darling Downs, Queensland and Lake Callabonna in Australia.

**FOSSIL EVIDENCE**
The first *Diprotodon* fossils were found in a cave near Wellington, in New South Wales, early in the 1830s, and the next in Queensland about ten years later. Fossil finds showed that *Diprotodon's* feet turned inwards, and its footprints indicated they were covered with hair. Several finds indicated that death was brought on by drought. The mud of Lake Callabonna contained hundreds of *Diprotodon* that had perished together. Their bodies were fairly intact, but their heads had been crushed. More than one female skeleton has been found with dead infants in its pouch.

PREHISTORIC ANIMAL
QUATERNARY

**HOW BIG IS IT?**

**LARGE MARSUPIAL**
*Diprotodon* was about the size of a modern hippopotamus. It was also the largest known marsupial that ever lived on Earth.

TIMELINE (millions of years ago)

540 | 505 | 438 | 408 | 360 | 280 | 248 | 208 | 146 | 65 | 1.8

# QUATERNARY (PLEISTOCENE)

# Glyptodon

• ORDER • Cingulata • FAMILY • Glyptodontidae • GENUS & SPECIES • Various species within the genus *Glyptodon*

## VITAL STATISTICS

| | |
|---|---|
| FOSSIL LOCATION | North and South America |
| DIET | Herbivorous |
| PRONUNCIATION | GLIP-toe-don |
| WEIGHT | 1000kg (1 ton) |
| LENGTH | 1.8m (6ft) |
| HEIGHT | Unknown |
| MEANING OF NAME | 'Grooved or carved tooth' |

*Glyptodon* was a large armoured mammal, closely related to the modern armadillos. The evident slowness of *Glyptodon* was not an enormous disadvantage for an animal so heavily armoured from head to tail. It had a bony shell covering its back, which was composed of over 1000 individual 2.5cm (1in) thick osteoderms.

**HEAD**
*Glyptodon* was unable to withdraw its head into its shell. Its head was instead protected by a bony cap on the top of its skull.

## WHERE IN THE WORLD?

Abundant fossils of glyptodonts have been found in North and South America. They have helped clarify the transfer of faunas between these continents in the Pleistocene when Central America formed.

## FOSSIL EVIDENCE

Fossils of *Glyptodon* are common in the Pleistocene sediments of South America. The genus goes back to more than 1 million years ago and extends until about 10,000 years ago. Their fossils have been reported since at least the 1820s and are still plentiful. But human experience of these animals goes back much further than that; people had made it to the Americas thousands of years before *Glyptodon* became extinct and they would have been familiar with the living animal. The group to which *Glyptodon* belongs, the glyptodonts, made it to North America towards the end of the Pleistocene Epoch.

PREHISTORIC ANIMAL

QUATERNARY

**HOW BIG IS IT?**

**SUPPORT**
*Glyptodon*'s body required extensive skeletal modifications for support of all of its bony armour, including short, massive limbs, a broad shoulder girdle, and fused vertebrae.

**TIMELINE (millions of years ago)**

| 540 | 505 | 438 | 408 | 360 | 280 | 248 | 208 | 146 | 65 | 1.8 |

QUATERNARY (PLEISTOCENE)

# Megaloceros

• ORDER • Artiodactyla • FAMILY • Cervidae • GENUS & SPECIES • Several species within the genus *Megaloceros*

## VITAL STATISTICS

| | |
|---|---|
| FOSSIL LOCATION | Europe, Asia |
| DIET | Herbivorous |
| PRONUNCIATION | MEG-uh-LAH-ser-us |
| WEIGHT | Varied |
| LENGTH | Varied |
| HEIGHT | 2m (6ft 6in) at the shoulder (Irish Elk); 65cm (26in) at the shoulder (Cretan species) |
| MEANING OF NAME | 'Great horn' |

The most well-known species of *Megaloceros* is the Irish Elk, which was very large and lived in open woods or meadows. Other species, from the Mediterranean, were much smaller.

**ANTLERS**
Potential mates would have been impressed by *Megaloceros*' magnificent antlers. Equally, any male rivals would have been discouraged.

### WHERE IN THE WORLD?

Remains have been found in Britain, Ireland, France and Crete, as well as in China and Japan.

## FOSSIL EVIDENCE

There are nine recognized species of *Megaloceros*, the earliest being *M. obscurus*, which lived in Europe during the early Pleistocene. Considerable variations in the physical size of *Megaloceros* species is reflected in differences between their antlers. For example, *M. obscurus* had long, crooked antlers, whereas the antlers of *M. savini*, found in France, were straight, with prongs. *M. pachyosteus*, which lived in China and Japan, sported long, curved antlers. The most splendid antlers of all belonged to the Irish Elk, or *M. giganteus*. Its antlers sprang from its head in broad, flat plates edged with pointed branches.

PREHISTORIC ANIMAL
QUATERNARY

**LEGS**
*Megaloceros* had strong legs which meant it could flee from danger.

### HOW BIG IS IT?

**TIMELINE** (millions of years ago)

| 540 | 505 | 438 | 408 | 360 | 280 | 248 | 208 | 146 | 65 | 1.8 |

# Glossary

**Amber**
A translucent or opaque yellow, fossil resin from coniferous trees.

**Amphibian**
A cold-blooded vertebrate animal that lives in the water during its early life and lives on land during its adult life.

**Ankylosaur**
An armoured, four-legged, herbivorous ornithischian dinosaur.

**Aquatic**
Living in or often found in water. Strictly speaking, dinosaurs only lived on land. However, some dinosaur-like reptiles, such as Libonectes, lived in water.

**Archosaurs**
A group of reptiles with teeth set in sockets, a pointed snout, and a narrow skull. This group gave rise to the dinosaurs, pterosaurs, crocodiles, and birds.

**Arthropod**
An invertebrate with an exoskeleton, segmented body and jointed limbs.

**Biped**
An animal that walks on two legs. Humans are bipedal and some types of dinosaur, such as *Tyrannosaurus rex*, were also bipedal.

**Camouflage**
Many creatures use special colours or patterns to help them blend into their surroundings and hide from prey or predators. This is called camouflage. We do not know exactly what colour the dinosaurs were, but many dinosaurs probably had camouflage patterns on their skin to help them hide.

**Carnivore**
An animal that eats meat.

**Carnosaur**
A large theropod dinosaur alive during the Jurassic and Cretaceous Periods.

**Cenozoic**
The Cenozoic Era is the most recent of the three classic geological eras and covers the period from 65.5 million years ago to the present.

**Cephalopod**
A marine mollusk with large eyes, tentacles and a large head.

**Coelurosaur**
A theropod closely related to birds.

**Cold-blooded**
Dinosaurs and other reptiles are cold-blooded. This means that they do not produce their own heat but take on the same temperature as their surroundings.

**Counterbalance**
To balance one weight against another. Many dinosaurs had long heavy tails to counterbalance the weight of their bodies.

**Cynodont**
A group of mammal-like reptiles that lived in the late Permian to early Jurassic Period. Although they were reptiles, they were not dinosaurs. The group probably included the ancestors of modern mammals.

**Epoch**
A division of a geologic period; it is the smallest division of geologic time, lasting several million years.

**Era**
Two or more geological periods comprise an Era, which is hundreds of millions of years in duration.

**Exoskeleton**
Some animals have a skeleton on the outside, instead of the inside, of their bodies. Exoskeleton is a tough body armour made of a type of protein called chitin. Animals with exoskeletons include insects, crabs and trilobites.

**Fossil**
A mineralized remain of a prehistoric plant or animal preserved in the earth's crust. Although the body eventually disappears, the impression of the creature is preserved in the rock forever. Most of our knowledge of dinosaurs comes from fossils.

**Genus**
A group of related or similar organisms containing one or more species. A group of similar genera (the plural of genus) forms a family. In the scientific name of an organism, the first name is its genus (for example, people are Homo sapiens - our genus is Homo).

**Giraffid**
A hoofed, herbivorous mammal with long forelegs and a long neck.

**Glyptodont**
A large mammal with a thick, bony shell; an ancient armadillo.

**Herbivore**
An animal that eats plants.

**Ichthyosaur**
A marine reptile from the Mesozoic Period. Although these reptiles spent their whole lives in water, they breathed air, so they were more like modern whales than fish.

**Incisor teeth**
The chisel-shaped cutting teeth found at the front of the mouth.

**Invertebrate**
Any creature without a backbone, such as an insect or a jellyfish.

**Lizard**
A long-bodied reptile, usually with two pairs of legs and a tail that becomes narrower towards the tip. A dinosaur is a type of lizard.

**Mammal**
A warm-blooded, hairy, vertebrate animal that gives birth to live young and feeds its young on milk.

**Marsupial**
A mammal with an external pouch for its young. Kangaroos and wallabies are present day examples.

**Molar teeth**
The rounded or flattened teeth used for crushing, chewing or grinding.

**Mollusc**
An invertebrate animal with a soft body protected by a hard shell.

**Nodosaur**
A type of ankylosaur with no tail club and a pear-shaped head.

**Order**
A group of related or similar organisms. An order contains one or more families. A group of similar orders forms a class.

**Ornithischian**
An herbivorous, bird-hipped dinosaur with hoof-like claws.

**Ornithopod**
An ornithischian dinosaur that walked upright on its hind limbs.

**Paleontologist**
A scientist who studies fossil plants and animals.

**Pangaea**
The supercontinent formed at the end of the Paleozoic Period, consisting of all of Earth's landmasses.

**Panthalassa**
The super-ocean that existed on Earth during the time of the super-continent Pangaea.

**Petrification**
The process in which an organic tissue turns to stone.

**Plankton**
Tiny plants or animals that live in water and drift along in the currents.

**Predator**
An animal that hunts and kills other creatures.

**Prehistoric**
Before recorded history. This can refer to millions of years ago, as in the case of dinosaurs, or a few thousand years ago.

**Prosauropod**
An early saurischian herbivorous dinosaur, that lived before the sauropods during the Late Triassic and Early Jurassic Periods.

**Pterosaur**
An ancient flying reptile with skin-covered wings.

**Reptile**
A cold-blooded, scaly vertebrate animal that lays eggs or gives birth on land.

**Saurischian**
A lizard-hipped dinosaur.

**Sauropod**
A large, four-legged herbivorous dinosaur.

**Stegosaurid**
An armour-plated dinosaur.

**Tetanuran**
A theropod dinosaur with a stiff tail.

**Tetrapod**
A vertebrate animal with four limbs.

**Theropod**
A predatory, carnivorous dinosaur that walked upright on its hind limbs and had grasping hands with clawed fingers.

**Vertebrae**
The linked bones forming the spine of a vertebrate animal.

**Vertebrate**
Any creature with a backbone, including mammals, birds, fish and reptiles.

# Non-Avian Dinosaur Fossil Sites

**HOW TO USE THIS MAP**

Fossils are found all over the Earth, from pole to pole, and even in the sea. This map only shows the most significant non-bird dinosaur fossil localities, spanning the entire Mesozoic Era. Some sites are special because of the abundance of fossils preserved while others are unique for the incredible mode of preservation or the rarity of the occurrence. Fossil localities spotlighted include the spectrum of fossil remains, such as bones, tracks, eggs, and soft tissues.

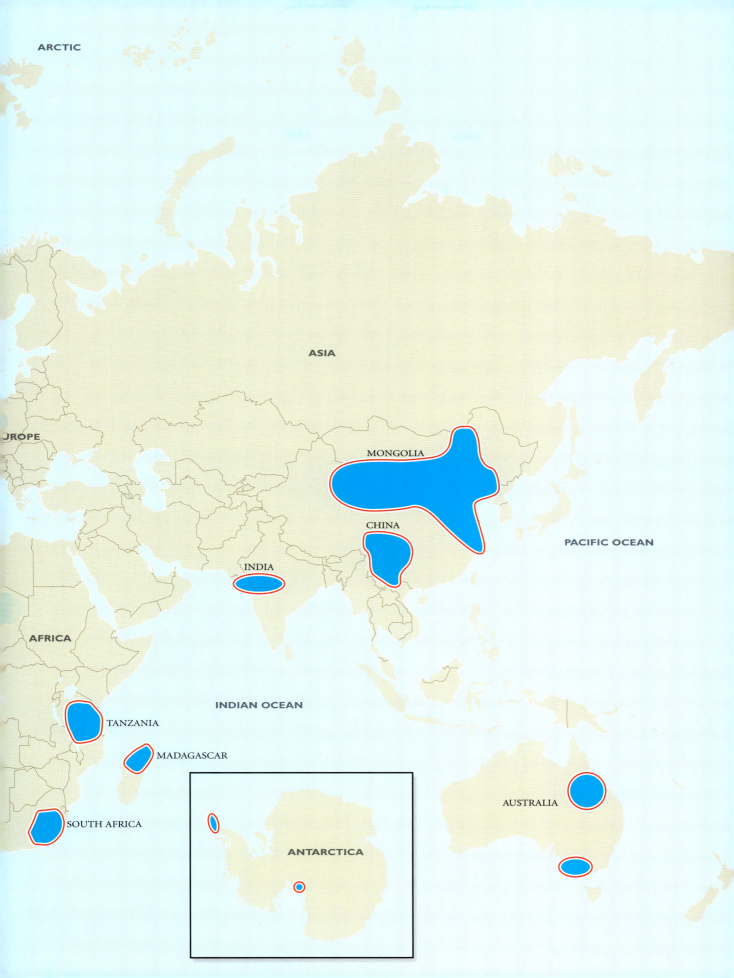

# NAME INDEX

*Utah Raptor*

# Name Index

**A**
Abelisaurus 230, 272, 286, 343
Abrictosaurus 85
Acanthostega 55
Acrocanthosaurus 198–201
Adasaurus 231
Aeolosaurus 232
Afrovenator 168
Alamosaurus 233
Albertosaurus 234, 353
Alectosaurus 235
Alioramus 236, 371
Allosaurus 97, 105, 125, 134, 136, 144–7, 148, 152, 155, 161, 171, 198
Alvarezsaurus 237
Amargasaurus 202–5
Amebelodon 412
Ammonite 16, 262
Ammosaurus 86
Anatotitan 172, 238
Anchiceratops 239
Anchisaurus 86, 87
Andrewsarchus 398–9
Ankylosaurus 312–15, 318, 353
Anomolocaris 12
Anserimimus 240
Antarctosaurus 241
Apatosaurus 141, 151
Appalachiosaurus 259
Aralosaurus 242
Archaeopteryx 58, 142
Archaeornithomimus 243
Argentavis 416–17
Argentinosaurus 223, 403
Aristosuchus 213
Arrhinoceratops 244
Arsinoitherium 408
Arthropleura 25
Atlascopcosaurus 169
Aublysodon 245
Avaceratops 246

**B**
Bactrosaurus 216
Bagaceratops 247
Bahariasaurus 228
Barapasaurus 88
Baryonyx 206–7, 213, 222
Basilosaurus 400–3
Becklespinax 170
Beipiaosaurus 282
Bison alticornis 374
Borhyaena 385
Borogovia 248
Borophagus 419
Bothriospondylus 124
Brachiosaurus 8, 102, 131, 136, 148–51
Brachyceratops 249
Brachylophosaurus 250
Brontosaurus 141
Brontotherium 404–7

**C**
Cacops 27
Camarasaurus 141
Camptosaurus 125
Carcharocles carcharias 390
Carcharocles megalodon 390
Carcharodontosaurus 224–7, 228, 336
Carnotaurus 340–3
Centrosaurus 251
Ceratosaurus 106, 136, 147, 152–5
Cetiosauriscus 119
Cetiosaurus 118
Chasmosaurus 252
Chialingosaurus 126
Chilantaisaurus 171
Chirostenotes 253
Cladoselache 17
Coelacanth 22

Coelodonta 435
Coelophysis 68–71
Coelurosauravus 42
Coelurus 127
Coloradia 82
Coloradisaurus 54, 82
Colossochelys 423
Compsognathus 143
Conchoraptor 254
Coryphodon 393
Corythosaurus 172, 255
Cryolophosaurus 96–7
Cynognathus 49, 72–5

**D**
Dacentrurus 138–9
Daeodon 413
Datousaurus 100
Deinocheirus 344–5
Deinogalerix 414
Deinonychus 184, 208–11
Deinosuchus 325
Desmatosuchus 83
Diadectes 28
Diatryma 386, 387
Diceratus 256
Dicraeosaurus 128, 131
Dilophosaurus 9, 11, 51, 94, 96
Dimetrodon 36–9, 103
Dimorphodon 110
Diplocaulus 40–1
Diplodocus 119, 129, 147, 151, 158, 205, 283
Diprotodon 436
Doedicurus 424–5
Dracopelta 130
Dracorex hogwartsia 361
Dravidosaurus 257
Dromaeosaurus 258, 286
Dryosaurus 131, 151, 165
Dryptosaurus 259
Dsungaripterus 166
Dunkleosteus 20–1

**E**
Edaphosaurus 36
Edmontonia 289, 346–9
Edmontosaurus 238, 260, 300
Einiosaurus 261
Elaphrosaurus 131
Elasmosaurus 262
Elmisaurus 263
Elvisaurus 97
Emausaurus 9–10, 89
Eogyrinus 23
Eohippus 394
Eoraptor 43
Erlikosaurus 264
Eryops 29
Erythrosuchus 44

Euhelopus 132
Euoplocephalus 265, 326
Euparkeria 45
Eurypterid 15, 26
Euskelosaurus 46
Eusthenoptero 18
Eustreptospondylus 112–13

**F**
Fulgurotherium 172

**G**
Gallimimus 327
Gargantuavis 271
Garudimimus 266
Gasosaurus 101, 117
Gastonia 167
Gastornis 386–9
Giganotosaurus 193, 336
Gigantoraptor 357
Gillicus arcatus 324
Gilmoreosaurus 173
Glyptodon 437
Goyocephale 267
Gracilisuchus 9, 66
Gryposaurus 294

**H**
Hadrosaurus 215, 268
Hallucigenia 13
Haplocanthosaurus 133
Harpymimus 174
Henodus 47
Herrerasaurus 57, 76–9
Hesperornis 328
Heterodontosaurus 85, 92
Homalocephale 269
Homalodotherium 415
Homotherium 434
Huayangosaurus 108–9, 120
Hylaeosaurus 175
Hylonomus 24
Hypacrosaurus 270
Hyperodapedon 48
Hypselosaurus 271
Hypsilophodon 9, 107, 194
Hypsilophus 9
Hyracotherium 387, 394

**I**
Ianthasaurus 36
Ichthyostega 19, 55
Iguanodon 125, 173, 179, 194, 212–15, 216, 355
Indosuchus 272
Ingenia 254, 273

**J**
Janjucetus 409
Jaxartosaurus 274, 285

# NAME INDEX

**K**
Kaijiangosaurus 101
Kannemeyeria 49, 72
Kentrosaurus 156–7
Kotasaurus 90
Kronosaurus 195

**L**
Lagosuchus 50
Lambeosaurus 329
Lapparentosaurus 102
Leaellynasaura 176
Leptoceratops 275, 280
Lesothosaurus 107
Lexovisaurus 120–1
Libonectes 330
Liliensternus 51
Liopleurodon 122
Loricatosaurus 120
Lotosaurus 52
Lufengosaurus 91
Lycorhinus 92
Lystrosaurus 80–1

**M**
Magyarosaurus 276
Maiasaura 350–3
Majungasaurus 277
Mammalodon 409
Mandschurosaurus 173, 278
Massospondylus 71
Materpiscis attenboroughi 21
Megaloceros 438
Megalosaurus 94, 103, 111
Megapnosaurus 93
Megatherium 421
Melanosaurus 53
Mesonyx 395
Mesosaurus 30
Metriacanthosaurus 103
Microceratus 279
Minmi 188–9
Moloch horridus 9
Mononykus 331
Montanoceratops 280, 353
Moropus 413, 418
Mosasaur 310, 332
Moschops 31
Mussaurus 54, 82
Muttaburrasaurus 177

**N**
Nanchangosaurus 55
Nanotyrannus 281
Nanshiungosaurus 282
Nemegtosaurus 283, 287, 295
Neuquensaurus 284
Nipponosaurus 285
Noasaurus 286
Nodosaurus 316–17

Nothosaurus 56

**O**
Omasaurus 138
Omeisaurus 100, 104
Ophthalmosaurus 164
Opisthocoelicaudia 287
Ornithimimus 243
Ornitholestes 134, 140
Orodromeus 353
Othniella 134
Ouranosaurus 196
Oviraptor 253, 254, 354–7

**P**
Pachycephalosaurus 358–61
Pachyrhinosaurus 288
Palaeocastor 411
Panoplosaurus 289
Parasaurolophus 333
Parksosaurus 290
Pelicanimimus 178
Pelorosaurus 179
Pentaceratops 291
Piatnitzkysaurus 105
Pinacosaurus 292
Pisanosaurus 57
Platybelodon 391
Platyhystrix 32
Polacanthus 180
Postosuchus 67
Prenocephale 293, 309
Pristichampsus 396
Probactrosaurus 216
Proceratosaurus 106
Prosaurolophus 294
Protoavis 58
Protoceratops 49, 247, 273, 279, 362–5, 380
Protosuchus 71
Psittacosaurus 197
Pteranodon 366–9
Pterodactylus 367
Pterodaustro 217
Pterygotus 15
Pyrotherium 410

**Q**
Quaesitosaurus 295
Quetzalcoatlus 334

**R**
Regnosaurus 108
Rhabdodon 296
Rhoetosaurus 99
Riojasaurus 59

**S**
Saichania 318–21
Saltasaurus 335

Saltopus 60
Saurolophus 294, 297
Sauropelta 190–1
Saurornithoides 248, 298
Saurosuchus 76
Scelidosaurus 89, 95
Scutosaurus 33
Secernosaurus 299
Seismosaurus 158–9
Sellosaurus 61
Seymouria 34
Shansisuchus 62
Shantungosaurus 300
Shonisaurus 63
Shunosaurus 100, 104, 114–17
Silvisaurus 181
Sivatherium 422
Smilodon 426–9
Sphenacodon 36
Spinosaurus 103, 224, 228–9, 336
Staurikosaurus 84
Stegoceras 293, 301, 309
Stegosaurus 108, 139, 147, 151, 155, 156, 160–3
Stenopelix 182
Stomatosuchus 228
Struthiomimus 302
Struthiosaurus 322–3
Stygimoloch 303, 309
Styracosaurus 384
Suchomimus 222
Suchosaurus 206
Syndyoceras 420
Syntarsus 93
Szechuanosaurus 135

**T**
Talarurus 304
Tanystropheus 64
Tapejara 183
Tapinocephalidae 31
Tarbosaurus 236, 370–3
Tenontosaurus 184, 211, 296
Testudo atlas 423

Thecodontosaurus 65
Therizinosaurus 305
Thescelosaurus 290, 306
Thylacosmilus 392
Titanosaurus 307
Tochisaurus 248
Torvosaurus 147
Triceratops 246, 249, 256, 374–7
Trichobatrachus robustus 9
Trilobite 14
Troodon 275, 298, 301, 353, 378–9
Tropeognathus 185
Tsintosaurus 308
Tuojiangosaurus 123
Tylocephale 309
Tylosaurus 310
Tyrannosaurus rex 20, 122, 152, 193, 201, 222, 224, 234, 245, 275, 281, 312, 315, 336–9, 370

**U**
Uintatherium 397
Ultrasauros 136
Utahraptor 166, 218–21

**V**
Velociraptor 362, 380–3

**W**
Wannanosaurus 311
Wolly Mammoth 430–3
Wuerhosaurus 186

**X**
Xiaosaurus 107
Xiphactinus 324

**Y**
Yaleosaurus 87
Yangchuanosaurus 137
Yaverlandia 187
Youngina 35
Yunnanosaurus 98

**Z**
Zephyrosaurus 192

CRYOLOPHOSAURUS

# GENERAL INDEX

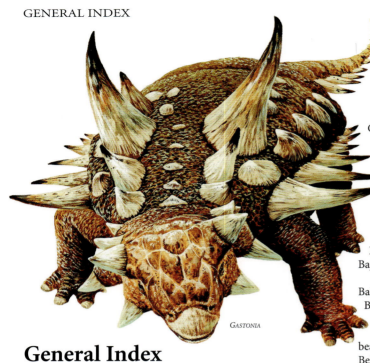
GASTONIA

# General Index

**A**
Abel, Roberto 343
abelisaurids 343
Academy of Natural Sciences, Philadelphia 215
Africa
  Devonian Period 22
  Early Cretaceous Period 165, 168, 196, 207, 222
  Early Jurassic Period 93
  Late Jurassic Period 128, 131, 144, 153, 156–7, 165
  Middle Cretaceous Period 225, 227
  Permian Period 30
  Tertiary Period 416
  *see also* Madagascar; Middle East; North Africa; South Africa; *individual countries*
Agate Springs Quarry, USA 413, 418
alligators 29
American Museum of Natural History 321, 383
American National Academy of Sciences 429
amphibians, definition 41
Andes mountains 416
ankylosaurs
  Early Cretaceous Period 167, 188, 189
  Jurassic Period 130
  Late Cretaceous Period 292, 304, 315, 316, 323, 326, 346, 349
Antarctica 72, 73, 81, 96, 176
Aral Sea, Kazakhstan 242
archosaurs 44, 45, 50–1, 55, 62, 75, 79, 83
Argentina
  Early Cretaceous Period 193, 203, 205, 217
  Jurassic Period 105, 164
  Late Cretaceous Period 230, 232, 237, 241, 272, 284, 286, 299, 307, 335, 341, 343
  Middle Cretaceous Period 223
  Neogene Period 385, 392
  Pleistocene Epoch 424
  Tertiary (Miocene) Period 415, 416
  Tertiary (Oligocene) Period 410
  Triassic Period 43, 50, 54, 57, 59, 65–6, 72, 73, 77, 79, 82, 84
Argile Plastique Formation, France 386
armadillos 424
Arthrodira 20
arthropods 12, 14, 15, 25
Asia
  Early Cretaceous Period 173, 174, 186, 197
  Late Cretaceous Period 231, 242–3, 274, 292, 297, 300
  Pleistocene Epoch 430, 433, 435
  Tertiary (Eocene) Period 395, 396, 400, 401
  Tertiary (Miocene) Period 413
  Tertiary (Pliocene) Period 422
  *see also* China; India; Mongolia Australia
  Cambrian Period 12
  Cretaceous Period 169, 172, 176–7, 188, 195
  Jurassic Period 99, 144
  Pleistocene Epoch 436
  Tertiary Period 409
  Triassic Period 49, 81
Austria 322

**B**
Bajo de la Carpa Formation, Argentina 237
Bakker, Robert 377
Barbour, George 420
Bayan Mandahu Formation, Mongolia 362
beavers 411
Belgium 212, 386
bennettitales 71
Bering Sea 412
Big Bend National Park, USA 334
bipedality
  Early Cretaceous Period 165, 218
  Early Jurassic Period 85, 93
  Late Cretaceous Period 230, 259, 294, 296–7, 299–300, 302, 311, 337
  Late Jurassic Period 127, 131
  Middle Jurassic Period 101, 105
  Triassic Period 45, 50, 51, 57
birds 11, 386–9, 416–17
bison 426
Black Hills Museum of Natural History, USA 201
Bolivia 312, 410
Brazil 84, 178, 183, 185
Brown, Barnum 294, 315, 377
Bungil Formation, Australia 188
Bunzel, Emanuel 322
Burgess Shale, Canada 12–13

**C**
Calizas de La Huérgina Formation, Spain 178
Cambrian Period 12–14
camels 407, 420
Canada
  Burgess Shale 12–13
  Cretaceous Period 234, 239, 244–5, 249–53, 255, 258, 263, 265, 270, 275, 280, 288–90, 294, 301–2, 306, 312, 326, 329, 333–4, 346, 351, 374, 379, 384
  Frasnian Escuminac Formation, Quebec 18
  Pangaea and 163
  Triassic Pardonet Formation 63
cannibalism 277
Canodon Asfalto Formation, Argentina 105
Carboniferous Period 19, 23–5, 41
carnosaurs 137
Carroll, Lewis 248
CAT scans 306, 361, 430
cats 392, 426, 434
cave paintings 433, 435
Cedar Mountain Formation, USA 167
ceratopsians 239, 244, 246–7, 249, 252
chalicotheres 418
chasmosaurines 239
cheetahs 434
chelae 15
Chengjiang, China 12
Chile 341
China
  Cambrian Period 12, 13
  Early Cretaceous Period 166, 171, 186, 197, 216
  Early Jurassic Period 91, 98
  Late Cretaceous Period 243, 245, 274, 278–9, 282, 292, 300, 305, 308, 354, 357, 365, 370, 380
  Late Jurassic Period 123, 126, 132, 135, 137, 166
  Middle Jurassic Period 100–1, 104, 107, 108, 109, 115, 117
  Pleistocene Epoch 438
  Tertiary (Eocene) Period 396
  Tertiary (Miocene) Period 412
  Triassic Period 52, 55, 56, 62, 64, 72, 81
Cleveland-Lloyd Dinosaur Quarry, USA 155
Cleveland Museum of Natural History, USA 20
climate change 75, 79, 291, 349, 430
Cloverly Formation, USA 191
coelurosaurs 106, 231, 259, 340
Colombia 195
conifers, earliest 71, 279
Continental Drift 30, 73
  *see also* land bridges; Laurasia; Pangaea
Cope, Edward Drinker 36, 141, 262, 386

444

# GENERAL INDEX

corals 75
Coues, Elliott 387
Cretaceous Period 11, 51, 396
    Early 165–222
    Late 230–384
    Middle 223–9
Cretaceous-Tertiary extinction 306, 332
Crete 438
crocodiles 9, 29, 44, 45, 66, 117, 228, 325, 396
Currie, Philip 357
cycads 71, 75, 123, 133, 239, 279, 306, 347, 375
cynodonts 72
Cynognathus Assemblage Zone, South Africa 44
Czech Republic 324

## D
daemonelix 411
Dakota Formation, USA 181
Dashanpu Formation, China 100, 115, 117
Dawson, Sir William 24
deer 420
Devonian Period 16–22
diapsids 35
dicynodonts 44
dinosaur, definition 10–11
    (film) 355
Dinosaur Provincial Park Formation, Canada 239, 251, 53, 384
Dinosaurland Fossil Museum, England 110
    (TV series) 355
Diplichnites cuithensis 25
Djadokhta Formation, Mongolia 354
dogs 419
dromaeosaurids 218, 231, 258, 380, 381

## E
Early Cretaceous Period 165–222
Early Jurassic Period 85–98
Echkar Formation, Africa 196
Edmonton Formation, Canada 346
Edson Quarry, USA 419
Egg Mountain, Montana 350, 351, 353
Egypt 228, 229, 400, 401, 408
elasmosaurs 330
elephants 391, 393, 410, 412 430, 433
elk 426, 438
Emu Bay Shale, Australia 12

endocasts 224
Ensenada Formation, Argentina 424
Eocene Epoch 393–407
Erickson, Bruce 377
etanurans 96
Ethiopia 408
etymology 9–10
Europe
    Early Cretaceous Period 175, 178–80, 182, 194, 207, 212
    Early Jurassic Period 89
    Late Cretaceous Period 271, 276, 296, 322, 324
    Late Jurassic Period 122, 130, 138, 142–3, 149, 161, 163
    Middle Jurassic Period 120
    Paleogene Period 386
    Permian Period 28, 34, 36, 37, 42
    Pleistocene Epoch 433, 435
    Tertiary (Eocene) Period 393, 396
    Tertiary (Miocene) Period 414
    Tertiary (Oligocene) Period 408
    Triassic Period 47, 51, 56, 61, 64
    *see also* Russia
eyes, earliest 14

## F
Falconer, Dr Hugh 423
feathers 28, 142, 143, 282, 302, 305, 354, 383
Felch, Marshall P. 147
fenestrae 113, 230, 246
ferns 71, 123, 176, 239, 279, 347, 375
'Fighting Dinosaurs' specimen 362, 380, 381
fish 324
fossilization process 8–9, 159
France 51, 120, 138, 143, 175, 271, 296, 322, 386, 396, 438
Frasnian Escuminac Formation, Quebec, Canada 18
frogs 9

## G
Galapagos Islands 423
gastralia 129
gastroliths
    Cretaceous Period 182, 197, 233, 241, 284, 307
    Jurassic Period 86, 87, 91,

95, 119, 124, 158, 159
    Triassic Period 53
Germany 51, 56, 61, 64, 89, 142, 143, 182, 212, 386, 396
gigantotherms 151
ginkgophyta 75
giraffes 420, 422
glyptodontids 425, 437
Gobi Desert
    Early Cretaceous Period 216
    Late Cretaceous Period 231, 235–6, 267, 283, 287, 292, 295, 298, 304, 319, 321, 331, 344, 362, 370
    Tertiary Period 398
gomphotheres 412
Goodwin, Mark 269
Great White Shark 390
Greenland 19
griffins 365
ground sloths 415, 421
gymnures 414

## H
hadrosaurs
    Early Cretaceous Period 173, 216
    Late Cretaceous Period 238, 242, 250, 255, 260, 268, 274, 278, 285, 297, 299, 322, 325, 329, 333, 370
Hall, Cephis 201
Hammer, Dr William 96
Harlan, Dr Richard 401
Hawkins, Benjamin Waterhouse 215
hedgehogs 414
Hell Creek Formation, USA 315, 374, 377
heteromorphids 16
Himalayan mountains 135, 423
hippopotami 304, 393, 436

Hopson, J.A. 85
Horner, Jack R. 350
horses 394, 410, 418
Horseshoe Canyon Formation, Canada 239, 244
horsetails 176
Huene, Friedrich von 172
human coexistence 11, 341, 425, 427, 431, 433, 435, 437
Hungary 307, 322
hyenas 419, 434
hypsilophodonts 172, 176, 296
hyraxes 394

## I
Ice Age 424, 435
ichtyodectids 324
ichthyosaurs 55, 63, 75, 122, 164
iguanodonts 296, 322
'index fossils' 14
India 49, 88, 90, 241, 257, 272, 307, 422–3
Indis 81
Indonesia 423
Ireland 438
Irish Elk 438
Ischigualasto Formation, Argentina 43, 50, 79
island gigantism 414
Israel 56, 64
Italy 56, 64, 414

## J
Japan 248, 262, 285, 373, 438
Jiangxi province, China 55
Joggins, Nova Scotia 24
Judith River Formation, USA 250, 258, 384
Junggar Basin, China 166 350
Jurassic Period 77, 99
    Early 85–98
    Late 122–66, 186, 194, 283
    Middle 98, 100–22

*Mosasaur*

# GENERAL INDEX

THYLACOSMILUS

## K
Karoo, South Africa 31, 35, 72, 80
Kazakhstan 242, 243, 274, 396
keratin 344
koala bears 436
Koch, Albert 400, 403
Kota Formation, India 88, 90

## L
La Amarga, Argentina 203, 205
La Brea Tar Pits 426
labyrinthodonts 41
lagosuchids 60
Lakes, Arthur 147
Lance Formation, USA 315
land bridges 168, 277, 298, 309, 371, 391, 412
Langdon, Doug 177
Late Cretaceous Period 230–384
Late Jurassic Period 122–66, 186, 194, 283
Late Triassic Period 47, 71, 82–4
Laurasia 126, 140
Lehman, Thomas M. 375
lepospondyls 41
Lesotho 92
life expectancy 339
Lightning Ridge, Australia 172
lions 429
lizards 316
Love, Sidney 201
Lufeng Formation, China 91
Lund, Peter Wilhelm 426
lycophytes 75
lycopods 24

## M
Madagascar 42, 102, 124, 277
mammals, evolution of 72, 75
mammoths 412, 426, 430
Manchuria, China 278
maniraptors 187
Maotianshan, China 13
marine iguanas 47
Marsh, Othniel Charles 141, 147, 374, 377
marsupials 392, 436
mass extinctions 11, 71, 75, 76, 79, 80, 81, 291, 306, 332, 349, 396
mastodons 426
McHenry, Colin 429
Mediterranean Sea 407
megacerops 404
megalosaurs 147
mesonychids 395
Mesozoic Era 328
meteor strike 291
Mexico 325
Middle Cretaceous Period 223–9
Middle East
    Tertiary Period 400, 401, 408
    Triassic Period 47, 64
Middle Jurassic Period 98, 100–22
Miocene Epoch 385, 392, 412–20
Mongolia
    Early Cretaceous Period 171, 174, 186, 197
    Late Cretaceous Period 231, 235–6, 240, 247–8, 254, 263–4, 266–7, 269, 273, 279, 283, 287, 292–3, 295, 298, 304–5, 309, 318–19, 321, 327, 331, 344, 354, 362, 370, 373, 380–1, 383
    Tertiary (Eocene) Period 398
    Tertiary (Oligocene) Period 408
    Triassic Period 81
monkeys 407
Morocco 225, 227, 228, 229
Morrison Formation, USA 129, 133, 134, 144, 147, 156
Museu da Lourinhã, Portugal 130

## N
Natural History Museum, London 175, 214–15
nectridians 41
Nemegt Formation, Mongolia 240, 254, 305, 370
Neogene Period 385, 390–2
Netherlands 56
New Zealand 240, 310
Niger 168, 196, 222, 225, 227
Niobrara Formation, USA 310, 366
nodosaurids 190, 191, 316–17, 307, 322, 346
Norell, Mark 383
North Africa
    Cretaceous Period 212, 224, 225, 227, 228, 229
    Jurassic Period 149
    Tertiary Period 408
    Triassic Period 47, 56
North America
    Carboniferous Period 24, 25
    Devonian Period 17
    Early Cretaceous Period 167, 174, 181, 184, 191, 192, 199, 201, 209, 211, 218–19
    Early Jurassic Period 86–7, 93, 94
    Late Cretaceous Period 233, 234, 238, 245–6, 249–50, 252, 255–6, 258–62, 265, 268, 270, 275, 280–1, 289, 293–4, 297, 301–3, 305–6, 310, 312, 315–16, 324–6, 329–30, 333–4, 336, 346–7, 351, 358, 366, 374, 379, 384
    Late Jurassic Period 125, 127, 129, 133–4, 136, 140–1, 144, 147, 149, 153, 160–1, 163
    Paleogene Period 386
    Permian Period 27–9, 32, 34, 36, 37, 40, 41
    Pleistocene Epoch 424, 425–7, 433, 437
    Silurian Period 26
    Tertiary (Eocene) Period 393, 394–7, 400–1, 404–5, 407
    Tertiary (Miocene) Period 412–20
    Tertiary (Pliocene) Period 421
    Triassic Period 58, 63, 64, 67–9, 83
    see also Canada
North American Inland Sea 328
North Carolina Museum of Natural Sciences, USA 306
North Sea 328
Norwegian-Greenland Sea 407

## O
Ogalalla Formation, USA 419
okapis 422
Oldman Formation, Canada 250
Oligocene Epoch 408–11
Ordovician Period 26
ornithischians 57, 85, 89, 95, 107
ornithocephala 322
ornithomimosaurs 174, 240, 243, 266, 327, 345
ornithopods 117, 172, 290
ornithosuchians 60
orthocones 14
ossicones 422
osteoderms 95, 130, 160, 167, 284, 313, 315, 424, 437
ostriches 327, 345
otic notch 27
oviraptorids 254, 263, 273, 357
Owen, Sir Richard 394, 401
Oxford University Museum, England 112

## P
pachycephalosaurs 182, 187, 267, 293, 303, 309, 311, 358, 359
Pakistan 400, 401, 423
Paleogene Period 386–9
paleontology, study of 11
paleotheres 394
Paleozoic Era 26, 40
palms 375
pampheres 425
Panama, Isthmus of 233
Pangaea 30, 48, 69, 71, 75, 156, 163
Pangaea 93
pareiasaurs 33
Patagonia 77, 164, 203, 335
Peabody Museum of Natural History, USA 127
pelycosaurs 39
Permian Period 14, 26–42
Permian-Triassic extinction 80
Petrified Forest, Arizona 69
pigs 407
Pipestone Creek, Canada 288
placoderms 21
placodonts 47
Planté, Gaston 386, 387
plateosaurs 46
Pleistocene Epoch 392, 423–38
plesiosaurs 122, 330
Pliocene Epoch 374, 392, 421–2

polacanthines 167
Portugal 130, 138, 144, 153, 161, 163, 179, 194, 207
Pritchard, George 409
prosauropods 53, 54, 59, 61, 65, 82, 86, 87, 90, 91, 98
pterosaurs 45, 50, 75, 110, 117, 166, 183, 185, 217, 334, 366, 369

## Q
Quaternary Period 423–38

## R
rats 407
Red Deer River, Canada 251, 302
reptilomorphs 34
rhinoceroses 288, 393, 394, 397, 404, 405, 407, 408, 418, 435
rhyncosaurs 48
Rio Limay Formation, Argentina 223
Rio Negro Province, Argentina 230, 232
Romania 276, 296, 322
'rostral organ' 22
Rozhdestvensky, Anatoly 216
Russell, Dale 228
Russia
    Cretaceous Period 197, 285
    Devonian Period 18
    Permian Period 31, 33
    Triassic Period 56, 81

## S
sacral shields 167, 180
Sahara Desert 168, 196, 222, 224
Sakhalin Island 285
Salta Province, Argentina 335
Santa Cruz Formation, Argentina 415
Santana Formation, Brazil 185
saurischians 57, 59
sauropodomorphs 86
sauropods
    Cretaceous Period 170, 198, 202, 233, 241, 271, 276, 282, 284, 287, 295, 307, 322
    Jurassic Period 88, 90, 99–100, 102, 104–5, 114, 117, 118–19, 128–9, 132–3
    Triassic Period 53, 59, 65
Science Museum of Minnesota, USA 377
sclerotic ring 297
Scollard Formation, Canada 315
scorpions 26
scutes 284
segnosaurs 235
Sereno, Paul 224, 225

Shackleford, J.B. 362
Shandong Province, China 132, 300
Shansi (Shanxi) province, China 62
sharks 17, 21, 122, 195, 390
Shaximiao Formation, China 101
Shinkehudug Formation, Mongolia 174
Siberia 430, 433
Sichuan Province, China 107, 109, 117, 123, 135, 137
Silurian Period 15, 26
sloths 415, 421
Smith, Professor Josh 229
Smithsonian Institution, USA 215
South Africa
    Early Jurassic Period 85, 92
    Permian Period 31, 33, 35
    Triassic Period 44, 45, 46, 49, 72, 73, 80, 81
South America
    Early Cretaceous Period 178, 183, 185, 193, 195, 203, 205, 217, 223
    Jurassic Period 105, 164
    Late Cretaceous Period 230, 232, 233, 237, 241, 284, 286, 299, 312, 335, 341, 343
    Late Triassic Period 82, 84
    Neogene Period 385, 392
    Permian Period 30
    Pleistocene Epoch 424, 425–7, 437
    Tertiary (Miocene) Period 415, 416
    Tertiary (Oligocene) Period 420
    Tertiary (Pliocene) Period 421
    Triassic Period 43, 46, 50, 54, 57, 59, 65–6, 72, 73, 76–7, 79
Spain 138, 163, 178, 271, 296
spinosaurs 171
squamates 332
squid 310
stegosaurs 89, 108, 109, 117, 120–1, 123, 137, 138–9, 156–7, 186, 257
Sternberg Museum of Natural History, USA 324
Stromer, Ernst 224, 225, 228, 229
suprascapula 290
Switzerland 56, 64
synapsids 31, 32, 33, 36

## T
Tamil Nadu Province, India 257

Tanzania 124, 128, 131, 144, 153, 156–7, 165
tapirs 393, 407, 418
telson 15
temnospondyls 32
Tendaguru beds, Tanzania 128, 157, 165
Tertiary Period
    Eocene Epoch 393–407
    Miocene Epoch 385, 392, 412–20
    Oligocene Epoch 416–17, 420
    Pliocene Epoch 374, 392, 421–2
testudinids 423
tetrapods 18–19, 23, 34, 41, 55
Thailand 197
thalattosaurs 75
thecodonts 62
therapsids 62
therizinosaurs 264, 282, 305
theropods
    Early Cretaceous Period 168, 170–1, 198, 208
    Jurassic Period 94, 96, 101, 105, 111–12, 117, 137, 140
    Late Cretaceous Period 230–1, 233, 234, 237, 243, 248, 253, 277, 286, 305, 315, 336, 340, 343, 354, 370
    Middle Cretaceous Period 229
    Triassic Period 51, 58, 66, 84
thorny devils 9
tigers 426, 429
titanosaurs 283, 284, 403
tortoises 423
trackways 221, 266
Triassic Pardonet Formation, Canada 63
Triassic Period 43–81
    Late 47, 71, 82–4
troodontids 248, 378
tumours 173, 250

Turgai Strait 328
Turkey 408
turtles 47, 117, 423
Two Medicine Formation, USA 249, 261
tyrannosaurs 233, 235, 236, 245, 259, 281, 357, 371

## U
uintatheriidae 397
unicorn 308
United Kingdom
    Carboniferous Period 23
    Devonian Period 18
    Early Cretaceous Period 170, 175, 179, 180, 187, 194, 206, 207, 212
    Jurassic Period 95, 103, 106, 110, 111, 112–13, 118–20, 124
    Permian Period 42
    Pleistocene Epoch 438
    Tertiary (Eocene) Period 394
    Triassic Period 60, 65
Upper Elliot Formation, Afric, 92
Uruguay 284

## W
Walker, William 206
Wandering Albatross 416
'wastebin taxon' 307
Wegener, Alfred 73
Wentz, Terry 201
Western Interior Sea 324
whales 400, 407, 409
Whiteaves, J.F. 12
wombats 436

## X
Xu Xing 357

## Y
Yunnan Province, China 91, 98

## Z
Zimbabwe 93

*Carnotaurus*

# Picture Credits

**Acknowledgments:**
The editor wishes to thank the following people for their help and support:
Brian Beatty, Fiona Brady, Eric Buffetaut, Charles Catton, Luis Chiappe, Samuel Ciurca, Kieron Connolly, Jack Conrad, Gilles Cuny, Terry Forshaw, Mary & Walter Mehling, Alvaro Mones, Chris Norris, Silvio Renesto, Cosimo Soltanto, Thomas Trombone, Tommy Tyrberg, Sarah Uttridge and Rich White.

**Picture Credits**
All Illustrations © **DeAgostini** except the following:
**Art-Tech:** 14, 21, 22, 26, 36b, 37, 40b, 41, 42, 66, 67, 68b, 69, 72b, 73, 76, 77t, 80b, 81, 94, 95, 96, 97b, 108b, 109, 110, 111, 112b, 113, 114b, 115, 120, 138b, 139, 140-143, 145, 149, 152b, 153, 156b, 157, 158b, 159, 160, 161b, 164, 193-197, 198b, 199, 203, 206 top and middle, 207b, 209, 212b, 213, 219, 222, 224b, 225, 228b, 229, 318b, 319, 323, 325-335, 341, 345, 347, 351, 355, 38b, 359, 363, 367, 371, 375, 379, 381, 384b, 385, 387, 390, 391, 392, 399, 400b, 401, 404b, 405, 416b, 417, 423, 424b, 425, 427, 430b, 431, 434.
**Amber books Ltd:** 108 main and insets, 121, 167, 188, 189b, 191, 313, 317, 337
**Dorling Kindersley:** 305
**Natural History Museum, London:** 272 (DeAgostini), 277 (Andrey Atuchin)

**Photographs:**
**Alamy:** 38/39 (Mervyn Rees), 78/79 (Ronald Karpilo), 162/163 (Peter Casolino), 200/201 (Jill Stephenson), 314/315 (PSI), 342/343 (MAF), 360/361 (Amberstock)
**Corbis:** 146/147 (Richard Cummins), 150/151 (Stephanie Pilick/EPA), 204/205 (Louie Psihoyos), 210/211 (Louie Psihoyos), 220/221 (Kimimasa Mayama), 226/227 (Photopress Washington/Sygma), 338/339 (Paul A. Souders), 348/349 (Richard T. Nowitz), 356/357 (Louie Psihoyos), 364/365 (Kevin Schafer), 368/369 (Jonathan Blair), 402/403 (Mike Nelson/EPA)
**Getty Images:** 70/71 (Louie Psihoyos), 352/353 (Ken Lucas), 372/373 (Martin Bernetti/AFP), 432/433 (Frederic J. Brown/AFP)
**Natural History Museum, London:** 74/75, 116/117, 214/215, 272 (De Agostini), 277 (Andrey Atuchin), 376/377, 388/389, 406/407, 428/429
**Photoshot:** 320/321 (Andrea & Antonella Ferrari/NHPA)
**Rex Features:** 382/383 (Solent News)
**Still Pictures:** 154/155 (Fototoro)